AF719332
WITHDRAWN
WITHDRAWN

Rozzell/Wagner
Biocatalytic Production of Amino Acids and Derivatives

Hanser Titles in Biotechnology

Biotechnology Focus 1:
Fundamentals-Applications-Information
Edited by Robert K. Finn and Paul Präve

Biotechnology Focus 2:
Fundamentals-Applications-Information
Edited by Robert K. Finn and Paul Präve

Biotechnology Focus 3:
Fundamentals-Applications-Information
Edited by Robert K. Finn and Paul Präve

Large Scale Cell Culture Technology
Edited by Bjorn K. Lydersen

Plant Cell and Tissue Culture in Liquid Systems
Gregory F. Payne, Venkataraman Bringi, Christopher L. Prince, Michael L. Shuler

Bioprocess Monitoring and Control
Edited by Marie-Noëlle Pons

Biocatalytic Production of Amino Acids and Derivatives
Edited by J. David Rozzell and Fritz Wagner

Diffusion In and Through Polymers:
Principles and Applications
Wolf R. Vieth

Membrane Systems: Analysis and Design
Wolf R. Vieth

Protein Purification:
Design and Scale up of Downstream Processing
Scott M. Wheelwright

Biocatalytic Production of Amino Acids and Derivatives

Edited by
J. David Rozzell and Fritz Wagner

With 115 Illustrations and 68 Tables

Hanser Publishers, Munich Vienna New York Barcelona

The Editors:
Dr. J. David Rozzell, Vice President, Research and Development, Exogene, Monrovia, CA 91016, USA
Prof. Dr. Fritz Wagner, Institut für Biochemie und Biotechnologie der TU Braunschweig, Braunschweig, Germany

TP
248.65
.A43
B56
1992

Distributed in USA and in Canada by
Oxford University Press
200 Madison Avenue
New York, NY 10016

Distributed in all other countries by
Carl Hanser Verlag
Kolbergerstraße 22
D-8000 München 80

The use of general descriptive names, trademarks, etc., in this publication, even if the former are not especially identified, is not to be taken as a sign that such names, as understood by the Trade Marks and Merchandise Marks Act, may accordingly be used freely by anyone.

While the advice and information in this book are believed to be true and accurate at the date of going to press, neither the authors nor the editors nor the publisher can accept any legal responsibility for any errors or omissions that may be made. The publisher makes no warranty, express or implied, with respect to the material contained herein.

Library of Congress Cataloging-in-Publication Data
Biocatalytic production of amino acids and derivatives / J. David Rozzell, Fritz Wagner (editors).
p. cm.
Includes bibliographical references and index.
ISBN 3-446-15699-2 (Carl Hanser Verlag). – ISBN 0-19-520982-6 (Oxford University Press : cloth)
1. Amino acids – Biotechnology. 2. Amino acids – Synthesis. I. Rozzell, J. David. II. Wagner, Fritz, Dr. rer. nat.
TP248.65.A43B56 1992
660'.63 – dc20 92-26917

Die Deutsche Bibliothek – CIP-Einheitsaufnahme
Biocatalytic production of amino acids and derivatives / J. David Rozzell ; Fritz Wagner (ed.). – Munich ; Vienna ; New York ; Barcelona : Hanser, 1992
(Hanser titles in biotechnology)
ISBN 3-446-15699-2
NE: Rozzell, J. David [Hrsg.]

ISBN 3-446-15699-2 Carl Hanser Verlag, Munich Vienna New York Barcelona
ISBN 0-19-520982-6 Oxford University Press

All rights reserved. No part of this book may be reproduced or transmitted in any form or by any means, electronic or mechanical, including photocopying or by any information storage and retrieval system, without permission from the publisher.

© 1992 Carl Hanser Verlag, Munich Vienna New York Barcelona
Printed and bound in Germany by Kösel, Kempten

INTRODUCTION

Amino acids are the building blocks of proteins, and as such form a group of critically important chemical compounds. There are 20 naturally occurring amino acids; 19 of these are optically active, having the L- configuration. Fourteen of the L-amino acids are essential for nutrition in the diet of humans. The combination of these essential amino acids in formulations for parenteral feeding solutions has led to a large and important pharmaceutical market. Several amino acids of the D- configuration also occur in nature: D-alanine, D-glutamic acid, and D-2-aminoadipic acid are three such examples found as components in the cell walls of microorganisms or in certain secondary metabolites. The unnatural D-amino acids D-phenylglycine and *p*-hydroxy-D-phenylglycine have developed large applications as side chains for β-lactam antibiotics. The annual sales volumes for L-lysine and DL-methionine have grown to several hundred million dollars annually as nutritional additives to fortify animal feed, and L-tryptophan and L-threonine have similar potential applications. New peptide drugs, synthesized from amino acid building blocks, are being increasingly submitted for biological testing. Given the ubiquitous presence of amino acids throughout nature, it is not surprising that important markets for them have developed in agricultural and pharmaceutical products for nutrition and drugs such as those listed above. What may be surprising, however, is the development and growth of significant markets for amino acids, both naturally occurring and non-naturally occurring, in other market areas.

Worldwide markets for amino acids total approximately two billion dollars annually. For example, L-phenylalanine and L-aspartic acid have developed an annual demand of nearly five thousand metric tons each as components in the manufacture of the high-intensity sweetener aspartame. L-Glutamic acid has a world market greater than one billion dollars for use as a flavor enhancer in foods; DL-alanine, glycine, and L-aspartic acid are also used as seasonings in smaller amounts. Glycine is manufactured in moderate quantities for use as an intermediate in the production of chelators and surfactants. Numerous D- and L-amino acids are used as homochiral intermediates in fine chemical manufacture, and dipeptides are also gaining in importance as chemical and pharmaceutical building blocks.

This book in intended to provide examples of recently developed methods for the production of amino acids and derivatives. The chapters fall into three sections: processes to produce various types of amino acids; processes to produce derivatives of amino acids and amino acid analogs; and process considerations for the use of biocatalysts in amino acid production. The reader will find much new technology integrated into these chapters: the development of membrane reactors, the use of enzymes in nonaqueous solvents, the application of recombinant DNA techniques for the cloning and expression of genes, newly discovered enzymes and microbial strains, and immobilized biocatalytic systems. It is our hope that these applications to the production of amino acids are only a preview of future developments in industrial biotechnology.

J. David Rozzell
May 29, 1992

CONTRIBUTORS

David M. Anderson
ChemGen Corporation
2501 Research Blvd.
Rockville, MD 20850, USA

W. H. J. Boesten
DSM Research
Bio-organic Chemistry Section
P.O. Box 18
6160 MD GELEEN
The Netherlands

Q. B. Broxterman
DSM Research
Bio-organic Chemistry Section
P.O. Box 18
6160 MD GELEEN
The Netherlands

C. Bull
W. R. Grace & Co.-Conn.
Washington Research Center
7379 Route 32
Columbia, MD 21044, USA

Gary J. Calton
SRCHEM, Inc.
5331 Landing Road
Elkridge, MD 21227, USA

Steven P. Crump
Coors Biotech Products
400 Rolling Hills Lane
P. O. Box 4136
Winchester, KY 40391, USA

D. R. Durham
W. R. Grace & Co.-Conn.
Washington Research Center
7379 Route 32
Columbia, MD 21044, USA

Nobuyoshi Esaki
Institute for Chemical Research
Kyoto University
Uji, Kyoto-Fu 611, Japan

Erwin Flaschel
Institute of Chemical Engineering
Swiss Federal Institute of
Technology
CH-1015 Lausanne, Switzerland
Present address:
Fermentation Technology
Technical Department
University of Bielefeld
P.O.Box 8640
D-4800 Bielefeld 1, FRG

A. Gross
Present address:
Opta Food Ingredients, Inc.
64 Sidney St.
Cambridge, MA 02139, USA

H. F. M. Hermes
DSM Research
Bio-organic Chemistry Section
P.O. Box 18
6160 MD GELEEN
The Netherlands

Humg-Yu Hsiao
ChemGen Corporation
2501 Research Blvd.
Rockville, MD 20850, USA

J. Kamphuis
DSM Research
Bio-organic Chemistry Section
P.O. Box 18
6160 MD GELEEN
The Netherlands

B. Kaptein
DSM Research
Bio-organic Chemistry Section
P.O. Box 18
6160 MD GELEEN
The Netherlands

Steven F. Karel
Department of Chemical Engineering
Princeton University
Princeton, NJ 08544, USA

Karsten Krohn
Institute of Organic Chemistry
Technical University of Braunschweig
Hagenring 40
D-3300 Braunschweig, FRG

R. J. Kupper
W. R. Grace & Co.-Conn.
Washington Research Center
7379 Route 32
Columbia, MD 21044, USA

Alexey L. Margolin
Marion Merrell Dow Research Inst.
Indianapolis Center
9550 N. Zionsville Rd.
Indianapolis, IN 46268, USA

E. M. Meijer
DSM Chemicals
Specialty Chemicals Group
Mauritslaan 49
6129 EL URMOND
The Netherlands

Ralf Müller
Institute of Biochemistry and Biotechnology A
Technical University of Braunschweig
Konstantin-Uhde-Straße 5
D-3300 Braunschweig, FRG

Kiyotaka Oyama
Chemical Research Center
Tosoh Corporation
4560 Kaisei, Shin-Nanyo
Yamaguchi 746, Japan

Markus Pietzsch
Institute of Biochemistry and Biotechnology A
Technical University of Braunschweig
Konstantin-Uhde-Straße 5
D-3300 Braunschweig, FRG

J. David Rozzell
Exogene Corporation
222 E. Huntingdon Drive, Suite 111
Monrovia, CA 91016, USA

H. E. Schoemaker
DSM Research
Bio-organic Chemistry Section
P.O. Box 18
6160 MD GELEEN
The Netherlands

Martin Siemann
Institute of Biochemistry and Biotechnology A
Technical University of Braunschweig
Konstantin-Uhde-Straße 5
D-3300 Braunschweig, FRG

Kenji Soda
Institute for Chemical Research
Kyoto University
Uji, Kyoto-Fu 611, Japan

Christoph Syldatk
Institute of Biochemistry and Biotechnology A
Technical University of Braunschweig
Konstantin-Uhde-Straße 5
D-3300 Braunschweig, FRG

Fritz Wagner
Institute of Biochemistry and Biotechnology A
Technical University of Braunschweig
Konstantin-Uhde-Straße 5
D-3300 Braunschweig, FRG

J. F. Walter
W. R. Grace & Co.-Conn.
Washington Research Center
7379 Route 32
Columbia, MD 21044, USA

CONTENTS

PART I

BIOCATALYTIC PRODUCTION OF AMINO ACIDS

THE ENZYMATIC PRODUCTION OF L-ASPARTIC ACID

Gary J. Calton

CONTENTS

1.1 INTRODUCTION

The production of L-aspartic acid from fumaric acid via an enzyme reaction now represents the highest productivity reaction known for an enzyme system. Methods of immobilization have been developed which allow the reuse of the enzyme to the extent that over 220,000 pounds of product per pound of enzyme can be produced.

Aspartic acid use in Japan has been large historically due to its use as a food flavoring. Estimates of the use of aspartic acid for this purpose are in the range of 100 metric tons per year. The historic use of this chemical in the U.S. has been in enteral nutrition. Although aspartic acid is used in parenteral nutrition, the percentage of aspartic acid in the 15 amino acid mix is in the 0.5 to 2% range in parenteral products. In the enteral products, aspartic acid may be as high as 15% of the total amino acid content. The introduction of the artificial sweetener, aspartame, in 1981, provided additional large markets specifically for L-aspartic acid. In 1986, the production of aspartame was estimated to consume 5,000 metric tons of aspartic acid per year. With the expansion of the aspartame market due to the introduction of its use into frozen desserts and fruit beverages, the market has once again begun its upward growth curve.

The market for aspartic acid has also begun growing in Europe, where aspartame is no longer under patent. In the United States at the present time, there is very little requirement for aspartic acid due to the captive production of aspartic acid by the NutraSweet Co., which meets all of its own requirements. That situation may change rapidly in the near future with the expiration of the U.S. patent covering the use of aspartame.

1.2 PRODUCTION BY FERMENTATION

Aspartic acid has been made by fermentation and chemical synthesis since the early 1950s in Japan. A national goal of improving the nutritional status of the Japanese population provided the impetus for the development of methods for the production of all the nutritionally beneficial amino acids by Japanese industry. Fermentation, using *E. coli* or *Proteus vulgaris* which were high in aspartase, has been used in Japan, China, Europe and the United States. The chemical synthesis of L-aspartic acid is quite uneconomical compared to the microbial synthesis. The largest plant ever announced for the production of aspartic acid by fermentation was the 7500 t (metric ton) capacity of Orsan in 1984. This plant, using excess fermentation capacity and presumably an *E. coli* for enzymatic activity was never known to have started production due to competition from Pfizer and Ajinomoto. When NutraSweet began producing all of its own requirements, this plant along with others was shut down. The possibility of any of the large fermentation companies moving into the aspartic acid business is always present as there is a large excess of low-quality fermentation capacity available worldwide.

The standard method for production of aspartic acid by fermentation is to add ammonium fumarate to a microbial fermentation which has reached stationary phase (Figure 1.1). The osmotic shock of the addition of the ammonium fumarate rapidly lyses the cells and the ammonium fumarate is rapidly converted to L-aspartic acid by the aspartase contained in the cells. Purification of the aspartic acid is relatively easy since the cell debris can be centrifuged away and the high concentration of aspartic acid allows precipitation by the addition of acid. Decolorizing the aspartic acid by passage of the liquor through a bed of charcoal will provide a very high quality aspartic acid, which may be improved by recrystallization. Assuming a density of 60 g/L of *E. coli* and the addition of 133 g/L of fumaric acid as its ammonium salt, the conversion would be 99.4% at pH 8.5. Calculation of the productivity of this method gives an estimate of 2.2 pounds of product per pound of *E. coli*. Even though this is an astronomical fermentation yield when compared to any of the immobilized cell methods (described below) which have been commercialized, the fermentation route is as outmoded as the synthetic route was when fermentation first began to be practiced.

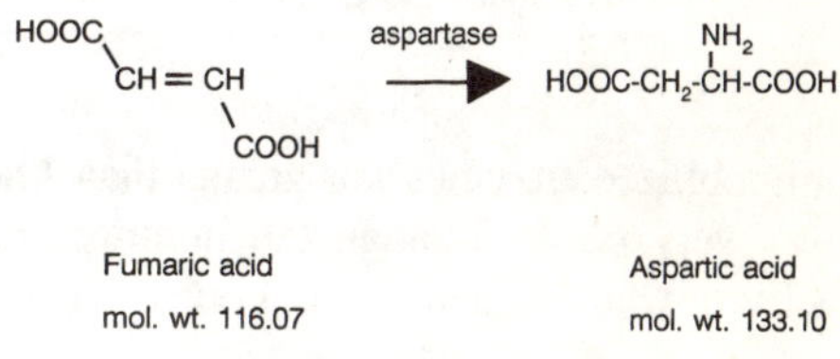

Figure 1.1. Conversion of Fumaric Acid to Aspartic Acid.

1.3 PRODUCTION BY IMMOBILIZED ENZYMES

The introduction of immobilized enzymes did not affect the production of aspartic acid greatly because the stability of aspartase was quite low and the extraction of the enzyme was not economically practical. Tosa and co-workers screened a number of enzyme immobilization procedures which had been used successfully for the immobilization of L-amino acid acylase for the immobilization of aspartase with results which ranged from 0.7 to 7% of initial activity (Table 1.1). The effect of the monomer concentration on the immobilization of aspartase was also evaluated and the highest yield of activity was found with 0.75 g of acrylamide, 40 mg of *N,N'*-methylenebisacrylamide and 80 mg of *E. coli*. It was determined by the inventors that the pH optimum was approximately 8 to 8.5. The effect of temperature on the reaction rate showed a maximum at 37 °C and a rapid drop-off to less than 20% of activity at 50 °C. The effect of Mn^{2+} on the immobilized enzyme was nearly twice that of the Mn^{2+} on the native enzyme. The heat stability of both the native and the

TABLE 1.1 Immobilization of Aspartase

Method	Aspartase available for binding (units)	Activity of immobilized preparation (μmol/h)	Yield (%)
TEAE-cellulose	4,800	336	7.0
DEAE-cellulose	4,800	276	5.8
ECTEOLA-cellulose	4,800	35	0.7
DEAE-Sephadex A-25	4,800	158	3.3
Silica gel	8,400	10	0.1
Calcium phosphate gel	4,800	230	4.8
CM-cellulose azide	700	0	0
PAB-cellulose diazonium salt	700	0	0
Polyacrylamide gel	1,440	417	29.0

immobilized enzymes was greater than 1 hour at 37 °C; however, at 50 °C the stability was very poor, although the immobilized enzyme still retained nearly 10% of its activity when heated for an hour (Figure 1.2) [1].

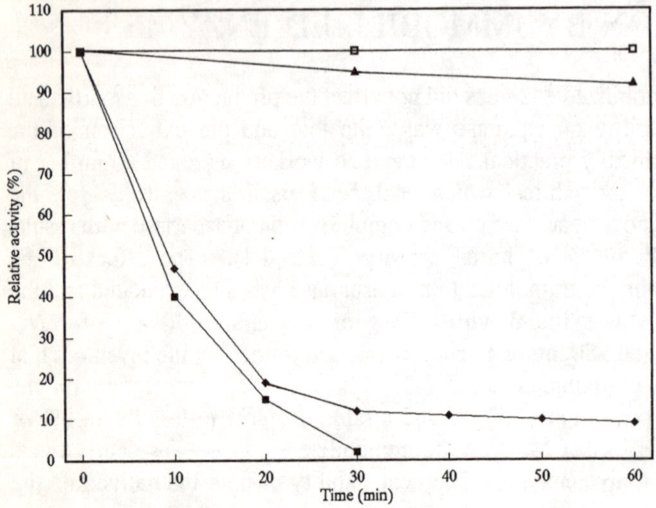

Figure 1.2. Heat Stability of Native and Polyacrylamide-Immobilized Aspartase. Key: –■– native aspartase, 37 °C; –◆– immobilized aspartase, 37 °C; –▲– native aspartase, 50 °C; –□– immobilized aspartase, 50 °C.

1.3.1 Immobilization on Polyacrylamide

A significant advance in the state of the art was introduced by Chibata and co-workers in 1974 when they successfully immobilized an *E. coli* high in aspartase activity. This process was described in a U.S. patent issued to I. Chibata, T. Tosa, and T. Sato, all of Tanabe Seiyaku. The successful production of aspartic acid was commercially exploited by the inclusion of high quantities of aspartic acid in health beverages containing quantities of amino acids, which are extremely popular in Japan. The early enzyme immobilization studies were conducted with a polyacrylamide matrix in which *E. coli* 11303 was immobilized. *E. coli* 11303 had been used previously for several years in the batch production of aspartic acid. The inventors immobilized 4.8 g of *E. coli* 11303 in 9 g of acrylamide, 480 mg of *N,N'*-methylenebisacrylamide, and 6 ml of 5% β-dimethylaminopropionitrile. This provided 120 ml of an immobilized preparation of *E. coli* which could be placed in a column over which a 1 M solution of ammonium fumarate could be passed, at pH 8.5, to provide L-aspartic acid in virtually 100% yield. The reported flow rate was 0.13 space velocities (SV) per hour [2]. The addition of a metal ion such as Ca^{2+}, Mg^{2+}, Mn^{2+} or Sr^{2+} was found to prolong the lifetime of the catalyst from 20 days to 50 days, in the case of Mn^{2+}. Variations in the structure of the acrylamide were investigated as well as the immobilization of other organisms containing high levels of aspartase, such as *Pseudomonas aeruginosa* [2].

The immobilization of *E. coli* 11303 cells in polyacrylamide gave a slight broadening of the optimum pH. The optimum temperature showed a significant increase over that of the native enzyme, however, moving to nearly 50 °C. The operational stability was better at 37 °C than at either 45 °C or 50 °C (Figure 1.3).

If 1 mM Ca^{2+}, Mg^{2+}, or Mn^{2+} was added, there was no loss of activity after 16 days when the immobilized cell column was run at a flow rate of 0.08 SV, whereas the immobilized cell preparation began decreasing dramatically in activity at day 3 without the addition of one of the divalent metals. It was found that the maximum flow rate for production of >99% aspartic acid with this preparation was 0.8 SV (Figure 1.4). In this system, the stability of the immobilized *E. coli* cells containing aspartase was directly related to the concentration of the fumaric acid substrate. At 0.5 SV/hr, no decay in the column operating activity was seen over a 40-day period, whereas at 0.8 SV/hr, the activity had dropped to approximately 82% and at 1.2 SV the activity had fallen from 80% initially to 62% (Figure 1.5) [3].

1.3.1.1 Cross-Linking

Additional studies on methods of maximizing the enzymatic activity were carried out using various agents to cross-link or to entrap the cells. Chibata and co-workers found that all of the cross-linking agents tested resulted in decreased yields of activity

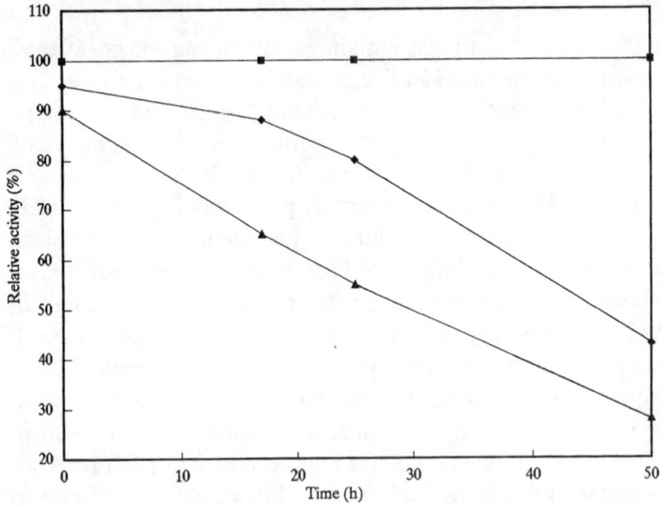

Figure 1.3. Operational Stability of Polyacrylamide-Immobilized *E. coli*. Key: –■– 37 °C; –◆– 45 °C; –▲– 50 °C.

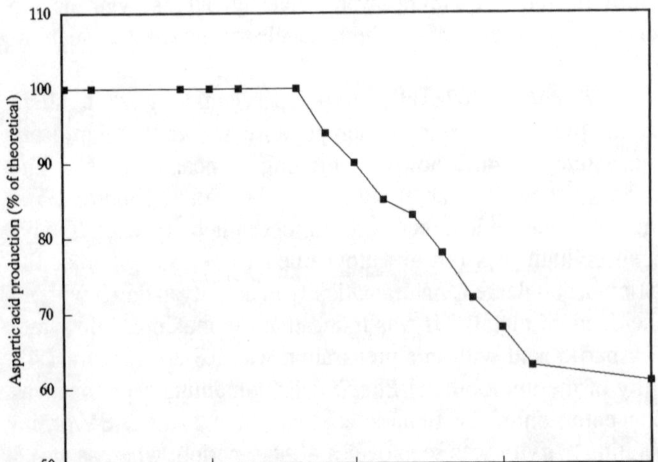

Figure 1.4. Production of Aspartic Acid as a Function of Flow Rate.

varying between 100 and 30% (Table 1.2). An evaluation of the activity of a number of microbes that contained aspartase levels near that of *E. coli* 11303 was also carried out. *Bacterium succinium, Proteus vulgaris, Pseudomonas aeruginosa,* and *Serratia marcescens* were evaluated; *E. coli* had the highest activity level, although the *Pseudomonas* had the highest yield of activity. It was also found that rinsing the immobilized cells with a solution of sodium phosphate buffer containing 0.5 M NaCl,

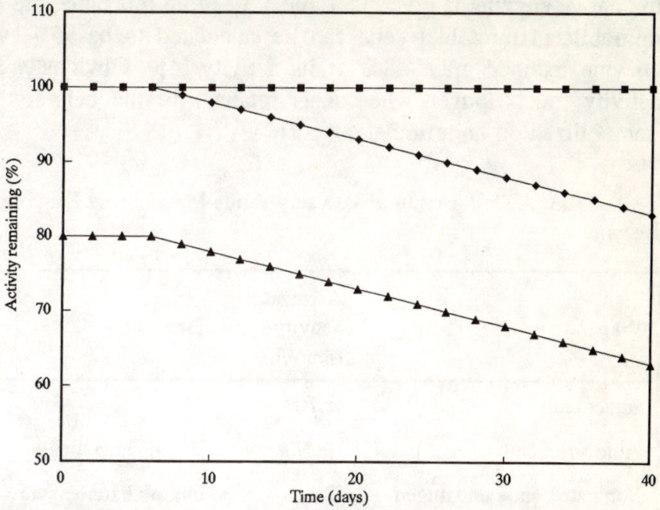

Figure 1.5. Stability of Polyacrylamide-Immobilized *E. coli* at Various Flow Rates. Key: –■– SV = 0.5; –◆– SV = 0.8; –▲– SV = 1.2.

TABLE 1.2 Immobilization of *E. coli* Cells with Acrylamides and Bifunctional Agents

Reagent	Aspartase activity	Yield (%)
Acrylamide	1320	72.9
Glutaraldehyde	620	34.2
Toluene diisocyanate	0	0
Toluene diisocyanate + hexamethylenediamine	0	0
N,N'-Methylenebisacrylamide	1220	67
N,N'-Propylenebisacrylamide	1104	60.7
Diacrylamide dimethyl ether	1048	57.6
1,2-Diacrylamide ethylene glycol	1136	62.4
N,N'-Diallyl tartardiamide	1320	72.5
Ethylene urea bisacrylamide	128	7.0
1,3,5-Triacryloyl hexahydro-*s*-triazine	128	7.0

pH 8.5, increased the activity by a factor of 5. The heat stability of the immobilized cells was maximized by the addition of 1 mM Mg^{2+} [4].

A comparison of immobilization in polyacrylamide of the enzyme, the cells and of the native enzyme is given in Table 1.3. From this table, the retained activity of the immobilized microbial cells can be calculated to be 95%, while the immobilized enzyme retained only 4.8% of its activity [4]. Obviously, the enzyme retains its activity much longer when it is retained in the cells rather than isolated and immobilized on an artificial support.

TABLE 1.3 A Comparison of Polyacrylamide-Immobilized Enzyme, Cells, and Native Enzyme

Preparation	Aspartase activity (μmol/h)	Treatment	Aspartase activity (μmol/h)
Intact cells	1,700		
Autolyzed cells	11,290	Homogenized cells	12,780
Sonicated and centrifuged	2,160	Immobilized enzyme	620
Immobilized cells	1,310	Activated immobilized cells	12,200

1.3.1.2 Kinetics

The kinetics of the batch vs. column showed a significant variation with the concentration of the substrate. This was thought to be due to the change in the diffusibility of fumaric and L-aspartic acid in the polyacrylamide gel matrix [5].

A column which allowed control of the temperature by recirculating water through an outer jacket was constructed of ten sections of 116 ml each. When samples were taken from each of the sections at various flow rates, the immobilized-cell column was found to follow zero-order kinetics between 0.1 and 1.0 M concentrations of ammonium fumarate. The half-life of the polyacrylamide-immobilized *E. coli* was found to be 120 days at 37 °C when a 1 M concentration of ammonium fumarate, containing 1 mM Mg^{2+}, was passed through the column at 37 °C for 180 days. The rate of formation of L-aspartic acid was independent of the column diameter in this system [6].

1.3.1.3 Stability

The immobilization yield of the native enzyme was examined after treatment with detergents (Table 1.4), to determine whether the complete release of the enzyme from the cell wall was the reason for low activity upon immobilization in polyacrylamide. The reason for the stability of this aspartase activity was presumed to be that the

TABLE 1.4 A Comparison of Polyacrylamide-Immobilized Enzyme, Cells, and Native Enzyme after Detergent Treatment

Preparation	Aspartase activity (μmol/h) Native	Aspartase activity (μmol/h) Immobilized	Yield (%) Post-Immob.	Stability (half-life, days)
Deoxycholate-treated cells				
Autolyzed cells	25,680	20,180	79	9
Ppt. fraction	11,080	15,220	137	13
Soluble fraction	10,550	3,140	30	8
Triton-treated cells				
Autolyzed cells	24,420	21,840	89	9
Ppt. fraction	12,050	16,970	141	9
Soluble fraction	11,320	6,190	55	5

enzyme was bound to cellular particles or to membranes within the cells, since the isolated aspartase was not nearly as stable when immobilized via covalent binding to CNBr-activated Sepharose™ (2.3% activity), ionic binding to TEAE cellulose (1%) or hydrophobic binding to aminopentyl Sepharose™ (1.3%) [7].

1.3.1.4 Improvements

A minor reference on a major improvement in the method of culturing *E. coli* was mentioned in passing in a discussion of the lack of transferable activity to polyacrylamide immobilization. By culturing in a medium which had corn steep liquor as its main nitrogen source, Nishida *et al.* had consistently obtained 20,000–25,000 units of aspartase activity per gram of cells. When Meast was included at 2% in the culture medium, the activity increased to over 100,000 units of activity per gram of cells. Immobilization of either the 25,000 U/g cells or the 100,000 U/g cells in polyacrylamide gave an immobilized activity of 19,000 units per gram of cells. In comparison with their earlier work, the immobilization yield was 18.7% whereas it had been 72.9%. This difference in immobilized activity was due solely to the ability to increase the activity of the cells, but an inability to retain the increased activity when the cells were immobilized in polyacrylamide [7].

The engineering group at Tanabe continued to work on improving the performance of their system. They found that the half-life of the catalyst using the polyacrylamide method of immobilization was approximately 300 days at 26 °C; 220 days at 30 °C; 130 days at 33 °C; 110 days at 35 °C, 90 days at 37 °C, and less than 10 days at temperatures of 40 °C and above. Useful activity was obtained at temperatures as high as 60 °C with the activity decreasing to less than 1% after 12 days. Operating procedures were set to obtain the desired conversion when the initial enzyme activity was higher than required for the productivity of the plant. The industrial use of temperature control to attain maximum plant production was put in place for the polyacrylamide-immobilized cells [8].

1.3.2 Immobilization on κ-Carrageenan

Having exhausted the potential with polyacrylamide but still wishing to improve the process, the group at Tanabe began evaluation of other methods of entrapping the *E. coli* cells. They evaluated κ-carrageenan and locust bean gum initially, where they found that the activity yield after immobilization was considerably higher for either of these systems than it was for polyacrylamide (Table 1.5). They also found that they could cross-link the immobilized cell preparation chemically, especially with glutaraldehyde and hexamethylenediamine, and increase the half-life substantially while significantly decreasing the retained activity. The half-life in these systems was strikingly prolonged as compared to the polyacrylamide system which had a half-life of 120 days at 37 °C and a relative productivity of 100. κ-Carrageenan showed marked improvement in the immobilization of these new higher activity *E. coli*, more than doubling the activity over that obtained in polyacrylamide with the low activity *E. coli*. However, when κ-carrageenan was cross-linked with both hexamethylenediamine and glutaraldehyde, the half-life increased to 680 days with a relative productivity of 1498. κ-Carrageenan and locust bean gum together cross-linked with both hexamethylenediamine and glutaraldehyde gave a half-life of 650 days or a relative productivity of 1214 [9].

The characteristics of the κ-carrageenan–immobilized catalyst were significantly improved when compared to those of the polyacrylamide-immobilized catalyst. Most exceptional were the lifetime and the fact that enhanced flow rates could be obtained. The pH optimum of the κ-carrageenan–immobilized enzyme was at 9.0. The native enzyme had a pH optimum of 9.5. The optimum temperature for the native enzyme was at 50 °C, while the κ-carrageenan–immobilized microbial cells without cross-linking showed an optimum at 55 °C. The cells which had been crosslinked with glutaraldehyde showed a broad optimum from 55 to 60 °C. The cells which had been cross-linked with glutaraldehyde and hexamethylenediamine showed an even broader optimum, from 55 to 65 °C.

TABLE 1.5 A Comparison of Polyacrylamide-, Carrageenan-, and Carrageenan/Locust Gum-Immobilized Cells and the Effects of Cross-Linking

Preparation	Treatment	Aspartase activity (μmol/h) Immobilized	Yield (%) Post-immob.	Stability (half-life, days)	Relative productivity
E. coli	none	101,710			
Polyacrylamide	none	19,000	19	120	100
κ-Carrageenan	none	56,340	56	70	174
κ-Carrageenan	ga*	37,460	37	240	397
κ-Carrageenan	ga*/hex*	49,400	49	680	1,498
κ-Carrageenan/ locust bean gum	none	54,310	54	40	96
κ-Carrageenan/ locust bean gum	ga*/hex*	46,360	46	650	1,214

* ga = glutaraldehyde; hex = hexamethylenediamine

1.3.2.1 Stability

The heat stability was assessed by keeping the κ-carrageenan–immobilized cells at an appropriate temperature for 30 minutes, after which the cells were cooled and the aspartase activity was determined. The temperature stability for the native enzyme showed a decrease at 28 °C which continued rapidly until the activity was at 0 at 55 °C. The microbial cells began to decrease in activity at 45 °C and dropped to below 10% activity at 55 °C. The cells which had been immobilized with κ-carrageenan, cross-linked with glutaraldehyde or cross-linked with glutaraldehyde and hexamethylenediamine showed similar stabilities. Each began dropping in activity at 55 °C and fell to approximately 20% activity at 65 °C (Figure 1.6).

The effect of flow rate on the κ-carrageenan–immobilized *E. coli* was also investigated. The κ-carrageenan–immobilized microbial cells without cross-linking gave 100% conversion of a 1 M solution of ammonium fumarate at 2.04 SV/hr. The cells which had been cross-linked with glutaraldehyde gave 100% conversion at 1.20 SV/hr, and the cells which had been crosslinked with glutaraldehyde and hexamethylenediamine gave 100% conversion at 1.75 SV/h (Figure 1.7).

The effect of the pH of the substrate on the operational stability of the κ-carrageenan–immobilized *E. coli* was assessed as shown in Table 1.6; the maximum stability was achieved at pH 7.0 with cells which had been immobilized and cross-linked with glutaraldehyde and hexamethylenediamine [10].

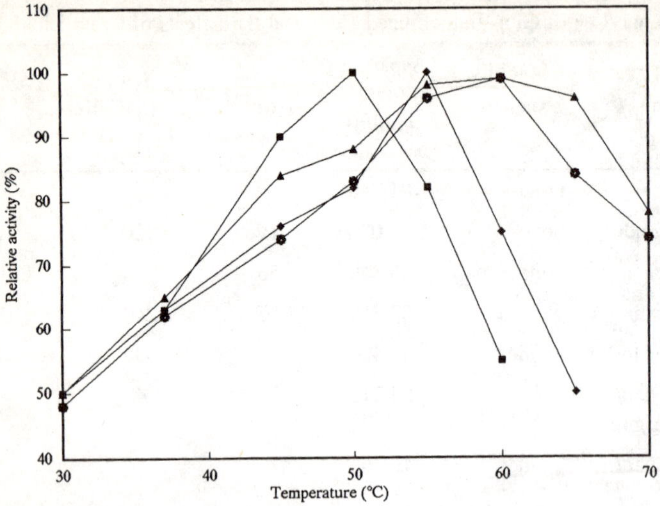

Figure 1.6. Effect of Temperature on the Activity of Various Enzymes and Immobilized Cell Treatment. Key: –■– native enzyme; –◆– polyacrylamide-immobilized cells; –⊕– glutaraldehyde-treated, polyacrylamide-immobilized cells; –▲– glutaraldehyde/hexamethylenediamine-treated, polyacrylamide-immobilized cells.

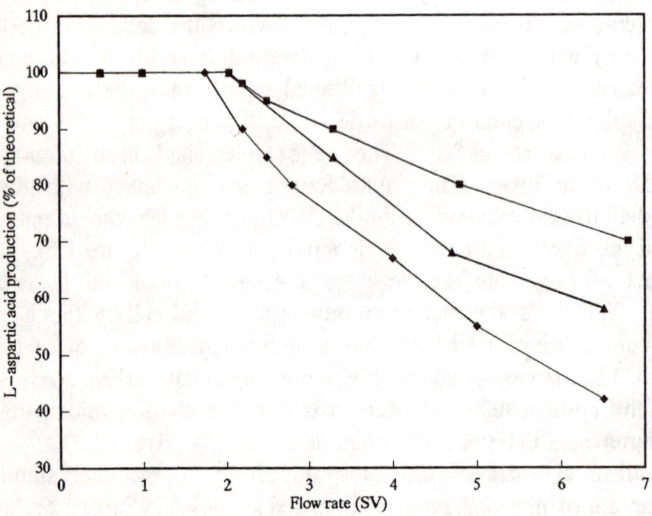

Figure 1.7. Effect of Flow Rate on Various Cell Preparations. Key: –■– polyacrylamide-immobilized cells; –◆– glutaraldehyde-treated, polyacrylamide-immobilized cells; –▲– glutaraldehyde/hexamethylenediamine-treated polyacrylamide-immobilized cells.

TABLE 1.6 The Effect of the pH of the Substrate on the Operational Stability of the κ-Carrageenan–Immobilized *E. coli*[*]

Substrate pH	Stability, *E. coli* with: κ-carrageenan	κ-carrageenan glutaraldehyde	κ-carrageenan glutaraldehyde/ hexamethylenediamine
7.0	95	238	693
8.0	93	240	675
8.5	92	237	682
9.0	54	132	391
9.5	27	71	136
10.0	11	21	48

[*] Ammonium fumarate, 1.5 M, was passed through a 12-ml column of the indicated preparation at 0.8 SV/h at 37 °C.

1.3.2.2 Cross-Linking

If the preparations of κ-carrageenan were homogenized and treated with Triton X-100 to activate the cells, it was found that the activity of the uncross-linked cells was 88% of the original value. κ-Carrageenan–immobilized, glutaraldehyde–cross-linked cells treated in this manner retained only 62% of their activity, and κ-carrageenan-immobilized cells cross-linked with both glutaraldehyde and hexamethylenediamine retained 81% of their activity. Undoubtedly, the particle size and diffusional constraints account for part of this discrepancy in the activity, whereas the cell disruption by the surfactant does not appear to have been a viable alternative due to stability losses [10].

1.3.2.3 Further Improvements

Workers at Tanabe were also able to select strains having higher levels of aspartase activity [11]. The activity of these strains of *E. coli* was approximately 7 times that of the parent strain of *E. coli* 11303; however, the fumarase activity of some strains was too high for them to be of use. When a new strain of *E. coli*, designated EAPc-7, was immobilized in κ-carrageenan, the fumarase activity was quite stable and accounted for 2% of the total fumaric acid disappearance. The effect of pH on the EAPc-7 cells was investigated and it was found that lowering the pH to 4.9 was most effective for eliminating fumarase activity without affecting aspartase activity. This effect was then combined with heat treatment. After 30 minutes at 45 °C at pH 4.9, the fumarase activity was reduced to 15% of its original activity. The side reaction of fumarase activity in these cells could be eliminated adjusting the pH to 4.9 and raising the temperature of the reaction to 45 °C for 1 hour. The half-life of the selected *E. coli* in this immobilized system was 126 days at 37 °C (Table 1.7). No data was given for the use of cross-linking agents such as glutaraldehyde and hexamethylenediamine with this strain [12].

TABLE 1.7 Comparison of the Aspartic Acid Productivity of κ-Carrageenan–Immobilized *E. coli* Strains

Preparation	Aspartase activity (μmol/h/g cells)	Half-life (37 °C, days)	Relative productivity (%)
E. coli 11303	56,300	70	100
Untreated EAPc-7	203,060	70	361
EAPc-7, 45 °C, 1 hr, pH 4.9	192,140	126	614

Finally, scientists at Tanabe worked on the production of higher levels of aspartase in *E. coli* and were able to clone the aspartase gene (*asp*A) of *Escherichia coli* into a multi-copy vector, pBR322, thus overproducing aspartase. The level of aspartase was reported to be 30-fold higher than in the wild strain. However, the industrial use of the genetically engineered strain was not possible because the plasmid was not stable [13].

1.3.3 Immobilization on Polyurethane

My group at W. R. Grace & Co. was the first to use polyurethane for immobilization of cells. We incorporated whole cells of *E. coli* in a hydrophilic polyurethane foam for the production of aspartic acid. Chibata and co-workers had used toluene diisocyanate as a cross-linking agent and found no retention of activity. Others at Grace had been using a hydrophilic polyurethane foam for the immobilization of enzymes for a number of years. This polyurethane foam was invented by Louis L. Wood [14] and its use in immobilizing enzymes was heavily patented although not highly published by scientists at W. R. Grace & Co. Materials which are nearly identical to it (Toyo Rubber Co.) have been widely used by Japanese workers for immobilization of enzymes and microbial cells [15]. The value of the hydrophilic polyurethane in the production of L-aspartic acid was not superior to that of the Tanabe system, however.

When *E. coli* cells high in aspartase activity (68,000 U/g) were immobilized in polyurethane prepolymer (Hypol™) in equal weights followed by curing, it was found that the foam both entrapped and covalently bound the microbial cells. The effect of the isocyanate content varied in relation to the number of milliequivalents of isocyanate in the prepolymer (Table 1.8). This system had extremely high retention of activity (98%). The effect of pH paralled that seen in the case of κ-carrageenan, being 20 to 50% lower at pH 9.0 than at pH 8.3. The use of 1.5 M ammonium fumarate as opposed to 1.0 M lowered the average rate approximately 8%. Complete conversion of a 1.0 M solution of ammonium fumarate to aspartic acid was accomplished at 3.77 SV/h [16]. Unfortunately, the half-life of this system was only 37 days. This would place its productivity at 3.58 times as high as the polyacrylamide system, but its half-life is 0.31 times that of the polyacrylamide system. The major advantage was the ease of immobilization on a small scale; however, no large-scale methods were developed.

TABLE 1.8 The Effect of Isocyanate Content on the Aspartase Activity of Polyurethane-Immobilized *E. coli* Cells

Prepolymer	Isocyanate content (meq/g)	Aspartase activity (units/g *E. coli*)
Free cells		68,000
A	2.16	48,966
B	1.63	61,035
C	2.48	46,035
D	1.79	68,017

Other investigators at W. R. Grace & Co. improved on the system of isolation of aspartic acid by using maleic acid to lower the pH and obtain the crystallization of aspartic acid. The mother liquors from the crystallization could then be sent through a well-known isomerization process to prepare fumaric acid which could then be recycled through the aspartic acid column. The use of this method of aspartic acid recovery would increase the material utilization and decrease the disposal problems; however it has never been used in commercial production [17].

1.3.4 Immobilization on Polyazetidine

The most advantageous system for the production of aspartic acid appears to be the one developed by us at Purification Engineering, Inc., using polyazetidine for immobilization of an *E. coli* high in aspartase activity. The *E. coli* was immobilized by mixing equal parts of aqueous polyazetidine solution (Hercules Polycup™ 172, 12% water) (Figure 1.8) and *E. coli* ATCC 11303, stirring vigorously to insure a homogeneous dispersion and then pouring the materials onto an appropriate matrix. These materials were prepared as membranes, or distributed onto various supports including molecular sieve beads, Amberlite™ IRA 45 ion exchange resin, or Amberlite™ IRA 938 ion exchange resin (Table 1.9). IRA 938 gave superior results in respect to the number of moles of aspartic acid per hour per kilogram of *E. coli* produced, with the exception of a film which was difficult to place in a column.

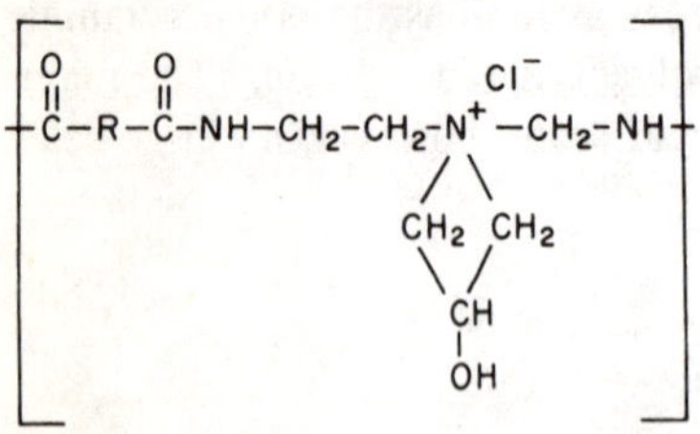

Figure 1.8. Idealized Polyazetidine Prepolymer. Typically, R = $-(CH_2)_4-$

Due to the resin characteristics, a total of 12.75 moles of aspartic acid per hour per liter of catalyst bed could be obtained. The catalyst maintained the characteristic properties of enzymatic activity of free or immobilized *E. coli*. The catalyst gave complete conversion of the 1.5 M ammonium fumarate at 5.5 SV/h (Figure 1.9).

TABLE 1.9 Aspartase Activity of Polyazetidine-Immobilized *E. coli*

Support	Aspartase activity $[mol \cdot h^{-1} \cdot (l\ catalyst\ bed)^{-1}]$	Aspartase activity $[mol \cdot h^{-1} \cdot (kg\ E.\ coli\ wet\ wt)^{-1}]$
Film	31.9	55.0
Molecular sieve	0.6	11.5
IRA 45 ion exch. beads	3.8	32.5
(ditto, 2nd expt.)	3.9	26.7
IRA 938 ion exch. beads	12.75	55.9
None	–	57.6

Even at 8 SV/h there was no decrease in activity, thus showing that the catalyst was stable, regardless of the flow rate of the substrate solution. This flow rate is significantly higher than the 2–3 SV/h for the other systems reported in the literature.

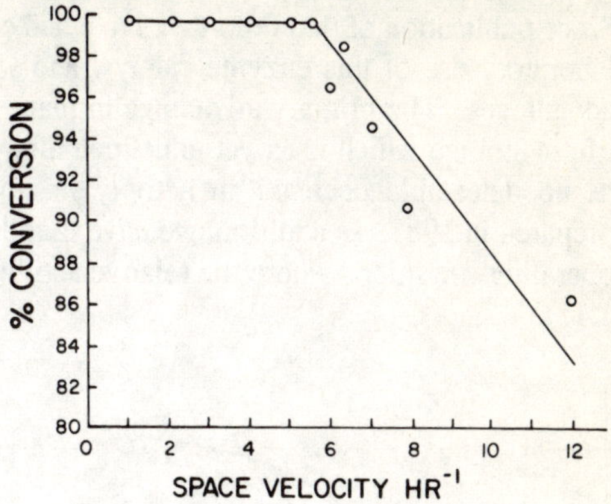

Figure 1.9. Effect of Flow Rate on Polyazetidine-Immobilized *E. coli* Activity.

1.3.4.1 Processing and Scale-up

The catalyst preparation was easy to scale up, and several hundred liters of catalyst were prepared by the same process as that used for laboratory preparation. The polyazetidine prepolymer was cured by drying with a gentle flow of 25%-humidity air. Particles of approximately 0.3 mm diameter were obtained, which were quite hard. These beads, when immersed in a substrate solution, swelled only slightly (approximately 10% of the total bed volume).

The flow characteristics of this material were extremely good, as a 15- × 122-cm chromatography column with a 0.31-cm orifice generated a pressure of only 19 psi at 60 SV/h. The substrate was, of course, not completely converted to aspartic acid and most of the pressure obtained was due to the size of the orifice in the inlet and outlet. Approximately 97% of the free cell activity was immobilized and retained after 10 washings. The half-life was estimated to be 611 days (Figure 1.10) [18].

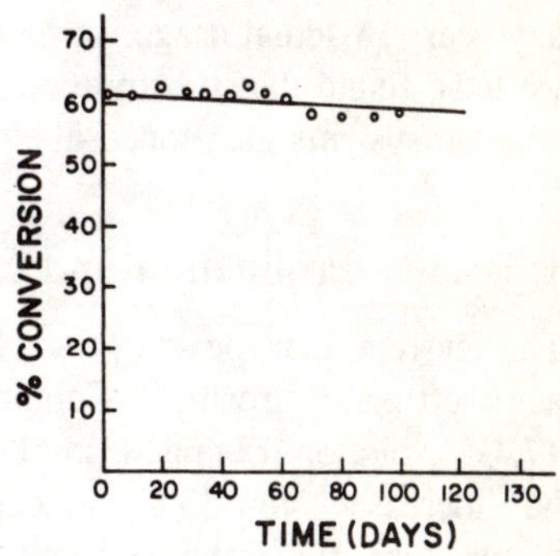

Figure 1.10. Stability of Polyazetidine-Immobilized *E. coli* Activity.

1.3.4.2 Stability

Since publication of that report we have had considerable experience in both plant and laboratory use of this enzyme catalyst and we have found that it has several unique advantages. The primary advantage in plant operation is that the material has a half-life in storage which is as yet undefined after 8 years of storage at room temperature, as no detectable decrease in activity has been noted since the material was first prepared in 1981. In addition, we have found a 3-month half-life at 55 °C under plant operating conditions where the relative activity is twice that obtained at 25 °C (Figure 1.11).

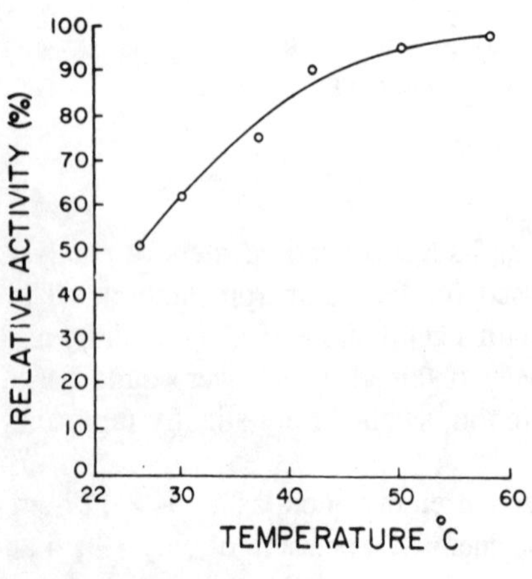

Figure 1.11. Effect of Temperature on Polyazetidine-Immobilized *E. coli.*

The stability in storage and use allows us to operate a plant on a campaign system, with no loss of activity of the stored catalyst during the storage period and no down time since the material is stored in converted substrate solution. A column of 208 liters will produce 4 million pounds of aspartic acid per year on a 24-hour day, 330-day year. In actual usage we have seen a column half-life in excess of 600 days, and we have found that substrate concentrations (1.5 M) which were deleterious to other enzyme systems mentioned above were not a problem.

1.3.4.3 Production and Yields

The system can be recycled by the addition of fumaric acid to the aspartic acid effluent stream. Fumaric acid is near its limit of solubility at the 17.4% solution obtained in 1.5 M concentration. However, the solution can be increased to 33% concentration of aspartic acid by the addition of ammonium fumarate and fumaric acid to the aspartic acid effluent stream (19.8%) and passage through the catalyst column an additional time. No decrease in stability or activity is seen; however, increased yields are obtained due to decreased losses in crystallization with a 33% solution of aspartic acid.

Plant shut-down requires no extra care. The cessation of substrate flow across the column puts the catalyst in storage conditions, thus allowing the greatest possible latitude in the production of aspartic acid. It should be noted that under these conditions 1 lb of enzyme will produce in excess of 220,000 lb of aspartic acid, thus establishing this process as the most productive enzyme process in the world at this time.

1.4 REFERENCES

1. Tosa, T., Sato, T., Mori, T., Matsuo, Y., Chibata, I., *Biotechnol. Bioeng. 15*, 69 (1973).
2. Chibata, I., Tosa, T., Sato, T., U.S. Pat. 3,791,926 (1974).
3. Tosa, T., Sato, T., Mori, T., Chibata, I., *Appl. Microbiol. 27*, 886 (1974).
4. Chibata, I., Tosa, T., Sato, T., *Appl. Microbiol. 27*, 878 (1974).
5. Takamatsu, S., Yamashita, K., Sumi, A., *J. Ferm. Technol. 58*, 129 (1980).
6. Sato, T., Mori, T., Tosa, T., Chibata, I., Furui, M., Yamashita, K., Sumi, A., *Biotechnol. Bioeng. 17*, 1797 (1975).
7. Tosa, T., Sato, T., Nishida, Y., Chibata, I., *Biochim. Biophys. Acta 483*, 193 (1977).
8. Takamatsu, S., Ueba, M., Yamashita, K., *J. Chem. Eng. Jpn. 17*, 647 (1984).
9. Nishida, Y., Sato, T., Tosa, T., Chibata, I., *Enzyme Microb. Technol. 1*, 95 (1979).
10. Sato, T., Nishida, Y., Tosa, T., Chibata, I., *Biochim. Biophys. Acta 570*, 179 (1979).
11. Nishimura, N., Kisumi, M., *Appl. Environ. Microbiol. 48*, 1072 (1984).
12. Umemura, I., Takamatsu, S., Sata, T., Tosa, T., Chibata, I., *Appl. Microbiol. Biotechnol. 20*, 291 (1984).
13. Komatsubara, S., Taniguchi, T., Kisumi, M., *J. Biotechnol. 3*, 281 (1986).
14. Wood, L. L., Hartdegen, F. J., Hahn, P. A., U.S. Pat. 3,929,547 (1979).
15. Fukushima, S., Nagai, T., Fujita, K., Tanaka, A. Fukui, S., *Biotechnol. Bioeng. 20*, 1465 (1978).
16. Fusee, M. C., Swann, W. E., Calton, G. J., *Appl. Environ. Microbiol. 42*, 672 (1981).
17. Sherwin, M. B., Blouin, J. J., U.S. Pat. 4,560,653 (1985).
18. Wood, L. L., Calton, G. J., *Bio/Technol. 2*, 1081 (1984).

ENZYMATIC METHOD FOR L-SERINE PRODUCTION

David M. Anderson and Humg-Yu Hsiao

CONTENTS

2.1 BACKGROUND

The worldwide markets for L-serine have been relatively small and the purchase price has been relatively high compared to several other amino acids. L-Serine is used commercially in therapeutic parenteral solutions, other pharmaceutical applications, and in some cosmetics. The establishment of much more efficient production methods for L-serine could lower the cost and expand the market. For example, potential applications for L-serine include use as an optically active precursor for the enzymatic synthesis of other amino acids including L-tryptophan [1, 2], L-cysteine [3, 4], and L-tyrosine [5, 6].

A number of methods have been developed for the production of L-amino acids. They include extraction from hydrolyzed protein, microbial fermentation using an inexpensive carbon source, fermentative bioconversion, enzymatic synthesis from chemically produced precursors, and straight chemical synthesis with or without additional enzymatic steps to provide the optical resolution. All of these methods have either been used commercially or investigated as production methods for L-serine.

The oldest method, extraction from protein hydrolysates, is still used today to supply L-serine. For example, hair is an efficient protein source used for the production of L-cysteine, and L-serine is one of the by-products of that process. Fermentation is the method of choice for production of many amino acids when high titer can be obtained from inexpensive carbon feedstocks. However, high level L-serine production by fermentation has apparently not been established [7]. In typical L-serine fermentation results, one report describes an *Arthrobacter* strain capable of producing about 5 g/l of L-serine from *n*-alkanes, but the addition of 5–10 g/l of L-threonine was used in the medium [8]. *Brevibacterium* or *Corynebacterium* mutants resistant to sulfaguanidine were reported to produce about 2 g/l of L-serine in two- to seven-day fermentations [7].

There are numerous reports of L-serine production through the fermentative bioconversion of glycine, or glycine plus an additional source of carbon such as methanol, glyoxylic acid or formic acid [9]. Metabolically, glycine is usually synthesized from L-serine with the release of a one-carbon compound in the form of N^5,N^{10}-methylenetetrahydrofolate. In the instances where methanol, glyoxylic acid or formic acid are added to the bioconversions, these compounds provide a source of one-carbon groups to shift the equilibrium toward L-serine synthesis from glycine. In *Sarcina albida*, glycine synthase [10] was shown to efficiently degrade glycine to formaldehyde and CO_2. Thus, glycine can be used as the sole precursor to L-serine in *Sarcina albida* bioconversions by also serving as the one-carbon source.

The common problems with the glycine bioconversion method for L-serine production are the toxicity of glycine to the growth of the microorganism and the rapid degradation of the L-serine that is formed. This results in a low yield of L-serine based on glycine and low L-serine titers in the fermentation broth. For example, *Pseudomonas* 3ab produced only 4.7 g/l L-serine from glycine and methanol [11].

These problems were partially solved by a temperature-sensitive mutant of *Pseudomonas* M-31 [12, 13]. After growing the mutant to high cell density at 30 °C, glycine was added and the temperature was raised to 38–40 °C. An L-serine titer of 6.89 g/l from 12 g/l of glycine was reported, with a yield based on glycine of about 70%. The authors suggested that at 38–40 °C the enzyme activity responsible for L-serine degradation in the mutant was greatly reduced. In another example, a *Sarcina albida* mutant that was both defective in L-serine degradation and glycine-resistant was able to produce 21 g/l L-serine from glycine [14]. However, an elaborate recovery scheme was required to isolate the L-serine from the high concentration of residual glycine at the end of the bioconversion. Based on the state of the art for L-serine production by fermentation or bioconversion, it appeared that new and improved stereospecific L-serine production methods were needed.

We have taken a different approach to the use of glycine as a feedstock for L-serine production similar to a previously described threonine aldolase method [15]. Rather than use a fermentative bioconversion method, an enzyme reactor based on the enzyme *serine hydroxymethyltransferase* (*SHMT*) was developed [16, 17]. The basic reaction scheme is as follows:

$$\textbf{formaldehyde + tetrahydrofolate (FH}_4\textbf{)} \longrightarrow \textit{N}^5\textbf{,}\textit{N}^{10}\textbf{-methylenetetrahydrofolate (m-FH}_4\textbf{)} \quad (1)$$

$$\textit{N}^5\textbf{,}\textit{N}^{10}\textbf{-methylenetetrahydrofolate + glycine} \longrightarrow \textbf{L-serine + FH}_4 \quad (2)$$

The net reaction is thus the synthesis of L-serine from formaldehyde and glycine. The first step, (1), is a nonenzymatic reaction with a favorable equilibrium that has been extensively studied by Kallen and Jencks [18]. The second step, (2), is catalyzed by the SHMT, which contains a bound pyridoxal phosphate cofactor. In order to make this enzyme reactor method practical, several criteria had to be met, including:

- a source of inexpensive SHMT catalyst;
- a minimal requirement for tetrahydrofolate and pyridoxal phosphate cofactors;
- prevention of formaldehyde inactivation of enzyme;
- high rate of L-serine synthesis;
- high percent conversion of glycine; and
- high final titers of L-serine (to simplify the purification).

The description of and rationale behind many of the developments that allowed this enzyme reactor method to meet the desired criteria is the topic of this chapter.

2.2 BIOCATALYST DEVELOPMENT

The enzyme serine hydroxymethyltransferase (SHMT) from mammalian sources is well-studied [19]. Less has been published on the characteristics of microbial SHMT enzymes, but their characteristics are similar. In order to develop an economical L-serine production process it was essential to keep the biocatalyst cost as low as possible. We chose to use the SHMT enzyme from *E. coli* encoded by the *gly*A gene [20]. Purified SHMT has a relatively low turnover number (specific activity 2.4 U/mg), thus gene copy number amplification by cloning on a stabilized plasmid was used to increase expression and obtain acceptable enzyme activity in crude extracts. For further cost reduction, a *Klebsiella* strain capable of growth to high density on inexpensive carbon substrates was used as host for the amplified gene. Also, an autolysis mechanism was built into the strain to allow simplified recovery of cell-free SHMT with a minimum number of unit operations. The detailed aspects of the biocatalyst production strain are described below.

2.2.1 Strain Development

2.2.1.1 SHMT Gene Cloning

The *gly*A gene of *E. coli* that codes for SHMT was cloned into pBR322 on an *Eco*RI-*Sal*I restriction fragment, then subcloned as a *Sma*I, *Bcl*I fragment onto a high copy number plasmid to yield pGX139 [2, 21, 22]. The published DNA sequence for *gly*A [23] was in agreement with our sequence data.

The expression of SHMT in *E. coli* cells with pGX139 was found to be significantly elevated over cells without plasmid pGX139, especially in cultures grown to stationary phase in a minimal salts medium. Therefore, it was not necessary to further manipulate the *gly*A gene by changing the promoter or regulatory mechanisms to increase expression. However, the pGX139 plasmid was further modified as shown in Figure 2.1.

2.2.1.2 Plasmid Stabilization

First, an improved plasmid stabilization method was needed. Plasmid pGX139 was maintained in the laboratory strains through selection for the β-lactamase gene that provides ampicillin resistance. This selection mechanism is unsuitable for fermentation since extracellular β-lactamase enzyme rapidly depletes the antibiotic in the medium. Plasmids derived from ColE1, such as pGX139, are lost at a rate of 1% per generation [24] without positive selection. Additional negative selection pressure of high-level gene expression from the plasmid can produce greater instability. Plasmid stabilization could have been achieved through use of a gene that codes for another antibiotic resistance such as tetracycline or insertion of a *par* locus [24], but these methods would have increased the fermentation medium cost or lowered expression by decreasing the plasmid copy number. Instead, a plasmid stabilization scheme was utilized that is based on nutritional selection and the tryptophan operon. The complete tryptophan operon was placed on plasmid pGX139, giving plasmid pGX2236

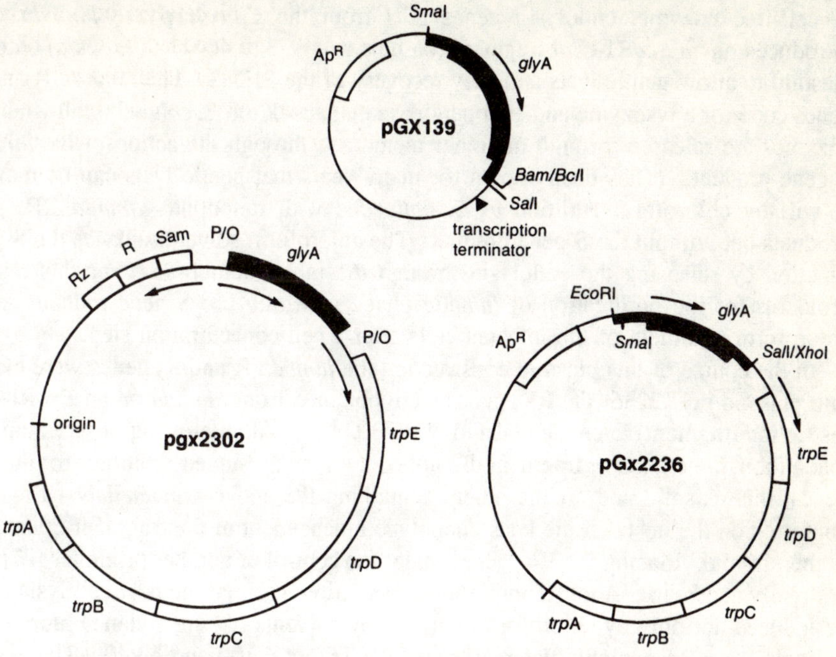

Figure 2.1. Plasmids Relevant to the Construction of SHMT Production Strains. Construction details of plasmids can be found in [2], [21], and [22].

[2, 22] that can be stably maintained in a *trp* mutant host. The regulation of tryptophan synthesis is very tight; thus a plasmid-borne multicopy operon is not sufficient to cause the overproduction and secretion of tryptophan [25]. Therefore, in media without tryptophan there is no cross-feeding and only the *trp* mutant cells that maintain the plasmid are able to grow. Because of the relative instability of tryptophan compared to other amino acids, even complex media can be easily formulated without tryptophan. For example, acid hydrolysates of protein do not contain tryptophan. In the case of SHMT production, however, a clean minimal salts medium was formulated so that fermentation broth containing enzyme released from cells could be used directly in the bioreactors. The use of the tryptophan operon to stabilize plasmids in tryptophan auxotrophs has proven to be a generally useful technique for production strains [26].

2.2.1.3 Autolysis

In the development of serine bioreactors, it was quickly learned that the SHMT enzyme was quite stable and that the cell-free extract was a good catalyst. Also, avoiding whole cells in the bioreactor reduced the solution viscosity and greatly simplified the subsequent L-serine purification. In order to facilitate the preparation of cell-free enzyme, endolysin genes [27] from the *E. coli* phage *lambda* were introduced on an *Eco*RI-*Cla*I fragment (base pairs 44973 to 46441 of *lambda*) onto the plasmid to allow gentle lysis and easy recovery of the SHMT. The *lambda* R and R_z genes code for a lysozyme and endopeptidase that attack the *E. coli* cell wall when the enzymes are released through the inner membrane through the action of the *lambda* S gene product. It has been known for many years that gentle lysis can be induced at will by chloroform addition to *E. coli* cells with functional *lambda* RR_z gene products but without the S gene product. The chloroform solvent replaces the S gene function by releasing the endolysin through the inner membrane. For this reason, protocols for the purification of *lambda* phage generally use S gene mutants and a chloroform treatment of the infected cells after a cell concentration step.

In the course of this plasmid construction the *lambda* R and R_z genes were cloned into plasmid pGX2236 (the RR_z genes were obtained from *lambda* on an *Eco*RI-*Cla*I restriction fragment) to create plasmid pGX2302 [22]. An interesting observation was that although the *lambda* fragment did not contain the presumed promoter for the RR_z genes, this plasmid and all the others containing the above *lambda* DNA fragment conferred on the host cell the lysis phenotype independent of the fragment orientation in the plasmid. Placing the RR_z genes under the control of another promoter [27] was not really necessary. Another important observation was that the host cell lysis could be induced not only by chloroform, but also by solvents like methylene chloride and a whole series of nonionic detergents such as Triton X-100 and NP40. This gave a range of choices to initiate cell lysis since it may not be desirable to use a volatile and toxic compound such as chloroform under some conditions. The *lambda* phage endolysin genes also functioned well in the *Klebsiella aerogenes* biocatalyst production host (see below).

After the biocatalyst fermentation, cell lysis was induced directly in the fermentation broth either by adding nonionic detergent or simply by allowing the cells to mix for an additional twenty-four hours. The silicone antifoam used in the fermentation broth may have a slight solvent effect that contributes to the initiation of the endolysin function. The broth with the lysed cells was filtered or passed through a Westphalia-type low-speed continuous centrifuge to remove the cell debris, and then diluted approximately one:four into the bioreactor. The SHMT lysate prepared by this method was much cleaner and was obtained in higher yield than similar preparations where the cells were lysed mechanically with a French Press or Matin-Gaulin apparatus. For example, the supernatants from mechanically lysed cells had much higher nucleic acid content and about ten times more solids (cell debris). The insoluble cell material after endolysin-generated lysis had better sedimentation characteristics than did mechanical lysates, presumably because the remaining cell fragments were much more intact. The use of the endolysin genes provided a better biocatalyst and reduced the number of unit operations for biocatalyst preparation.

2.2.1.4 Host Strain

A number of Gram-negative strains were evaluated as possible hosts for SHMT production. *Klebsiella aerogenes* (previously called *Aerobacter aerogenes*) was chosen because it has favorable growth characteristics. The strain is a soil microbe that grows to high density and will use inexpensive carbon sources such as starch and molasses. Also, a mutant (LV025) with reduced serine dehydratase activity was available [28]. Although serine dehydratase is much less stable than SHMT and serine dehydratase activity could have been inactivated before use of SHMT-containing lysates, the mutant strain eliminated the need for this step. In order to use the tryptophan operon stabilization mechanism with this strain, LV025 was treated with the mutagen nitrosoguanidine, and tryptophan auxotrophic strains were selected. One mutant, GX1705, was found to be defective in tryptophan synthase with an undetectable reversion frequency. Strain GX1705 retained good growth characteristics and was chosen as the host strain for plasmids and the biocatalyst fermentation development program.

Plasmid pGX2302 was placed into *Klebsiella aerogenes* GX1705 by transformation or by mobilization from *E. coli* by F factor [29]. In the latter case, the resulting strain (GX3036) is considered exempt from the NIH Guidelines for Recombinant DNA Research [30] under Section III-D-3 because the *Klebsiella* transformant was produced by natural means.

2.2.2 Fermentation for SHMT Production with Strain GX3036

As described above, a minimal salts medium was chosen for fermentation of GX3036 for the production of SHMT. This provided the selection for plasmid stabilization due to the lack of tryptophan. The lack of complex components also facilitated the direct use of fermentation broth in the bioreactor after the release of SHMT from the cells by autolysis without causing any serious problems for the purification of L-serine. A typical fermentation for the production of SHMT with GX3036 is shown in Figure 2.2. In this case a cell density of approximately 25 grams dry cell weight per liter was obtained with a specific activity of 900 units (β-phenylserine assay [19]) per gram dry cell weight. Based on the specific activity of purified SHMT there was approximately 4 g/l SHMT, representing 13.5% of the dry cell weight or 27% of the total protein. The dry cell weight in this fermentation frequently reaches 30 g/l with an equivalent increase in enzyme concentration.

An interesting aspect of this fermentation is that the SHMT specific activity increases throughout the cycle. It is not clear whether the SHMT expression is being induced or if another effect such as slower degradation of SHMT than other

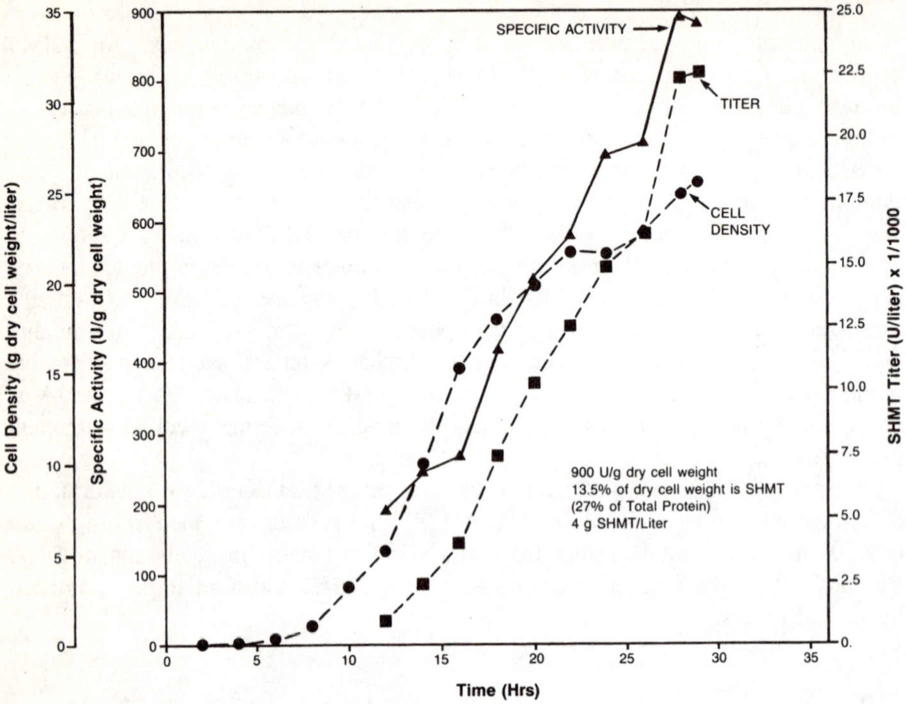

Figure 2.2. Fermentation of Strain GX3036 in 10 l in an NBS 14-l Fermentor with a Minimal Salts Medium and Starch Carbon Source [40]. SHMT activity was measured with the modified phenylserine assay [17].

proteins in late log or stationary phase cells is occurring. This phenomenon, although not clearly understood, is very convenient for preparation of high-titer SHMT for the L-serine bioreactor.

2.3 BIOREACTOR DEVELOPMENT

A well-mixed batch reactor was chosen as the optimal approach for the L-serine reactor because of the inherent economic and reaction kinetic advantages of this design. The stability of SHMT also makes this design practical. For example, a crude lysate with SHMT retained more than 70% of its initial activity after incubation at 37 °C for seven days [16]. Also, the reactivity of formaldehyde and its ability to inactivate the SHMT enzyme is best controlled in a batch reactor as opposed to immobilized enzyme columns. Mixing, along with the automated formaldehyde addition methods [17], managed the formaldehyde at a uniformly low level and virtually eliminated enzyme inactivation.

Several other important factors had to be kept in mind during the development of the bioreactor. First, glycine is a relatively expensive feedstock (roughly $4/kg today [31]); thus a high efficiency of utilization was very important for the economics of the bioreactor. Also, serine and glycine have similar solubilities in water or water-organic solvent mixtures [32]. A high final ratio of L-serine to glycine was necessary in order to obtain efficient separation of glycine and the L-serine product by simple fractional crystallization.

Many of the possible enzyme reactor production methods are made impractical because of the cost of enzyme cofactors that would have to be added. SHMT has been reported to function without tetrahydrofolate (FH_4) in the L-serine reaction [33], but too slowly to be practical. Because the exogenous addition of FH_4 cofactor was necessary in the L-serine bioreactor, and since FH_4 is expensive, it was also very important to obtain a high molar yield of L-serine to the FH_4 used. (Although not described here, a proprietary method for efficient chemical reduction of folate to tetrahydrofolate was also developed. It was possible to use the chemically produced FH_4 directly from the chemical reaction without purification to further decrease the costs.) SHMT also has a tightly bound pyridoxal 5'-phosphate (PLP) cofactor. It was found that the SHMT activity observed in extracts from the overproducing *Klebsiella* strains was greater when additional PLP was added to crude extracts. Presumably the PLP synthesis rate does not correlate with the significantly increased level of SHMT in these strains. A PLP concentration of 0.5 mM was routinely used in the bioreactors. Our estimates are that this would not make a significant contribution to the total cost.

Several of the key experimental results obtained during the development of the bioreactor are described along with the reaction conditions of an optimized bioreactor.

2.3.1 Reactor Characterization

2.3.1.1 Enzyme Characteristics

Two desired reactions occur in the bioreactor: the chemical reaction of formaldehyde with FH_4 to form methylene-FH_4 (m-FH_4), and the SHMT-catalyzed reaction to condense m-FH_4 and glycine to form L-serine. The pH optimum for the enzymatic L-serine synthesis reaction [16] was compared with the published data [18] on the chemical reaction as shown in Figure 2.3. At the optimal pH for the SHMT reaction, m-FH_4 formation is very slow. We have used a compromise pH between pH 7.5 and 6.5 for L-serine production.

The effect of glycine concentration on the initial rate of L-serine synthesis was examined in the presence of an excess concentration [34] of FH_4 as shown in

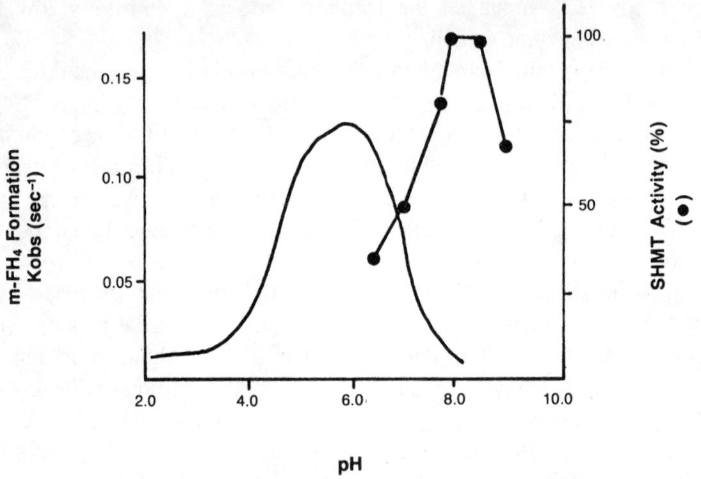

Figure 2.3. The Effect of pH on the Rate of m-FH_4 Synthesis and SHMT Initial Rates. Data on m-FH_4 formation (solid line) are from [13]. Serine formation was used to study the pH optimum of SHMT activity. The buffer used in the pH range from 6.5 to 8.0 was 0.1 M potassium phosphate. Tricine was used for the pH range from 8.5 to 9.0. The HCHO was preincubated with the reaction mixtures for 20 minutes prior to SHMT addition to allow m-FH_4 formation. The enzyme was then introduced to start the reaction.

Figure 2.4. These results show that there is no substrate inhibition. This allows the use of high concentrations of glycine in the reactor. Also, the K_m for glycine of 11.6 mM means there is no kinetic limitation because the glycine concentration is always greater than 0.25 M throughout the reaction. The reaction is not only kinetically favorable, but also thermodynamically favorable. The overall reaction has the equilibrium constant of 3.1×10^3 M^{-1} in favor of serine formation [35, 36].

2.3.1.2 Oxygen Sensitivity

The L-serine bioreactor is sensitive to oxygen because FH_4 is very sensitive to oxygen [18]. Methylene-FH_4 is less sensitive than FH_4 [16], but exclusion of oxygen is very important. Rather than attempt to measure oxygen directly in the bioreactors, the redox potential was monitored through a common pH and redox combination electrode. Table 2.1 demonstrates that FH_4 is stable as long as the redox potential is maintained at or below −330 mV. The redox potential was monitored in bioreactors prior to the addition of the FH_4 and continuously during the reaction. Redox potential was maintained below −330 mV by blanketing the reactor with nitrogen, and in small-scale bioreactors by adding a reducing agent such as 2-mercaptoethanol. The 2-mercaptoethanol feed pump was interfaced with the redox output from the

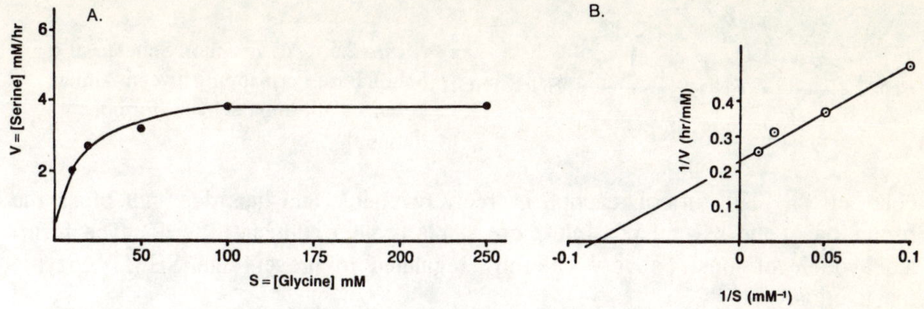

Figure 2.4. The Effect of Glycine Concentration on the Rate of Serine Synthesis. (A) The rate of serine formation detected by HPLC with crude enzyme extract (0.07 units/ml) in 0.1 M potassium phosphate buffer (pH 8.0) with 0.5 mM PLP and 8.8 mM FH_4 was used in this study. (B) Double reciprocal plot of A.

TABLE 2.1 The Effect of Redox Potential on FH_4 Stability

Redox Potential (mV)	FH_4 Concentration (mM) 0 h	7.5 h	22 h	48 h	70 h
−270	1.90	1.90	1.75	1.40	1.14
−300	1.96	2.00	1.72	1.60	1.40
−330	1.96	2.00	2.00	2.00	2.00

combination electrode in such a way that if the redox potential increased above a chosen value, the pump was automatically activated. In larger scale reactors (volumes greater than one liter) it was found that the desired redox potentials were easy to maintain with the use of nitrogen alone.

2.3.1.3 Formaldehyde Toxicity

Formaldehyde is very reactive toward amines [37] as well as the side chains of many amino acids after incorporation into protein [38, 39]. It is not surprising that SHMT is inactivated quickly in the presence of fairly low concentrations of formaldehyde [36]. The presence of high glycine and L-serine concentrations, as exist in the L-serine bioreactor, slows down the inactivation of SHMT but does not prevent it. Competitive formaldehyde reaction with the free amino groups of the amino acids in the well-known formol titration [37], as shown in Figure 2.5, is likely responsible for this

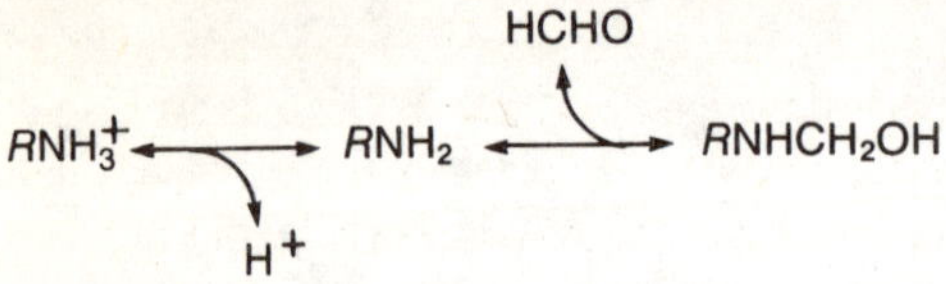

Figure 2.5. The Reaction Scheme of the Schiff Base Formation between Amino Groups and Formaldehyde (formol titration).

delay effect. The formol reaction is freely reversible and thus does not block the formation of the several possible more stable protein adducts [38, 39]. The formaldehyde level must be very carefully regulated to prevent the SHMT enzyme inactivation.

2.3.2 Automated Formaldehyde Control and Reactor Optimization

2.3.2.1 Formol Titration

The formol titration as shown in Figure 2.5 provided a means to monitor the available formaldehyde in the reactor. Formaldehyde reacts with the unprotonated form of the amine so that when the equilibrium existing between the protonated and unprotonated amines is shifted to the right (Figure 2.5) by formaldehyde addition, the pH of the solution goes down. This is clearly demonstrated in Table 2.2 where the effect of formaldehyde addition on pH in a 2 M glycine solution is tested. Addition of as little as 5 mM formaldehyde caused a clear decrease in the pH. The sensitivity of the pH change decreased with increasing initial pH, but at pH 6.5 there was a change of 0.06 pH unit for each increase of 1 mM in formaldehyde concentration up to at least 10 mM. Serine was not quite as sensitive in the formol titration reaction as glycine [17]. Therefore, the pH drop during the reaction reflects the accumulation of formaldehyde because the formaldehyde reaction with FH_4 to form m-FH_4 and the SHMT-catalyzed reaction to form L-serine do not involve protons and a change in pH.

TABLE 2.2 Formol Titration with 2 M Glycine

Initial pH	HCHO Conc. (mM)	Final pH	pH Change	HCHO Conc. per pH Change (mM/unit pH)
6.54	5	6.24	0.30	16.7
6.54	10	5.96	0.58	17.2
7.03	5	6.80	0.23	21.7
7.03	10	6.58	0.47	21.3
7.50	5	7.38	0.12	41.7
7.50	10	7.24	0.26	38.5

The formaldehyde-FH_4 reaction has a very favorable rate and favorable equilibrium [18]. This is demonstrated in Table 2.3 where it can be seen that the addition of 5 mM FH_4 to a formol titration similar to that described in Table 2.2 results in no pH change until the added formaldehyde exceeds the FH_4 concentration. This means that the formol reaction with the free amines, even at high amino acid concentration, does not interfere with the formaldehyde-FH_4 reaction. Since the SHMT enzyme also tolerates high amino acid concentrations, the bioreactor was operated at total amino acid concentrations of 3 M and greater. This simplified the development of the L-serine bioreactor and also allowed for efficient utilization of the enzyme and cofactors in the batch reactor configuration.

TABLE 2.3 Formol Titration with 2 M Glycine in the Presence of 5 mM FH_4

Initial pH		HCHO Conc. (mM)	Final pH	pH Change	HCHO Conc. per pH Change (mM/unit pH)
6.54	1st add	5	6.54	0	–
	2nd add	10	6.27	0.27	18.5
7.01	1st add	5	7.01	0	–
	2nd add	10	6.80	0.20	25.0
7.50	1st add	5	7.50	0	–
	2nd add	10	7.38	0.12	41.7

2.3.2.2 Feedback Control of Formaldehyde Addition

The pH was observed in a bioreactor set up with a constant formaldehyde feed rate as shown in Figure 2.6. The pH was maintained between pH 7.4 and 7.5 by automatic titration with 10% KOH solution. It can be seen that for the first five hours there is no change in pH, and the L-serine produced correlates exactly with the cumulative amount of formaldehyde added. However, after five hours the KOH titration started and the conversion of glycine to L-serine stopped. This result indicated that as the L-serine synthesis reaction proceeded toward equilibrium and began to slow down in the direction of net L-serine formation, the formaldehyde feed rate exceeded the L-serine formation rate. The unreacted formaldehyde began the formol titration, causing a pH decrease, and inactivated the enzyme, stopping the reaction altogether. This observation combined with the formol titration data in Tables 2 and 3 showed that the pH could be used as an indicator to regulate the formaldehyde feed rate for optimal bioconversion.

A bioreactor with the configuration shown in Figure 2.7, providing semiautomated formaldehyde control, was assembled. A pH controller was attached to the pumps

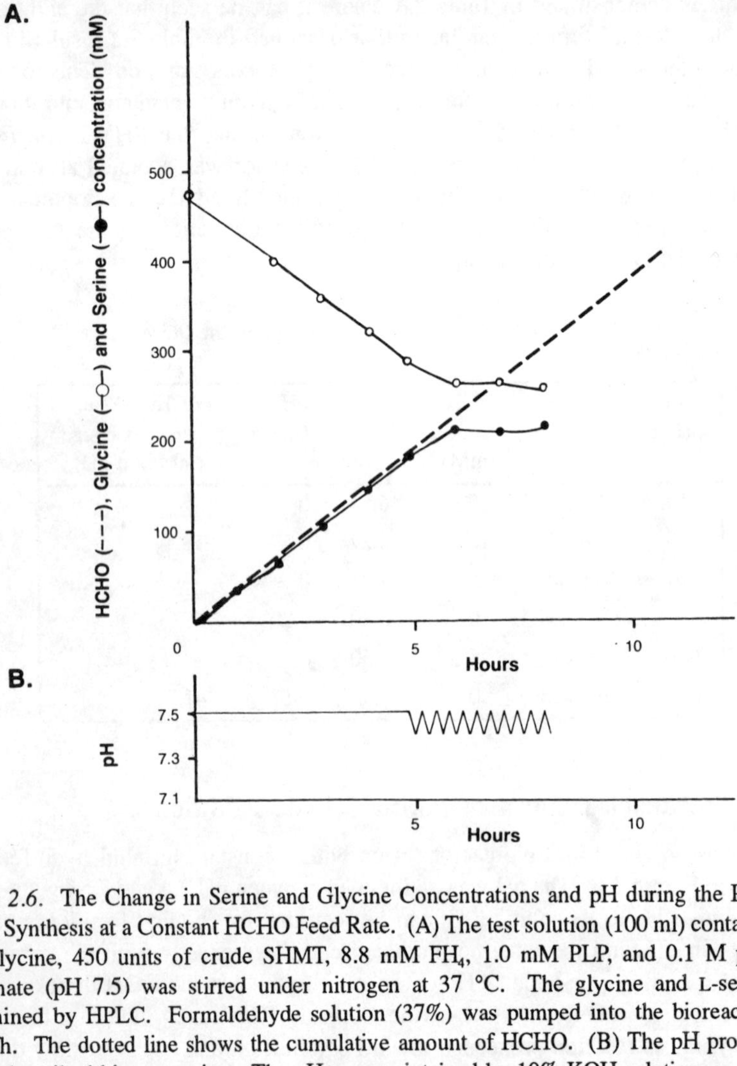

Figure 2.6. The Change in Serine and Glycine Concentrations and pH during the Process of Serine Synthesis at a Constant HCHO Feed Rate. (A) The test solution (100 ml) containing 470 mM glycine, 450 units of crude SHMT, 8.8 mM FH_4, 1.0 mM PLP, and 0.1 M potassium phosphate (pH 7.5) was stirred under nitrogen at 37 °C. The glycine and L-serine were determined by HPLC. Formaldehyde solution (37%) was pumped into the bioreactor at 3.7 mmol/h. The dotted line shows the cumulative amount of HCHO. (B) The pH profile of the above described bioconversion. The pH was maintained by 10% KOH solution.

adding formaldehyde and 10% KOH feed solutions in such a way that the formaldehyde pump was activated and the KOH pump was inactivated when the pH was within a small set range (*e.g.*, pH 7.48 to 7.42 or pH 6.48 to 6.42). The base pump was used to raise the initial pH to the desired value, but was disconnected after the pH stabilized. A small 150-ml reactor initialized with 3 M glycine and this control system

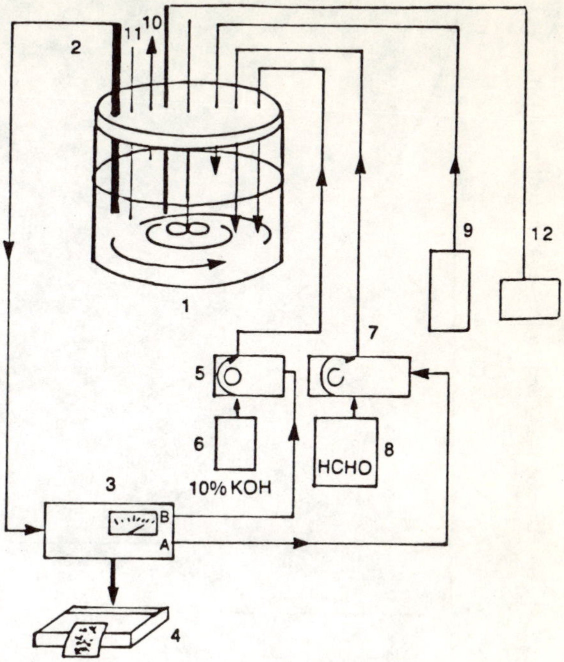

Figure 2.7. The Configuration of the Bioreactor System (see text).
(1) bioreactor; (2) pH electrode; (3) pH controller; (4) recorder; (5) pump; (6) 10% KOH; (7) pump; (8) 37% HCHO solution; (9) nitrogen tank; (10) nitrogen outlet; (11) sampling tube; (12) temperature control.

is shown in Figure 2.8. The feed rate of 37% (w/v) formaldehyde was initially adjusted to 1.7 ml/h. After about three hours the pH started to decrease, causing the formaldehyde pump to be turned off. Because of the reversible nature of the formol reaction and the favorable equilibrium of the L-serine formation, the excess formaldehyde was converted to L-serine, causing the pH to recover and the formaldehyde pump to be automatically reactivated. The formaldehyde feed rate was manually decreased whenever a pH inflection was observed, allowing the reaction to again proceed for several hours without formaldehyde accumulation and a pH drop. If the feed rate was not immediately adjusted as occurred at about 20 hours into the reaction, there was a period of cycling pH, but no serious accumulation of formaldehyde. Using this approach, the reactor is operated under a formaldehyde limitation. With the small 0.06 unit pH change allowed by the controller, the formol titration data in Table 2.2 indicates that only 1 mM to 3.3 mM available formaldehyde is allowed to accumulate at the time of the pH inflection. At the same time, it is possible to run the reactor at close to the maximum rate the amount of added enzyme will allow without the danger of inactivating the system. In the best case, very small adjustments to the formaldehyde feed rate and very sensitive detection of pH change are used in the formaldehyde auto-control system. This was achieved by using a custom-designed microprocessor-controlled formaldehyde feed pump in a 22-liter bioreactor.

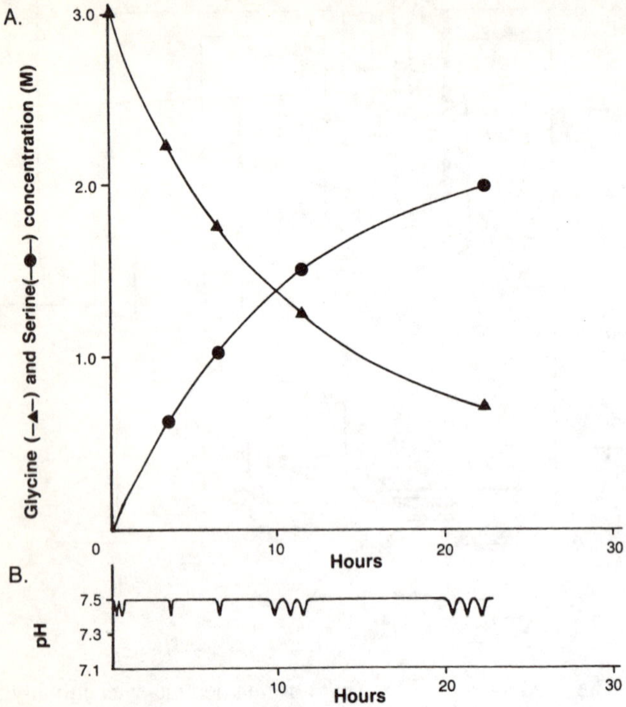

Figure 2.8. Change in Amino Acid Composition and pH Profile of the Serine Production Bioreactor, Which Employed an Automatic Feedback Control System for HCHO Addition. The reactor contained 30 mM 2-mercaptoethanol, 5 mM FH_4 and 300 units of SHMT. Other details are described in the text. (A) The glycine and serine concentrations were determined by HPLC. (B) The pH profile of the bioconversion.

2.3.2.3 Optimization of the Bioreactor

As shown in Figure 2.3, there must be a trade-off between the optimal pH for the chemical reaction of formaldehyde with FH_4 and the enzymatic SHMT reaction. When the formaldehyde feeding control described above is used, the loss of enzyme activity is minimal so that enzyme can be readily recycled into new bioreactors. On the other hand, FH_4 is also an expensive component (approximately $200/kg even with on-site synthesis from folate) and is more difficult to recycle because of the oxidation problem. Therefore, it was more cost-effective to use a pH closer to the optimum of the formaldehyde-FH_4 reaction than of the SHMT reaction. At a set point of pH 6.5, it was found that good overall productivity could be obtained with inclusion of 0.5 mM FH_4.

An example of a bioreaction run with 0.5 mM FH_4 and 0.5 mM PLP is shown in Figure 2.9. Independent experiments demonstrated that addition of 0.5 mM PLP caused about a twofold increase in SHMT activity observed in crude enzyme extracts from GX3036, and that higher PLP concentrations did not give further improvement [17]. The bioreactor example in Figure 2.9 demonstrates an 80% glycine to L-serine conversion in 25 hours with a productivity 10 g/l/h. When the reaction was

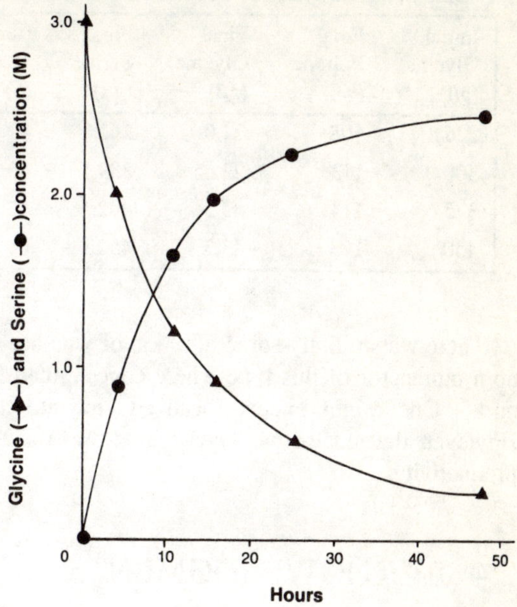

Figure 2.9. The Performance of a 100-ml Bioreactor with 0.5 mM PLP, 0.5 mM FH_4, 3 M Initial Glycine, 30 mM 2-Mercaptoethanol and 300 Units of SHMT at 37 °C at pH 6.5. The initial 37% formaldehyde feed rate was 1.7 ml/h. The glycine and L-serine concentrations were determined by HPLC.

continued for another 23 hours, the conversion was increased to 90%, but the productivity was reduced to 5.9 g/l/h. Thus, another trade-off must be made in running the bioreactor, where reduced productivity due to longer reaction times has to be balanced with the benefit of a higher ratio of final L-serine to glycine that makes the purification by crystallization more efficient.

Another factor important for efficient L-serine production is the initial concentration of glycine. Since glycine does not inhibit SHMT, as shown above (Figure 2.4), the effect of increasing the initial glycine concentration on the reactor productivity was examined, as shown in Table 2.4. Increasing the initial glycine concentration and thus the final L-serine titer increases the efficiency of use of the cofactors and enzyme, and reduces the size requirements and thus capital cost of the bioreactor. It can be seen in Table 2.4 that increasing the initial glycine does not adversely affect the productivity or molar conversion in the bioreactor, but does significantly improve cofactor utilization. The molar yield of L-serine molecules per FH_4 molecule increases from 4975 at 225 g/l initial glycine to 8611 at 450 g/l initial glycine. In the last example, where the initial glycine was 450 g/l, the initial condition is a suspension containing solid glycine. The final condition with 452.1 g/l L-serine is also a solid suspension. This result suggests that it should be possible to run continuous reactors with high efficiency where the glycine concentration is maintained below saturation and crystalline L-serine is continuously collected, for example, by circulation through a screen filter.

TABLE 2.4 The Effect of Glycine on the Performance of the Bioreactor

Initial Glycine (g/l)	Final Volume (%)	Final Glycine (g/l)	Final Serine (g/l)	React. Time (h)	Molar Conversion (%)	Productivity (g/l/h)
225	108	21.0	261.2	42	90	7.5
300	113	23.45	339.1	46	91	8.8
375	114	43.5	402.1	46	87	10.2
450	123	43.5	452.1	66	88	8.9

There was initially some question of whether it would be possible to readily scale up a bioreactor of this type where concentrated formaldehyde is being added at one port. One might expect localized enzyme inactivation at the site of addition. However, the reactor has been scaled up to a 22-liter volume without reduction in productivity.

2.4 ACKNOWLEDGMENTS

The L-serine process development was performed at Genex Corporation. Several individuals participated in this development work, including: Debbie Young who assisted in some aspects of the strain development; Tena Wei and Karen Campbell who worked on the bioreactor optimization; Huei-Hsiung Yang, Daleep Bhatia, and Matt Hanley who developed fermentation conditions for SHMT production and recovery; and Bill Watkins who worked on the scale-up of the reactor to 22-liter volumes and L-serine recovery in the pilot plant.

2.5 REFERENCES

1. Asai, Y. Y., Shimada, Y. M., Soda, U. K., U.S. Pat. 4,335,209 (1982).
2. Hamilton, B. K., Hsiao, H. Y., Swann, W. E., Anderson, D. M., Delente, J., *Trends in Biotechnol. 3*, 64 (1985).
3. Schlossmann, K., Lynen, F., *Biochem. Z. 328*, 591 (1957).
4. Hsiao, H. Y., Wei, T., *Biotechnol. Bioeng.30*, 875 (1987).
5. Yamada, H., Kumagai, H., *Appl. Environ. Microbiol. 19*, 249 (1979).
6. Lee, T. K., Hsiao, H. Y., *Enzyme Microb. Technol. 9*, 523 (1986).
7. Hirose, Y., Okada, H. In *Microbial Technology*, 2nd ed.; Peppler, H. J., Perlman, D., Eds.; Academic Press, New York, 1979; Vol. I, pp 211–237.
8. Kubota, K., Shiro, T., Jpn. Pat. 42/15102 (1967).
9. Kotani, Y., Araki, K., Nakayama, K., *J. Agric. Chem. Soc. Japan 48*, 131 (1974).

10. Masahiro, E., Kakimoto, T., Chibata, I., *Appl. Environ. Microbiol. 37*, 1053 (1979).
11. Keune, H., Sham, H., Wagner, F., *Eur. J. Appl. Microbiol. 2*, 176 (1976).
12. Morinaga, Y., Yamanaka, S., Takinami, K., *Agric. Biol. Chem. 45*, 1419 (1981).
13. Morinaga, Y., Yamanaka, S., Takinami, K., *Agric. Biol. Chem. 45*, 1425 (1981).
14. Omori, K., Kakimoto, T., Chibata, I., *Appl. Environ. Microbiol. 45*, 1722 (1983).
15. Yamade, H., U.S. Pat. 3,755,081 (1973).
16. Hsiao, H. Y., Wei, T., Campbell, K., *Biotechnol. Bioeng. 28*, 857 (1986).
17. Hsiao, H. Y., Wei, T., *Biotechnol. Bioeng. 28*, 1510 (1986).
18. Kallen, R. G., Jencks, W. P., *J. Biol. Chem. 241*, 5851 (1986).
19. Schirch, L., *Adv. Enzymol. 53*, 83 (1982).
20. Stauffer, G. V., Plamann, M. D., Stauffer, L. T., *Gene 14*, 63 (1981).
21. Anderson, D. M., Hsiao, H. Y., Somerville, R. L., Herrmann, K. M., U.K. Pat. 2,130,216 (1983).
22. Hsiao, H. Y., Wei, T., Anderson, D. M., PCT International Pat. Applic. WO 8605515 A1 (1986).
23. Plamann, M. D., Stauffer, L. T., Urbanowski, M. S., Stauffer, G. V., *Nucleic Acids Res. 11*, 2065 (1983).
24. Skogman, G., Nilsson, J., Gustafsson, P., *Gene 23*, 105 (1983).
25. Anderson, D. M., Herrmann, K. M., Somerville, R. L., U.S. Pat. 4,371,614 (1983).
26. Anderson, D. M., Eur. Pat. Applic. EP 178764 A1 (1986).
27. Young, R., *Mol. Gen. Genet. 182*, 326 (1981).
28. Vining, L. C., Magasanik, B., *J. Bacteriol. 146*, 647 (1981).
29. Kilbane, J., Malamy, M. H., *J. Mol. Biol. 143*, 73 (1980).
30. *Federal Register 51* (No. 88), 16958–16985 (1986).
31. *Chemical Marketing Reporter*, April 20, 1987, p 32.
32. *CRC Handbook of Biochemistry*; Sober, H. A., Ed.; Chemical Rubber Co., Cleveland, OH, 1970; pp 13–66.
33. Chen, M. S., Schirch, L., *J. Biol. Chem. 248*, 3631 (1973).
34. O'Connor, M. L., Hanson, R. S., *J. Bacteriol. 124*, 985 (1975).
35. Blakley, R. L., *Biochem. J. 77*, 459 (1960).
36. Alexander, N., Greenberg, D. M., *J. Biol. Chem. 220*, 775 (1956).
37. Kallen, R. G., Jencks, W. P., *J. Biol. Chem. 241*, 5864 (1966).
38. French, D., Edsall, J. T., *Adv. Protein Chem. 2*, 277 (1945).
39. Means, G. E., Feeney, R. E. *Chemical Modification of Proteins*; Holden-Day, San Francisco, 1971; pp 125–128.
40. Anderson, D. M., Hsiao, H. Y., Yang, H. H., Abstract 70, Div. Microb. Biochem. Technol.; 188th ACS Annual Meeting, Am. Chem. Soc., 1986.

BIOCATALYTIC PRODUCTION OF AMINO ACIDS BY TRANSAMINATION

Steven P. Crump and J. David Rozzell

CONTENTS

3.1 INTRODUCTION

Amino acids currently have application as additives to animal feed, nutritional supplements for human food, components in infusion solutions, and synthetic intermediates for the manufacture of pharmaceuticals and agricultural chemicals. L-Glutamic acid is used as a flavor enhancer for food with a world market estimated at greater than one billion dollars annually. L-Lysine and methionine are large-volume additives to animal feed, and L-tryptophan and L-threonine have similar potential applications. L-Phenylalanine and L-aspartic acid have very important markets as key components in the manufacture of the sweetener aspartame. Infusion solutions require 14 amino acids which are essential in human diets.

Methods developed for the production of amino acids involve fermentation, chemical synthesis, extraction from protein hydrolysates, and enzymatic conversion. Fermentation methods are especially effective in cases where very high yields and accumulations have been achieved (for example, L-glutamic acid and L-lysine). Synthetic chemical methods generally involve the initial formation of a racemic mixture, followed by the resolution of this mixture to yield the optically active product. The resolution may be accomplished either chemically, by fractional crystallization of diastereomeric salts of the amino acid, or, if desired, enzymatically using enzymes such as L-aminoacylase [1], amidase [2], or hydantoinase [3]. The undesired isomer can be re-racemized and then recycled through the process. A few amino acids can be recovered from protein hydrolysates, but, due to the often laborious purification required to produce an amino acid in high purity, this method is limited to those amino acids which are either easily recovered (such as L-tyrosine) or are present in certain protein sources in high levels (such as L-cysteine). In those cases where enzymatic conversions offer advantages, the primary determining factors are reduced capital investments, lower purification costs, and generally high extents of conversion. Such processes are useful for the production of both naturally occurring and unnatural amino acids, provided suitable precursors are available. Enzymatic processes are especially useful for the production of optically pure forms of unnatural amino acids which cannot be made by fermentation methods.

A number of biocatalytic routes have been proposed and developed for the production of specific L-amino acids. For example, Chibata and co-workers at Tanabe Seiyaku, as well as workers at several other companies, have developed a process for the production of L-aspartic acid from ammonium fumarate, catalyzed by the enzyme aspartase [4, 5, 6]. Genex Corporation commercialized a process for the production of L-phenylalanine by the phenylalanine ammonia lyase–catalyzed addition of ammonia to *trans*-cinnamic acid [7]. Serine hydroxymethyltransferase has been shown to be useful in the production of L-serine from glycine and formaldehyde [7]. Porcine aminoacylase has been used to resolve DL-mixtures of certain amino acids such as tryptophan, valine, and methionine [1], and has the potential of some breadth of scope as a resolution technology. Likewise, transaminases have similar broad potential, but as catalysts for the asymmetric synthesis of amino acids, D- and L-, both naturally occurring and unnatural [14, 15].

3.2 USE OF TRANSAMINASES IN BIOCATALYTIC REACTIONS

We have been studying aminotransferases (transaminases) as potentially useful biocatalysts for the production of a wide range of different amino acids. Transaminases catalyze the transfer of an amino group from an L-amino acid donor to a 2-keto acid acceptor. The amino group transfer is mediated by the cofactor pyridoxal phosphate, which is reversibly bound to the enzyme through a Schiff-base linkage to the ε-amino group of an active-site lysine [8, 9, 10]. A desired L-amino acid can be produced from a given 2-keto acid precursor using an inexpensive L-amino acid as the amino group donor (Figure 3.1). As a coproduct of this reaction, a second 2-keto acid is produced along with the desired L-amino acid.

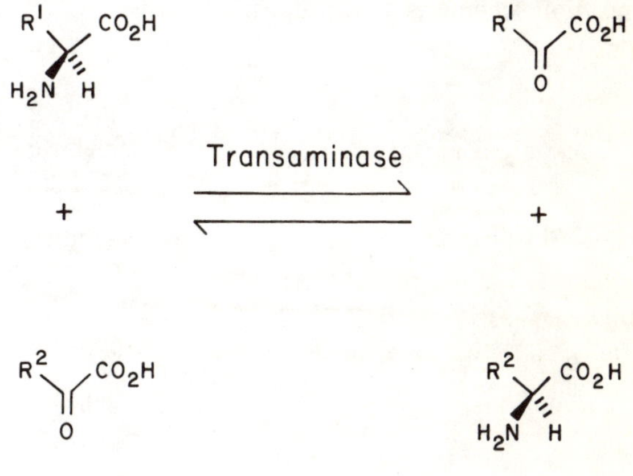

Figure 3.1. General Reaction Catalyzed by Transaminases.

Among the advantages of transaminases as biocatalysts for the production of optically pure amino acids are the following:

1. Transaminases have high stereoselectivity for a given enantiomer. Optically active L-amino acids (or D-) are produced stereoselectively; no unwanted D-isomer (or L-) is produced, and no resolution is required.
2. The catalytic rates of these enzyme-catalyzed reactions are generally relatively rapid.
3. Capital costs for such a biocatalytic process are low, and often existing chemical process equipment can be used.

4. A large number of the required 2-keto acid precursors are accessible through chemical synthesis, expanding the range of potential products.
5. Transaminases are potentially applicable to the production of a wide range of amino acids, because enzymes are available with a wide range of specificities, such as for amino acids with aromatic side chains, acidic side chains, branched alkyl side chains, etc.
6. In some cases, the 2-keto acid by-products may also have significant value (in addition to the L-amino acids).

One of the simplest examples of an efficient transamination process is the production of L-alanine and 2-ketoglutarate from the precursors L-glutamate and pyruvic acid. Porcine glutamic-pyruvic transaminase is available commercially as a partially purified preparation, and this enzyme was chosen as a convenient model system for study. We immobilized enzyme having a specific activity of 50 units per milligram (Lee Scientific, St. Louis, MO, USA) on porous aminopropyl glass using a water-soluble carbodiimide as coupling agent [13]. At a loading of 20 milligrams of total protein bound per gram of glass, the activity of the biocatalyst when assayed for the production of L-alanine was 400 units per gram of biocatalyst. The enzymatic activity retained after immobilization was 40%. The equilibrium constant for the reaction was 1.86 in favor of the desired products, L-alanine and 2-ketoglutarate. The reaction scheme is shown in Figure 3.2.

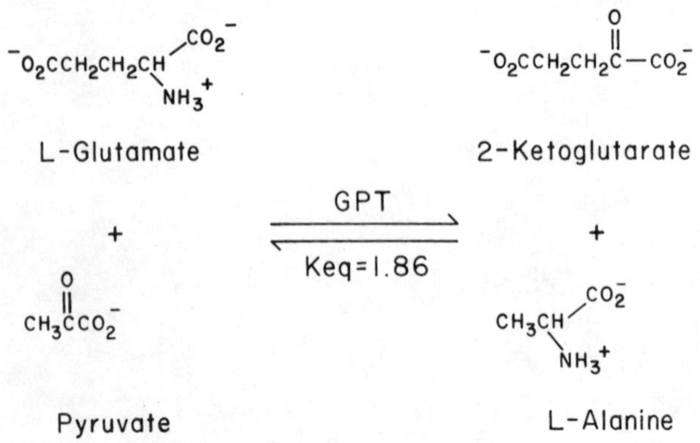

Figure 3.2. Scheme for the Production of L-Alanine and 2-Ketoglutarate by Transamination.

Using L-glutamate at a concentration of 200 mM and pyruvate at a concentration of 400 mM in the presence of pyridoxal phosphate at a concentration of 0.5 mM, one can achieve approximately 80% conversion of the limiting reagent, in this case L-glutamate. The amino acid L-alanine is produced along with the co-product 2-ketoglutarate. The productivity of the biocatalytic system when operated at a constant level of 80% conversion was found to be 200 grams of L-alanine per kilogram of biocatalyst per hour and 328 grams of 2-ketoglutarate per kilogram of biocatalyst per hour. Productivities of this magnitude compare favorably to those for other successful immobilized enzyme processes. The long-term stability of the system under continuous operating conditions was found to be greater than 200 days, as shown in Figure 3.3.

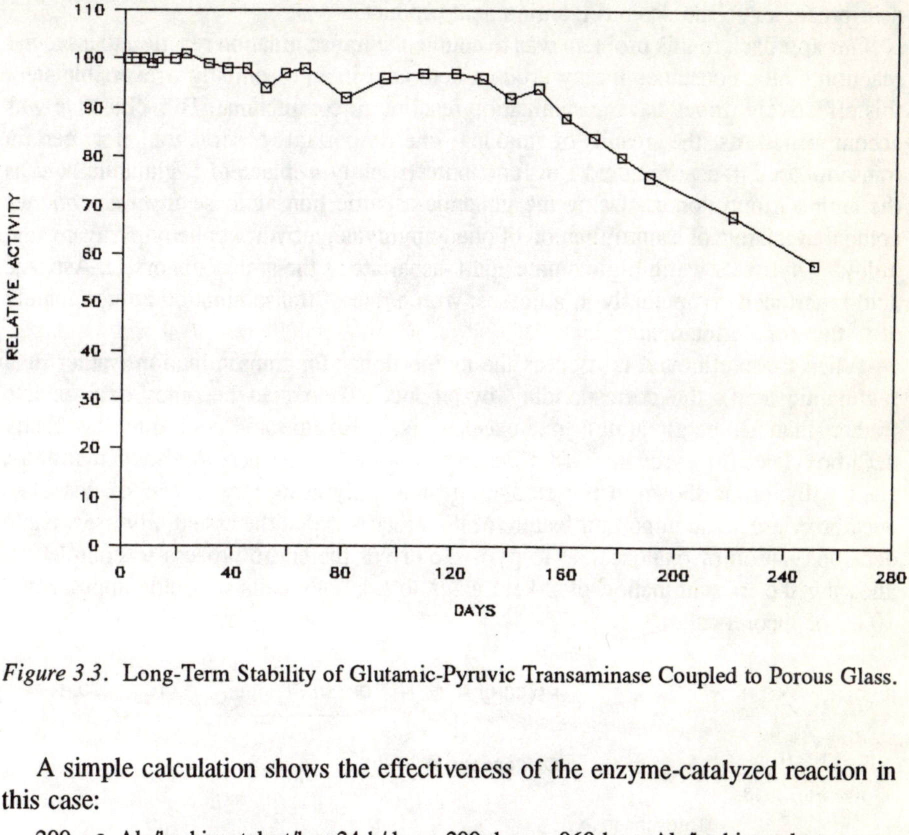

Figure 3.3. Long-Term Stability of Glutamic-Pyruvic Transaminase Coupled to Porous Glass.

A simple calculation shows the effectiveness of the enzyme-catalyzed reaction in this case:

200 g L-Ala/kg biocatalyst/h × 24 h/day x 200 days = 960 kg L-Ala/kg biocatalyst

Thus, one kilogram of biocatalyst, which contains only 20 grams of enzyme, is capable of producing about one ton of final product. Porcine glutamic-pyruvic transaminase thus proved to be a useful model for understanding the capabilities of transaminase enzymes in general for the production of optically pure amino acids.

3.3 DRIVING THE REACTION TO COMPLETION

There is one major disadvantage to the transamination technology as presented above: because the transamination involves an amino acid reacting with a 2-keto acid to generate products which consist of a 2-keto acid and an amino acid, the equilibrium constant of the general reaction would be expected to be near unity. As a result, the net conversion of substrates to products for the general transamination process will always be limited. The key to the development of an efficient transamination technology rests in overcoming the problem of incomplete conversion of the 2-keto acid precursor to the desired L-amino acid product.

Our approach to this problem was to couple the transamination reaction to a second reaction which consumes the by-product keto acid in an essentially irreversible step; this effectively drives the transamination reaction to completion. This objective was accomplished as the result of making one important observation: for certain transaminases, L-aspartic acid can function efficiently in place of L-glutamic acid as the amino group donor. Using the glutamic-aspartic transaminase from *E. coli*, we compared the rate of transamination of phenylpyruvate, *p*-hydroxyphenylpyruvate, and indoyl-3-pyruvate using L-glutamate and L-aspartate as the amino donors. L-Aspartic acid functioned competently in all cases, with a rate of transamination approximately 70% that for 2-glutamate.

When L-aspartic acid is used as the amino donor for transamination (rather than L-glutamic acid), the corresponding by-product 2-keto acid becomes oxaloacetate (rather than 2-ketoglutarate); oxaloacetate is a β-keto acid and may be easily decarboxylated to pyruvate. This decarboxylation can be accomplished thermally, chemically, or, as shown in Figure 3.4, enzymatically using the enzyme oxaloacetate decarboxylase. The important feature of the process is that the essentially irreversible decarboxylation of oxaloacetate to pyruvate drives the entire process to completion, allowing the transamination of 2-keto acids to L-amino acids in yields approaching 100% of theoretical.

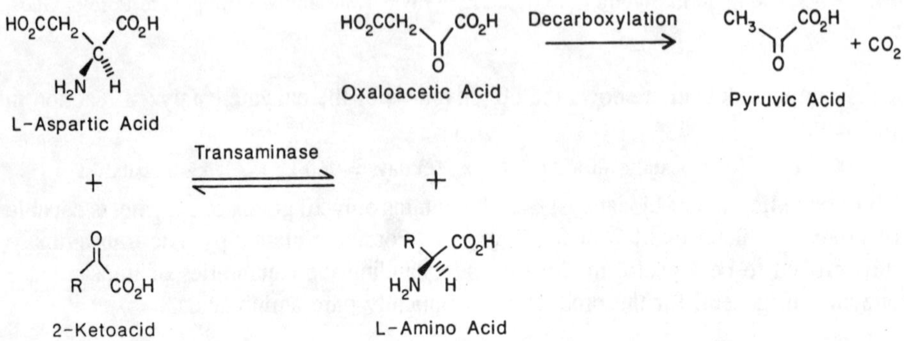

Figure 3.4. Coupled Two-Enzyme Reaction Scheme for the Production of L-Amino Acids by Transamination.

Initially, a model system in which 2-ketoglutarate was converted to L-glutamate in the presence of L-aspartic acid was used to study the coupled transamination-decarboxylation reaction. Both the commercially available porcine transaminase and the glutamic-aspartic transaminase isolated from *E. coli* were used. The decarboxylation reaction catalyzed by the enzyme oxaloacetate decarboxylase was examined using enzyme from four different sources—*Pseudomonas putida*, *Micrococcus luteus*, and two strains of *Azotobacter vinelandii*. The highest rates were obtained with the oxaloacetate decarboxylase isolated from *Pseudomonas*, a Mg^{2+}-requiring enzyme, and this enzyme was used for all subsequent work involving the coupled enzyme system.

Based on the work using the model reaction system, the production of other L-amino acids, both naturally occurring and unnatural, was investigated. One such compound is L-phenylalanine. This L-amino acid has seen rapid market growth as a result of its use as an ingredient in the manufacture of the noncaloric sweetener aspartame. L-Phenylalanine can be produced by the coupled enzyme-catalyzed reaction process just described, using phenylpyruvate as the starting 2-keto acid. An early experiment clearly shows the effect of decarboxylation in driving the reaction to completion. In this experiment, phenylpyruvate sodium salt and L-aspartate were incubated with *E. coli* transaminase at room temperature at pH 7.5 in both the presence and absence of oxaloacetate decarboxylase from *P. putida*. Magnesium ion, which is a cofactor for the decarboxylase, was also present in both reaction mixtures at a concentration of 6 mM. The transamination reaction was monitored by following the disappearance of phenylpyruvate. As shown in Figure 3.5, the reaction in which oxaloacetate decarboxylase was included in the mixture proceeded to completion much more rapidly than the reaction without the decarboxylase.

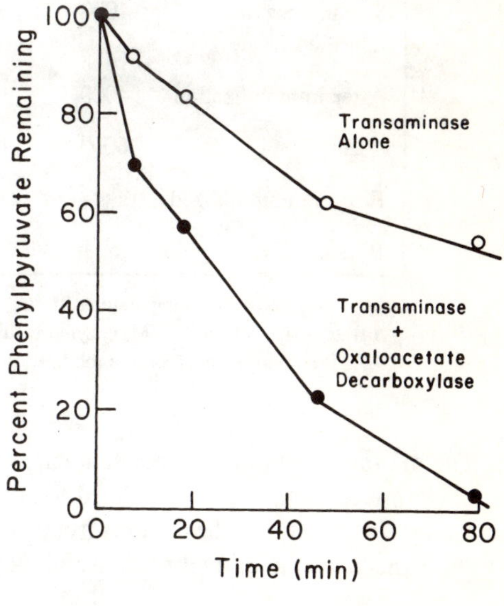

Figure 3.5. Effects of Oxaloacetate Decarboxylase on the Conversion of Phenylpyruvate to L-Phenylalanine by Transamination. $[Mg^{2+}] = 6$ mM; [Tris] = 50 mM; pH = 7.5.

3.4 PRODUCTION OF L-AMINO ACIDS USING IMMOBILIZED TRANSAMINASES

Continuous decarboxylation of oxaloacetate as it is formed is an important part of an efficient, high-yielding transamination process. This decarboxylation may be accomplished enzymatically as described above, or chemically using a metal ion such as Mg^{2+} in sufficient concentration. Immobilization of the enzyme allows continuous production of amino acid.

Immobilization of the *E. coli* aspartic transaminase can be accomplished by covalent attachment to a glutaraldehyde-activated PVC-silica support matrix which had been pretreated with a polyamine [15]. In the example described in Table 3.1, 4000 milliliters of cell lysate containing 61.6 grams of enzyme (activity of 5.2 million international units) was clarified by centrifugation at 13,000 × *g* for 30 minutes and recirculated through a preactivated support matrix for 1.5 hours. After washing, 57 grams or 93% of the enzyme remained bound to the support. Bound activity was 4.2 million units. The retained activity of the enzyme after immobilization was approximately 89%.

TABLE 3.1 Immobilization of Aspartic Transaminase from *E. coli*

	Volume (ml)	Protein conc. (mg/ml)	Total protein (g)	Total activity (MM IU)
Starting enzyme (clarified lysate)	4000	15.4	61.6	5.2
After immobilization	4000	0.3	1.6	0.3
Wash	6000	0.5	3.0	0.2
Enzyme immobilized			57.0	4.7
Bound activity				4.2

Assay conditions: inlet temperature, 21 °C; support weight, 680 g; phenylpyruvate concentration, 0.2 M; enzyme loading, 8.4%; L-aspartate concentration, 0.3 M; activity retained on immobilization, 89%; pH, 7.0.

The pH-rate profile for the reaction catalyzed by the *E. coli* aspartic transaminase was determined using the immobilized transaminase with phenylpyruvate as the keto acid and L-aspartate as the amino group donor (Figure 3.6). The transamination reaction displayed a fairly broad useful pH range; the immobilized transaminase

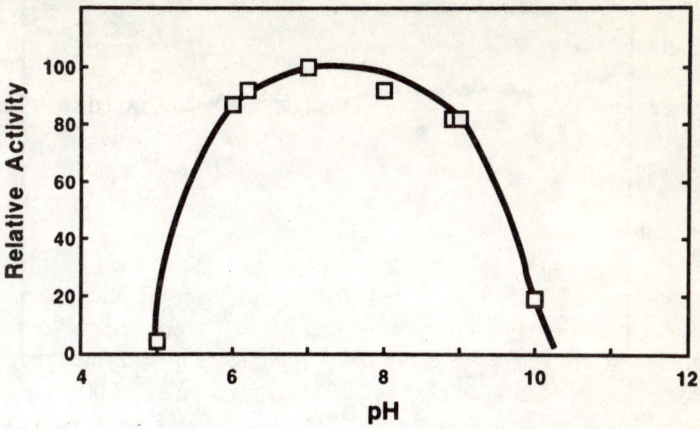

Figure 3.6. Plot of the pH-Rate Profile for Immobilized Aspartic Transaminase from *E. coli* (*asp*C gene product) with L-Aspartate and Phenylpyruvate as Substrates.

had a pH optimum of 7.0–7.2, but retained at least 80% of its maximum activity in the range of pH 6.0–8.5. At pH 5.0 and 10.0, activity fell to less than 20% of that measured at pH 7. A typical set of operating conditions maintains a pH of 7.2, temperature range of 37–40 °C, and reactants at the following concentrations:

Sodium phenylpyruvate	200 mM
L-Aspartate	300 mM
Pyridoxal phosphate	0.1 mM
$MgCl_2$	100 mM

Under these conditions, the productivity of the reactor for the production of L-phenylalanine at 90% conversion of phenylpyruvate is 102 grams of L-phenylalanine per kilogram total biocatalyst weight per hour, or approximately 50 grams of L-phenylalanine per liter per hour.

The long-term stability of the enzyme under these conditions was tested in the laboratory. For the immobilized *E. coli* aspartic transaminase used to catalyze the production of L-phenylalanine, approximately 10–12% of the initial activity was lost after 33 days of continuous operation. A typical profile of activity under continuous operating conditions as a function of time is shown in Figure 3.7.

One of the main advantages of the transamination system is its applicability to a range of other L-amino acids, including non-naturally occurring amino acids. For example, the *E. coli* transaminase (encoded by the *asp*C gene) will also transaminate efficiently the 2-keto acids corresponding to the L-amino acids *p*-fluoro-L-phenylalanine, L-tyrosine, *m*-hydroxy-L-phenylalanine, L-tryptophan, L-methionine, L-homophenylalanine, L-2-aminoadipic acid and a number of others (Table 3.2).

However, for the commercially interesting group of branched-chain L-amino acids L-leucine, L-isoleucine, and L-valine, the *asp*C gene product has little or no catalytic activity. To enable production of this group of L-amino acids, another transaminase

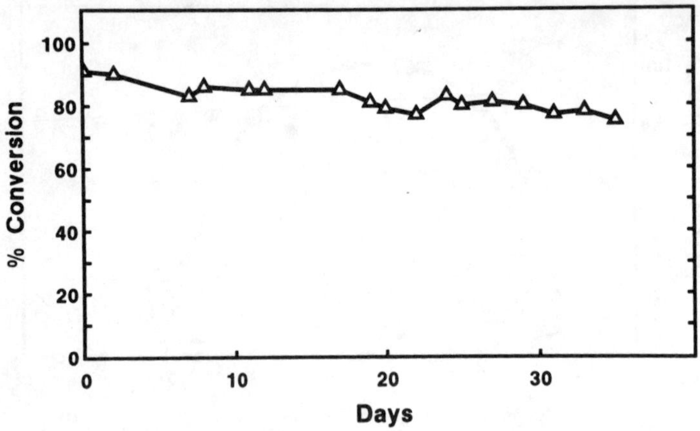

Figure 3.7. Long-Term Operational Stability of Immobilized Aspartic Transaminase from *E. coli.* Substrates were L-aspartate (300 mM), phenylpyruvate (200 mM), pyridoxal phosphate (0.1 mM) and $MgCl_2$ (100 mM).

TABLE 3.2 Relative Rates of Transamination of a Range of 2-Keto acids Catalyzed by Aspartic Transaminase from *E. coli*

L-Amino acid	Relative Rate
L-phenylalanine	100
L-tyrosine	130
L-tryptophan	150–190
L-2-aminoadipic acid	22
L-4-phenyl-2-aminobutanoic acid	18
L-*m*-tyrosine	~40
L-methionine	20–40
L-serine	3
L-cysteine	2–3

Amino group donor in all cases was L-aspartic acid.

from *E. coli* was identified, the so-called branched-chain amino acid transaminase or transaminase B, encoded by the *ilv*E gene.

After isolation of the *E. coli* branched-chain transaminase from wild-type *E. coli* K-12, we found that the keto acids 2-ketoisocaproic acid, 2-keto-3-methylbutanoic acid, and 2-ketoisovaleric acid could all be transaminated by the amino donor L-glutamic acid to L-leucine, L-isoleucine, and L-valine respectively. L-Aspartic acid proved to be completely inert as the amino donor for this particular transaminase, however; providing a driving force for effecting the transamination reaction in high yield was not possible through the direct replacement of L-glutamate by L-aspartate.

Despite the narrow substrate specificity of the donor amino acid, a scheme was devised to overcome this apparent problem, as outlined in Figure 3.8 using L-valine as an example. The conversion makes use of a coupled three-enzyme system. In the first reaction, the *ilv*E branched-chain transaminase catalyzes the reaction of L-glutamate with 2-ketoisovalerate to produce L-valine and 2-ketoglutarate. The second reaction, catalyzed by glutamic-oxaloacetic transaminase, converts L-aspartate and 2-ketoglutarate to oxaloacetate and L-glutamate. L-Aspartate is present in stoichiometric amounts relative to 2-ketoisovalerate and is used to continously convert the 2-ketoglutarate formed in the first step to L-glutamate, allowing continuous recycling of L-glutamate/2-ketoglutarate as the reaction proceeds. Finally, oxaloacetate decarboxylase is used to catalyze the essentially irreversible decarboxylation of oxaloacetate to pyruvate, thus driving the entire sequence of reactions to completion. The net reaction is the transamination of 2-ketoisovalerate to L-valine with L-aspartate, using 2-ketoglutarate as an intermediary amino transfer agent.

PRODUCTION OF L-VALINE

2-Ketoisovalerate + L-Glutamate ⇌ (ilv E) L-Valine + 2-Ketoglutarate

2-Ketoglutarate + L-Aspartate ⇌ (aspC) L-Glutamate + Oxaloacetate

Oxaloacetate → Pyruvate + CO_2

Net:

2-Ketoisovalerate + L-Aspartate → L-Valine + Pyruvate + CO_2

Figure 3.8. Coupled Enzyme System for the Production of Branched-Chain L-Amino Acids.

Using this type of reaction scheme, the branched-chain amino acids can also be produced efficiently. There is no detectable inhibition of the branched-chain transaminase by L-aspartate, and there is little cross-reactivity of the *ilv*E transaminase with pyruvate. The enzyme is stable, and can be immobilized with approximately 40% retention of activity. In laboratory-scale experiments, solutions containing 200–600 mM keto acid were transaminated to the corresponding branched-chain L-amino acid, with a concentration of L-glutamate between 50 mM and 100 mM and a 1.1 molar excess of L-aspartate. Yields obtained for the branched-chain amino acids L-valine and L-leucine are 90% based on starting 2-keto acid [16].

A special case of this three-enzyme coupled system is illustrated by the production of L-alanine, shown in Figure 3.9. In this system, the reaction mixture contains a high concentration of L-aspartate; the reaction is initiated with a small amount of pyruvate and L-glutamate. Glutamic-pyruvic transaminase catalyzes the formation of 2-ketoglutarate and L-alanine. The 2-ketoglutarate is recycled to L-glutamate by transamination by L-aspartate using glutamate transaminase. Again as before, the reaction is driven to completion by decarboxylation of oxaloacetate, which in this case also continuously supplies the starting 2-keto acid pyruvate to be transaminated to L-alanine. The net reaction is the decarboxylation of L-aspartate to L-alanine and CO_2.

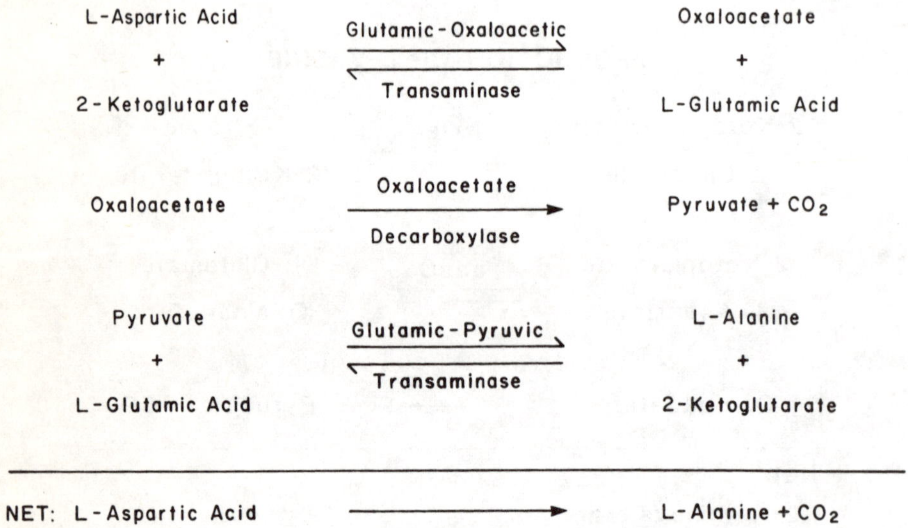

Figure 3.9. Coupled Enzyme System for the Production of L-Alanine.

An amino acid analysis of the product mixture of a reaction in which the starting concentration of L-aspartate was 0.42 M showed that the L-aspartate had been completely consumed, and the products were solely L-alanine plus a small amount of L-glutamic acid, produced by necessity as a result of its use as recycling agent.

In another variation, if it is the 2-keto acid that is the product of interest, the transamination system can be set up to produce these compounds as well. The reaction is described schematically in Figure 3.10. In this system, 2-ketoglutarate can function as the amino group *acceptor*, and a transaminase catalyzes the production of the desired 2-ketoacid plus L-glutamate as the L-amino acid by-product. In order to drive the reaction to completion, it is necessary to use L-glutamate decarboxylase to convert glutamate, again in an essentially irreversible reaction, to 4-aminobutyrate. This sequence of reactions can be carried out using either an L-amino acid or a DL-amino acid as the starting material. In the first case, 4-aminobutyric acid and the desired 2-keto acid are the products. In the latter case, the D-amino acid in the DL-starting material is neither a substrate nor an inhibitor for the transaminase. Thus, at the end of the reaction, the products are 4-aminobutyric acid plus the desired 2-keto acid and the corresponding D-amino acid in equimolar quantities.

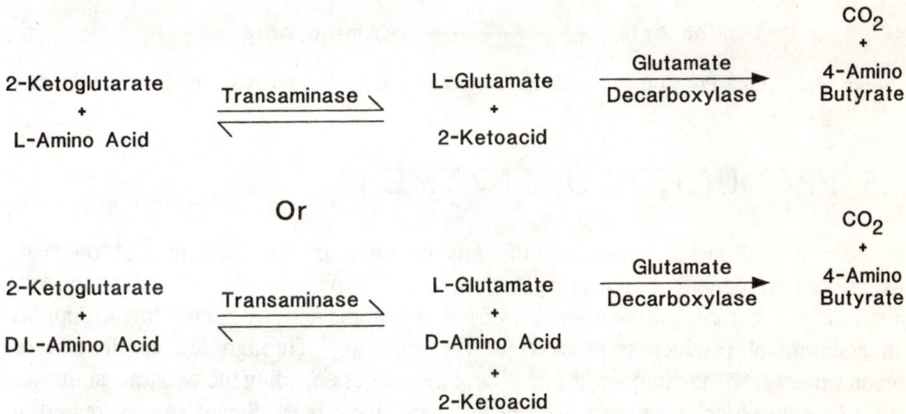

Figure 3.10. Production of 2-Keto acids and D-Amino Acids by Coupling of Transaminase to Glutamate Decarboxylase.

One final coupled enzyme reaction demonstrates the versatility of the transaminase system in biocatalysis. Using a racemic DL-amino acid mixture as the starting material, the enzyme D-amino acid oxidase will convert the D-amino acid in the mixture selectively to the corresponding 2-keto acid. The L-amino acid of the DL- pair is neither a substrate nor an inhibitor of D-amino acid oxidase. If a transaminase is present in the same reaction mixture, the 2-keto acid can be transaminated in the presence of L-aspartate to the corresponding L-amino acid. The entire reaction can be driven to completion as described before by decarboxylation of the oxaloacetate.

Thus, in a single pot, racemic DL-amino acids can be converted directly into optically active L-amino acids (Figure 3.11).

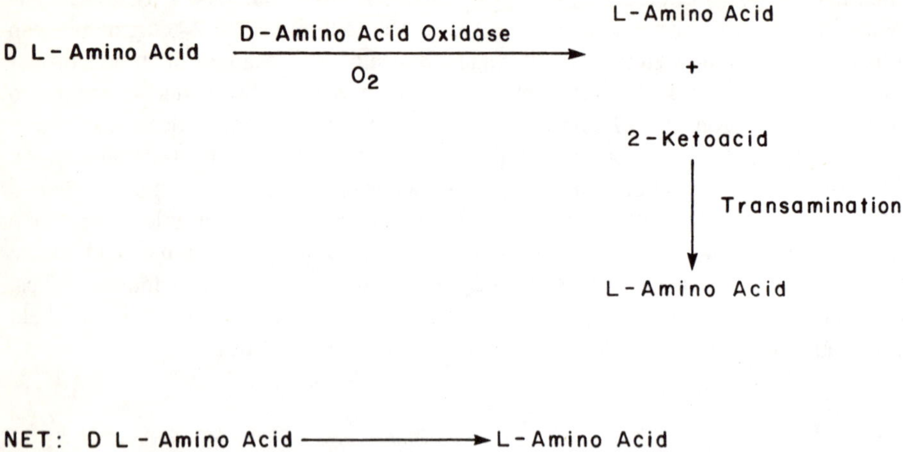

Figure 3.11. One-Pot Coupled Enzyme Conversion of DL-Amino Acids to L-Amino Acids.

3.5 PRODUCTION OF ENZYME

In order to facilitate the production of adequate amounts of transaminase at low cost, we turned to genetic engineering techniques. Transaminases are indeed widely distributed in nature, but the normal levels of enzymatic activity are too low to support an economical production process for the enzyme. Through the application of recombinant DNA techniques, the *asp*C and *ilv*E genes encoding the two transaminases from *E. coli* which have been used in this work have been cloned and expressed at high levels. The *asp*C gene product is a broad substrate range transaminase which has optimal activity on the substrates L-glutamate and L-aspartate, but also transaminates the keto acids corresponding to L-phenylalanine, *p*-fluoro-L-phenylalanine, L-tyrosine, L-homophenylalanine, L-methionine, L-2-aminoadipic acid, and many others. The *ilv*E gene product is a more specific transaminase which has optimal activity on the three branched-chain amino acids L-leucine, L-isoleucine, and L-valine. The genes have been inserted into the P_L expression vector in *E. coli* K-12, allowing expression of both genes at approximately 40% of the total soluble protein in the cell. Induction of expression can be achieved by raising the temperature of the culture from 30 °C to approximately 40 °C. Synthesis of the enzyme by the recombinant *E. coli* was rapid, reaching maximum levels approximately two hours after induction. Such high production levels have significantly reduced the cost of the enzyme and facilitated its isolation prior to immobilization. Figures 3.12 and 3.13 show pictures of polyacrylamide gels in which the crude cell lysates of recombinant *E. coli*, harboring plasmids containing either the *asp*C or *ilv*E genes, have been loaded before and after induction. The heavy bands on each gel have been identified as the *asp*C and *ilv*E gene products respectively, and the protein that is produced is in both a soluble and an active form. The crude enzyme is of sufficient activity that no purification of the enzyme is required prior to immobilization.

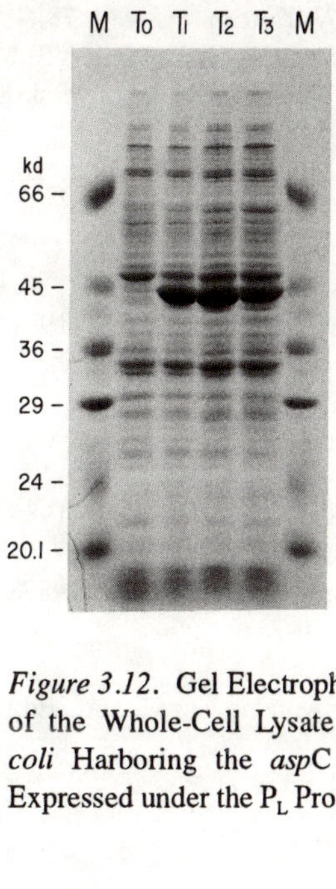

Figure 3.12. Gel Electrophoresis of the Whole-Cell Lysate of *E. coli* Harboring the *asp*C Gene Expressed under the P_L Promoter.

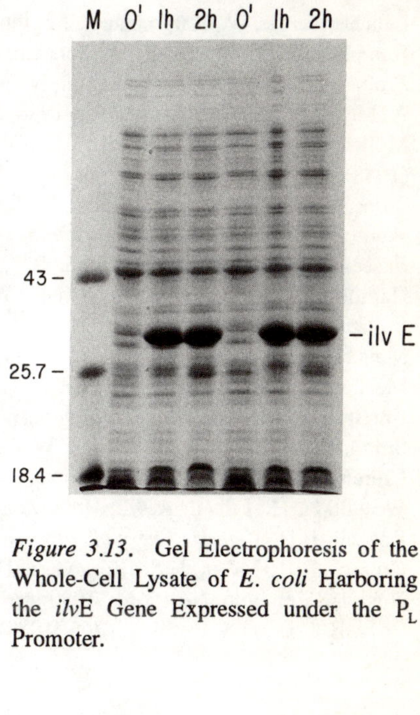

Figure 3.13. Gel Electrophoresis of the Whole-Cell Lysate of *E. coli* Harboring the *ilv*E Gene Expressed under the P_L Promoter.

3.6 SUMMARY

In summary, we have found that transaminases have broad applicability as useful, versatile biocatalysts for the production of L-amino acids, as well as certain 2-keto acids and D-amino acids. Genes encoding two transaminases from *E. coli* have been cloned and expressed at levels of 40% of the total soluble protein in the cell, facilitating the isolation of the protein, and the enzymes have been immobilized to produce a stable, active biocatalyst.

3.7 ACKNOWLEDGMENTS

We enthusiastically thank our co-workers Gwynneth Edwards, Jeff Heier, Jasbir Seehra, Paul Schendel, Tony Schuber and David Hill for their assistance in carrying out this work, and Genetics Institute, Inc., for sponsoring this research.

3.8 REFERENCES

1. Leuchtenberger, W., Karrenbaurer, M., Plocken, U., *Ann. N. Y. Acad. Sci. 434*, 78 (1984).
2. Kamphuis, J., Kloosterman, M., Schoemaker, H. E., Boesten, W. H. J., Meijer, E. M. In *Proc. 4th Eur. Congress Biotechnol.*; Neijssel, O.M., van der Meer, R. R., Luyben, K. C. A. M., Eds.; Elsevier, Amsterdam, 1984; Vol. 4, pp 331–348.
3. Möller, A., Syldatk, C., Schulze, M., Wagner, F., *Enzyme Microb. Technol. 10*, 618 (1988).
4. Tosa, T., Sato, T., Mori, Y., Matuo, Y., Chibata, I., *Biotechnol. Bioeng. 15*, 69 (1973).
5. Wood, L. L. Calton, G. J., *Biotechnology* 2, 1081 (1984).
6. Fusee, M.C., Swann, W. E., Calton, G. J., *Appl. Environ. Microbiol. 42*, 672 (1981).
7. Hamilton, B. K., Hsiao, H. Y., Swann, W. E., Anderson, D. M., Delente, J. J., *Trends Biotechnol. 3*, 64 (1985).
8. Meister, A., *Adv. Enzymol. 16*, 185 (1985).
9. Meister, A., *Ann. Rev. Biochem. 25*, 29 (1956)
10. Christen, P., Metzler, D. E. *Transaminases*; John Wiley and Sons, New York, 1985.
11. Smith, D. L., Ringe, D., Finlayson, W. C., Kirsch, J. F., *J. Mol. Biol. 191*, 301 (1986).
12. Cromm, C. N., Malcolm, B. A., Kirsch, J. F., *J. Am. Chem. Soc. 109*, 2222 (1987).
13. Weetall, H. H., Filbert, A. M., *Methods Enzymol. 34*, 59 (1974)
14. Rozzell, J. D. *Methods Enzymol. 136*, 479 (1987).
15. Crump, S. P., Meier, J. S., Rozzell, J. D. *Biocatalysis*; Abramowicz, D. A., Ed.; Van Nostrand Reinhold, New York, 1990; pp 115–133.
16. Rozzell, J. D., U.S. Pat. 4,826,766 (1989).

THE ENZYMATIC PREPARATION OF L-ALANINE

Gary J. Calton

CONTENTS

4.1 INTRODUCTION

L-Alanine is used in the food industry and in enteral and parenteral nutrition. The use of DL-alanine as a flavorant for pickled foods in Japan has involved large quantities of this amino acid. On the other hand, outside Japan very little use has been made of DL-alanine and, as a result, the world market has remained rather small. In 1980 there were approximately 2,000 metric tons of DL-alanine consumed, while there were only 250 tons of L-alanine consumed. One reason for this difference is the extremes in cost between the racemic and stereospecific forms, with an eightfold increase from the price of the racemic mixture to the price for L-alanine. Recently, there have been additional uses for L-alanine in two drugs, enalopril and captopril, both of which have achieved substantial markets in the United States (reported to total $900 million in 1986). In addition, although L-alanine is not used as a starting material in their chemical synthesis, there are several agricultural products which could take advantage of the L-alanine moiety to provide materials with higher specific activity and lower toxicity, including metalochlor and metalaxyl [1, 2].

The enzymatic production of L-alanine from L-aspartic acid is a long-established method: Meister and his co-workers [3, 4, 5] in 1951 showed that aspartate 4-decarboxylase (EC 4.1.1.12), in conjunction with small quantities of pyridoxal 5'-phosphate and 2-keto acids, produced high yields of L-alanine from aspartic acid (Figure 4.1). Aspartate 4-decarboxylase is an enzyme common to many organisms; the microbe used in this transformation by Meister, *Alcaligenes faecalis*, was obtained by an aspartic acid enrichment procedure on mud taken from the Boston Commons [3]. One gram of soil was placed in a culture medium of salts and 100 ml of 3.76×10^{-2} M aspartic acid. After twenty-four hours a 1-ml aliquot was placed in fresh medium. After one additional transfer the culture was serially diluted and placed on agar. A pure culture of *Alcaligenes faecalis* was obtained. L-Aspartate 4-decarboxylase is a pyridoxal phosphate–dependent enzyme, and α-keto acids act as allosteric activators of the enzyme [6].

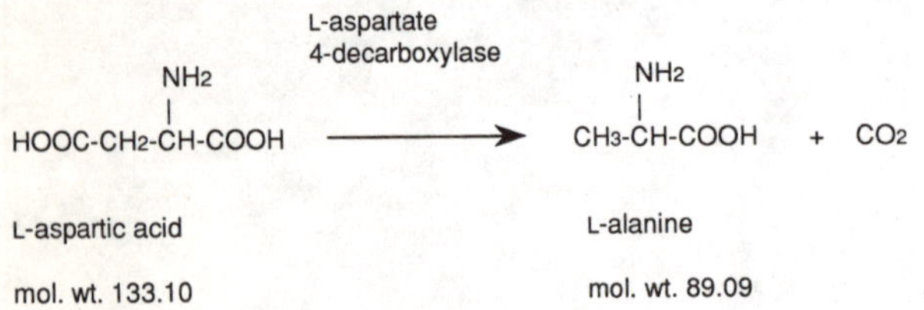

Figure 4.1. Production of L-Alanine from L-Aspartic Acid.

4.2 L-ALANINE PRODUCTION BY *PSEUDOMONAS DACUNHAE*

4.2.1 Batch Fermentations

Since Meister's discovery, a large number of other microorganisms having high levels of aspartate 4-decarboxylase have been obtained. These include *Desulfovibrio desulfuricans*, *Nocardia globerula*, *Acetobacter* sp., and *Acromobacter* sp. [7]. Very little difference was found in the total quantity of L-aspartate 4-decarboxylase between *Alcaligenes faecalis* and the other microbes evaluated [7].

The highest enzyme level was found in a strain of *Pseudomonas dacunhae*. The enzymatic production of L-alanine was reported in isolated yields of greater than 90% from L-aspartic acid when this organism was used, and a titer of 300 g/l of L-alanine could be obtained. No other amino acids were produced in this fermentation. Initially, the enzymatic reaction was carried out as a batch reaction by the addition of L-aspartic acid to a microbial culture broth containing whole cells of *P. dacunhae*. It was also shown that the addition of polyoxyethylene sorbitan monolaurate at 0.5% accelerated the production of L-alanine 2.54-fold. Enzyme activities, measured by the release of CO_2, as high as 3,910 μl CO_2/h/ml culture medium were achieved [7]. It was found that the formation of L-aspartic 4-decarboxylase reached its maximum concentration at the end of the logarithmic phase of growth at 30 °C. L-Aspartic acid could be added to the broth in excess of 70% and served as a buffer due to its low solubility (1%). When the aspartic acid had been completely transformed to L-alanine, the solution no longer contained aspartic acid as a solid. Isolation of the L-alanine was obtained by filtering the fermentation broth and passing the clarified liquor over a strong cation exchange resin. After washing the resin with water, pure L-alanine could be eluted with 5% aqueous ammonia, concentrated, and precipitated by the addition of methanol [8].

Fermentation optimization of *P. dacunhae* to achieve high levels of L-aspartate 4-decarboxylase led to evaluation of a number of dibasic, tribasic, and amino acids to induce production of the enzyme (Table 4.1). Optimum growth and activity were obtained with the addition of 2% sodium L-glutamate. Although the addition of L-glutamate was effective when it was added at any time during the fermentation, it was most effective when added during the early log phase of growth. In a typical experiment using the improved *P. dacunhae* grown under the optimum conditions (100 ml), an equal quantity of L-aspartic acid (100 g) and 7 ml of 28% ammonium hydroxide were added, followed by incubation at 37 °C. The typical course of the reaction is shown in Figure 4.2 [9].

TABLE 4.1 The Effect of Dibasic, Tribasic, and Amino Acids on the Production of L-Aspartate 4-Decarboxylase in *P. dacunhae*

Acid	Growth (mg/ml)	Enzyme activity (U/ml)
none	1.1	1.1
malonate	1.2	1.1
succinate	3.2	3.2
fumarate	4.1	4.5
maleate	1.1	1.1
malate	3.2	3.2
tartrate	1.9	1.0
citrate	3.0	3.3
isocitrate	3.7	4.1
cis-aconitate	3.8	4.6
L-alanine	3.9	4.7
L-aspartate	4.2	4.2
L-asparagine	3.8	3.8
L-glutamate	4.2	9.2
L-glutamine	3.8	7.2
glycine	4.1	7.4
L-leucine	3.6	6.5
L-proline	4.4	8.4
L-serine	3.3	0.3
L-threonine	4.2	6.3

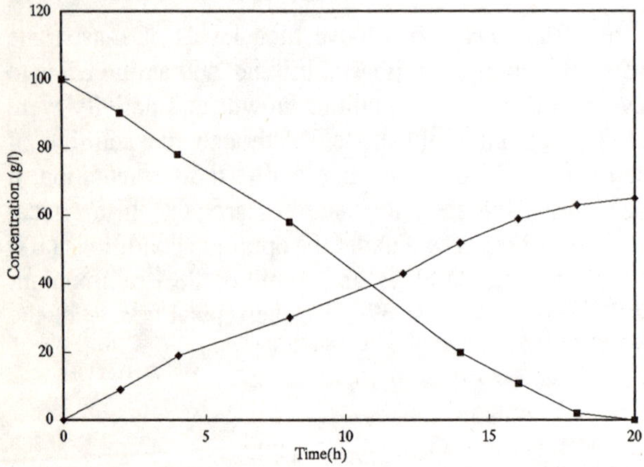

Figure 4.2. Typical Course of Conversion of L-Aspartic Acid to L-Alanine. Key: –■– L-aspartic acid; –◆– L-alanine.

The purification of *P. dacunhae* L-aspartate 4-decarboxylase to its crystalline form proceeded through ammonium sulfate precipitation, heat treatment, and chromatography on DEAE Sephadex™, Sephadex G-150™, and hydroxylapatite to provide crystals of the enzyme. The crystalline enzyme had an activity of 89 U/mg, where each unit was able to produce 1 μmol of carbon dioxide per minute. The crystalline form had 16 mol of pyridoxal phosphate as coenzyme per mol of enzyme and had a molecular weight of 800,000 daltons by gel filtration [9].

Although the production of 300 g/l of L-alanine represents an excellent yield by batch fermentation, the difficulties involved in commercial production included a high cost of capital equipment, an expensive purification, and a single use of the enzyme. A reduction in the cost of production by immobilization of the microbe was an obvious commercial goal.

4.2.2 Acrylamide-Immobilized Cells

The use of immobilized cells for the production of L-alanine was first accomplished by immobilizing *P. dacunhae* by polymerization in acrylamide. *N,N'*-methylenebisacrylamide and β-dimethylaminopropionitrile with potassium persulfate as catalyst were added to a suspension of *P. dacunhae* in physiological saline solution and allowed to stand at 25 °C for 10 minutes [10]. The cells were grown in a broth of 0.5% ammonium fumarate, 1% ammonium fumarate, 0.55% corn steep liquor, 1.8% peptone, and salts. This acrylamide polymer gel could be ground into small pieces of semirigid gel, but the flow properties are poor due to the irregular particle shape and size as well as its compressibility. The particles were placed in a 1.6-cm × 19-cm column and a 1 M solution of ammonium aspartate at pH 5.5 containing 10^{-4} M pyridoxal phosphate was passed through the column at 37 °C and converted to L-alanine at various flow rates, as shown in Figure 4.3. The maximum flow rate for achieving 100% conversion was 15 ml/h (0.2 SV/h). With the addition of pyridoxal phosphate, the column gave no decrease in activity for 21 days; however, without the addition of pyridoxal phosphate, the enzyme activity had decreased to 78% of the original activity in 3 days. It was found that the addition of Co^{2+} to the medium increased the operational stability of the catalyst from 36 days at 100% of original activity to an operational stability at 100% of 54 days (Figure 4.4).

One of the problems in running a reaction which produces carbon dioxide in a column is the occlusion of the catalyst surface by the gas produced. As an alternative to the column reaction, a batch process was carried out. The gel (24 g) containing the immobilized *P. dacunhae* was placed in a solution of ammonium L-aspartate (500 ml) and allowed to react under the conditions above. The reaction was 90% complete in 8 h and complete in less than 24 h (intermediate points were not reported) [10].

Bis(acrylamidomethyl) ether and β-dimethylaminopropionitrile were found to be not as effective as *N,N'*-propylenebisacrylamide and acrylamide for immobilizing

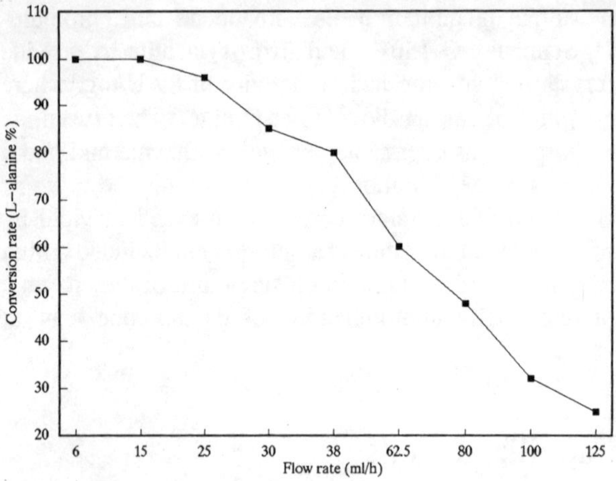

Figure 4.3. Conversion of L-Aspartic Acid to L-Alanine at Various Flow Rates.

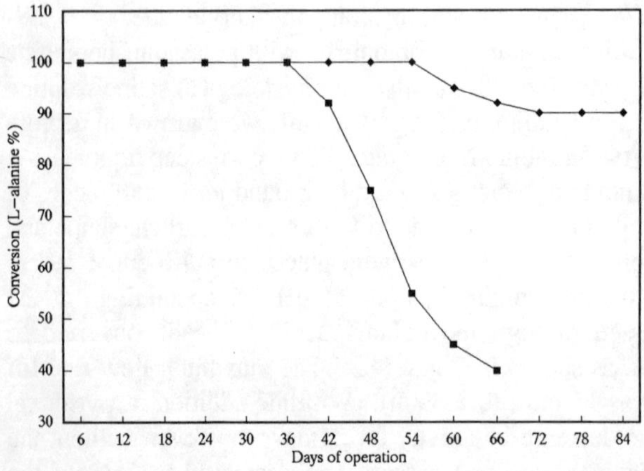

Figure 4.4. Conversion of L-Aspartic Acid to L-Alanine with and without Added Cobalt Ion. Key: –■– No Co^{2+}; –◆– Co^{2+} added.

P. dacunhae, as the reactivity in a stirred batch was 25% slower for equivalent amounts of immobilized cells. The same was true of *N,N'*-propylenebisacrylamide and *N,N,N',N'*-tetramethylethylenediamine. *N,N'*-methylenebisacrylamide and *N,N,N',N'*-tetramethyethylenediamine were found to be 17% less effective for immobilization. The acrylamide-based process apparently could not compete economically with the batch fermentation process since it does not appear to have been commercialized [10].

4.2.3 κ-Carrageenan–Immobilized Cells

The immobilized cell process was improved upon by the introduction of κ-carrageenan–immobilization of *P. dacunhae*. This immobilization was carried out by mixing a 50% (weight/volume) solution of *P. dacunhae* cells (concentrated by centrifugation) in a 5% κ-carrageenan solution, cooling to form a gel and hardening by incubating in 2% KCl solution at 5 °C [11]. Further stabilization of the polymer network was obtained by suspending the cubes cut from the gel in a solution of KCl. The activity of the immobilized cells was 89% that of the native cells (Table 4.2) after sonication. However, activation of the immobilized cells by standing in substrate solution at 37 °C did not increase the activity level to that of the sonicated cells.

TABLE 4.2 Effects of Immobilization and Activation on the L-Aspartate 4-Decarboxylase Activity of *P. dacunhae* Cells

Preparation	Activation time (h)	Activity (units)	Relative activity (%)
Intact cells	0	510	4
Sonicated cells	0	13500	100
Sonicated immobilized cells	0	12000	89
Immobilized cells	0	560	4
	20	3480	26
	44	8020	59
	68	7850	58
Sonicated glutaraldehyde-treated immobilized cells	0	10450	78
Glutaraldehyde-treated immobilized cells	0	2300	17
	20	7000	52
	44	6880	51

It was found that the L-aspartate 4-decarboxylase activity could be enhanced by the addition of pyridoxal 5'-phosphate, as indicated earlier by Meister and co-workers [3, 4, 5]. The effect of pyridoxal phosphate on the carrageenan-immobilized cells was to decrease the rate of the loss of activity of the cells and also to increase the activity

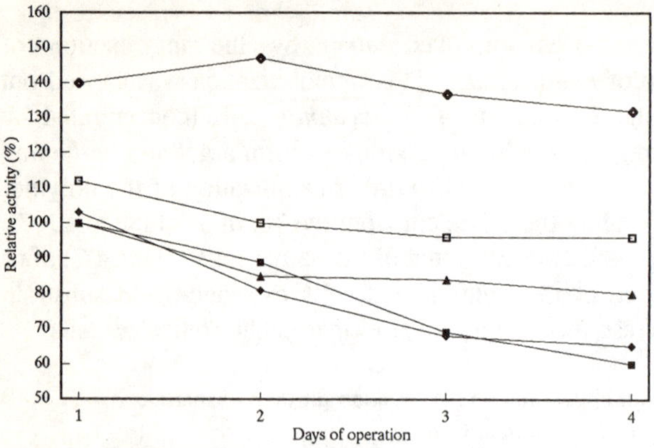

Figure 4.5. Effect of Pyridoxal 5'-Phosphate (PLP) on L-Aspartate 4-Decarboxylase Activity of Carrageenan-Immobilized *P. dacunhae*.
Key: –■– no added PLP;
–◆– 0.001 M PLP;
–▲– 0.01 M PLP;
–□– 0.1M PLP;
–◊– 1 M PLP.

of the immobilized cells (Figure 4.5). The pH optimum of both free and immobilized cells was between 5.5 and 6.0, but the optimum was slightly broader for the immobilized cells. Glutaraldehyde treatment of the immobilized cells did not affect the pH optimum. The optimum temperature for activity of the immobilized cells was 57 °C, while the free cells had an optimum at 53 °C. When 12.6 g of glutaraldehyde-treated carrageenan-immobilized cell preparation was placed in a 1.6-cm × 12-cm column and a 2 M ammonium L-aspartate solution at pH 6.2, containing 0.1 M pyridoxal phosphate, was passed upward through the column, a retention time of 6 h or more was required to obtain complete conversion. The limitation on this procedure is the carbon dioxide produced in the reaction, which makes a column difficult to run [11].

Chibata and his co-workers at Tanabe found that there was very little difference in the enzyme activity as a result of varying the growth phase in which the *P. dacunhae* cells were harvested. However, there was an effect on the operational stability of the cells which had been immobilized in κ-carrageenan and treated with glutaraldehyde [12]. When they investigated the effect of cross-linking the κ-carrageenan with hexamethylene diamine and glutaraldehyde, an increased catalyst lifetime was obtained. Various ratios of hexamethylenediamine and glutaraldehyde and treatment times were tried, with clear-cut results showing that glutaraldehyde at 0.1 M and hexamethylenediamine at 0.1 M were superior. The half-life under these conditions was 81 days if the treatment time was 30 minutes. This system retained 63% of the initial activity of the cells.

Superior results were obtained with L-lysine at 0.4 M and glutaraldehyde at 0.4 to 0.5 M, as the half-life was 102 and 103 days, respectively. All of this data was obtained in a continuous enzyme reactor since the column mode of operation was not practical. In this system, the retained enzyme activity was 72% that of the free cells. The effect of pH in this system was the same as that obtained for the κ-carrageenan-immobilized cells. No significant differences were seen in the pH profile regardless of the method of cross-linking. The cross-linking mixture of L-lysine and glutaraldehyde was superior in every respect to the use of hexamethylenediamine and glutaraldehyde [12], which had been so successful for this group in the production of aspartic acid.

4.3 BIOREACTOR DESIGN CONSIDERATIONS

4.3.1 Problems Due to Carbon Dioxide Evolution

The main problem in the efficient conversion of L-aspartic acid to L-alanine is not the catalyst stability or microbial enzyme activity level but rather the lack of a suitable bioreactor. In the conversion of aspartic acid to alanine, the production of carbon dioxide is stoichiometric and nearly 50 liters of CO_2 is evolved for each liter of reaction mixture after a 2 M addition of aspartate. Although a large quantity of CO_2 can be dissolved in the solution by allowing the pH to rise, the activity of the enzyme is decreased at these pH levels. The quantity of CO_2 evolved cannot be dissolved at constant pH near the optimum for the enzyme. The soft nature of carrageenan gels prevents the use of fluidized bed reactors or continuous stirred tank batch reactors with gentle agitation, since these gels will not withstand the abrasion which occurs in either of these reactor types. The tremendous evolution of CO_2 prevents column operation since CO_2 entrapped between the beads substantially decreases the available catalyst-substrate interaction. To overcome this problem, a novel type of bioreactor was required.

4.3.2 Pressurized Reactors

Furui and Yamashita [13] investigated pressurized reactor methods for the production of L-alanine. They found that the CO_2 was freely evolved from a stirred batch reactor when the pH was allowed to move as the reaction proceeded, but the enzymatic activity rapidly decreased under these conditions. When the pH was held constant at 6.2, however, only 15–16% of the CO_2 was evolved. In a pressurized system, the conversion of the L-aspartic acid to L-alanine was slightly more rapid than that obtained at atmospheric pressure.

The stability of the immobilized system in a pressurized column (1470 kPa-G) was equivalent to that of a similar column under atmospheric pressure. Thus, a pressurized column reactor at a constant pressure of 780 kPa/g gave maximum enzymatic production of L-alanine when a solution of 1.22 M ammonium aspartate at pH 6.2 containing 0.1 mM pyridoxal 5'-phosphate was incubated in a 720-ml closed vessel containing 90 g (wet weight) of immobilized cells. The substrate solution was recirculated continuously and was converted to L-alanine with a residence time of 2.5 h. The reaction rate was increased by the increased CO_2 pressure in the pressurized column as compared to an unpressurized batch reactor. No pH change was seen in the reaction medium. Also, no change in the catalyst decay rate was seen in the pressurized reactor, as compared to an unpressurized reactor, thus providing approximately a 50-day half-life [13].

4.3.3 Reactors Containing Both *E. coli* and *P. dacunhae*

Takamatsu and co-workers [14] attempted to continuously produce L-alanine from fumaric acid using two microorganisms in a single reactor (Figure 4.6). They found a significant reduction of yield due to side reactions. The side reactions of *E. coli* containing high levels of aspartase consisted primarily of the production of L-malic acid by fumarase from fumaric acid (16% of aspartase level) and alanine racemase (1.5% of aspartase level). Both of these side reactions could have been significant problems. It is extremely difficult to separate D-alanine from L-alanine. The formation of L-malic acid from fumaric acid is reversible; however, the possibility of a slower reaction rate existed. They found that the alanine racemase activity of the *E. coli* cells was more stable than the aspartase activity between pH 8 and 10, while the alanine racemase activity was less stable at pH 4 to 5. The fumarase activity was much more stable than the alanine racemase, however the use of 10 mM aspartic acid at pH 5 rather than 1 M ammonium fumarate resulted in the reduction of the fumarase to 10% of its original level. This meant that the side reactions in *E. coli* could be eliminated by heating the cells to 45 °C for one hour at pH 5 in aspartic acid. The alanine racemase activity was less stable than the L-aspartate 4-decarboxylase activity when the pH was changed from pH 7.

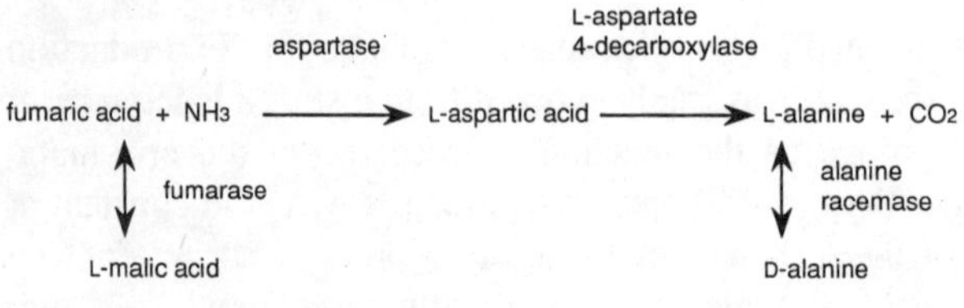

Figure 4.6. Production of L-Alanine from Fumaric Acid.

The use of pH adjustment alone was not sufficient to eliminate the fumarase activity in *P. dacunhae*. Side reactions of the *P. dacunhae* cells (8.0% fumarase, 0.5% alanine racemase) were eliminated by treating this culture at 30 °C for one hour at pH 4.75. The L-aspartate 4-decarboxylase activity could be stabilized to some degree during this treatment by the addition of 10 mM aspartic acid, 1 mM pyridoxal phosphate, 1 mM pyruvic acid, and 1 mM Co^{2+}.

Subsequent immobilization of these two microbes in κ-carrageenan allowed the production of L-alanine in a singe reactor without the production of L-malic acid and L-alanine (Figure 4.7). These investigators obtained 1 M L-alanine in 8 h in a nonpressurized batch reactor in which 5 g of immobilized *E. coli* and 10 g of immobilized *P. dacunhae* cells were mixed with 50 ml of 1 M ammonium fumarate, pH 6, containing 1 mM magnesium and 0.1 mM pyridoxal 5'-phosphate.

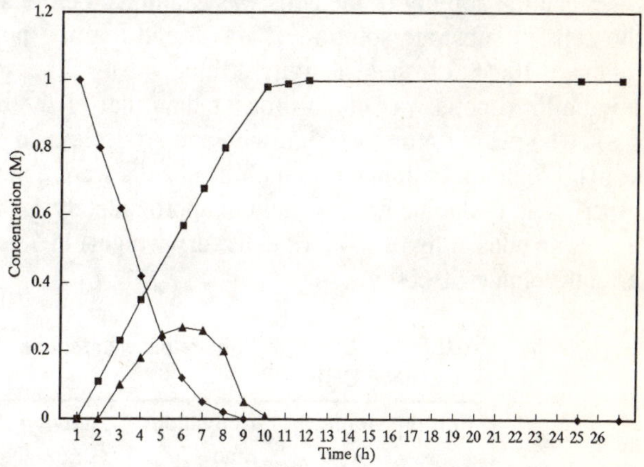

Figure 4.7. L-Alanine Production with Immobilized *E. coli* and *Pseudomonas dacunhae* in a Batch Reactor. Key: –■– L-alanine; –◆– fumarate; –▲– aspartic acid.

4.3.4 Membrane Reactors

The continuous production of L-alanine from ammonium fumarate in a two-stage membrane reactor was carried out by Jandel *et al.* [15]. To prevent side reactions of conversion of fumarate to malate by fumarase and the conversion of L-alanine to D-alanine by alanine racemase, the aspartase was isolated from *E. coli*. This 21-fold purification was accomplished by three consecutive liquid-liquid extractions using polyethylene glycol/dextran and two polyethylene glycol/salt systems after the manner of Kula [16]. The L-aspartate 4-decarboxylase was isolated from *P. dacunhae* by cell disruption, ammonium sulfate precipitation, heat treatment and DEAE Sephadex™ chromatography. The soluble enzymes were then used continuously in two enzyme membrane reactors. The aspartase system had a deactivation constant of 6.5%/day, while the aspartate 4-decarboxylase system was completely deactivated in 2 days. Due to the low half-life of the enzymes and their cost of isolation, this system is not viable commercially.

4.4 POLYURETHANE-IMMOBILIZED CELLS

A group at W. R. Grace and Co. immobilized whole cells of *P. dacunhae* in a hydrophilic polyurethane foam [17]. This system is extremely convenient for the immobilization of microbial cells. A cell paste (approximately 80% water) is mixed with the polyurethane prepolymer (Hypol™) [18] at room temperature and mixed well. The isocyanate reacts more rapidly with the amine groups on the cell than with the hydroxyl groups in the water; cells become covalently attached as well as entrapped by the secondary reactions of the water and the isocyanate groups.

When the activity of the cells was examined before and after activation by storing the cells in substrate solution, it was found that the polyurethane-immobilized cells retained 100% of their activity (Table 4.3). The temperature optimum of the immobilized cells was much broader than that of the nonimmobilized cells (Figure 4.8). The pH optimum was 5.5, with a rapid decline on either side. It was shown that at pH 5.5, the polyurethane-immobilized cells were able to convert 1.7 M ammonium aspartate to L-alanine in 98% yield in approximately 3.5 hours, when recirculated over a column consisting of 6.4 g of cells (dry weight) in 53 g of foam particles occupying a total volume of 600 ml.

TABLE 4.3 Effect of Polyurethane Immobilization on *P. dacunhae* Cells

Cell preparation	Activation time (h)	Activity (U/g cell dry wt)	Retention of activity (%)
Intact	24	8,025	
	120	51,061	100
Polyurethane-immobilized	24	46,549	91
	120	51,061	100

Unfortunately, the structural stability of the hydrophilic polyurethane is very low, and agitation of the immobilized catalyst to decrease CO_2 occlusion of the catalyst surface was not possible. The stability of the L-aspartate 4-decarboxylase activity was increased by the addition of pyridoxal phosphate, as shown by earlier workers in the field. The lifetime was also extended significantly by the addition of ketoglutaric acid (Table 4.4). The use of α-ketoglutarate not only improved the stability but also restored a great deal of the activity. The lifetime of the polyurethane system was 2.5% loss in 39 days when used in a semicontinuous manner in a batch reactor. Since

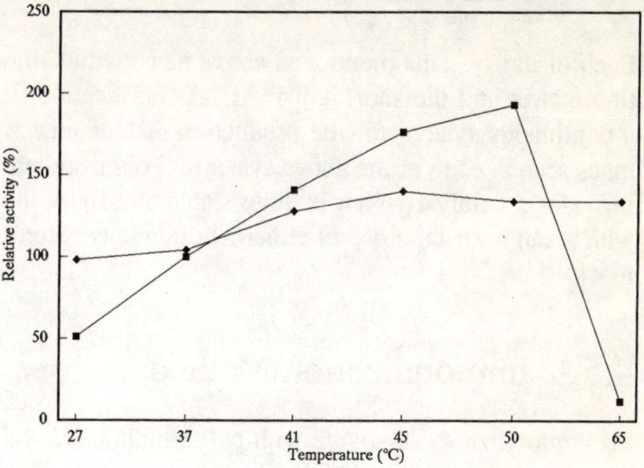

Figure 4.8. Effect of Temperature on *P. dacunhae* L-Aspartate 4-Decarboxylase Relative Reaction Rate. Key: –■– intact cells; –◆– polyurethane-immobilized cells.

TABLE 4.4 Stability of *P. dacunhae* L-Aspartate 4-Decarboxylase Activity before and after Polyurethane Immobilization.

Addition	Days of use[a]	Time to equil. (h)	Average conversion rate (mmol/l/min)	Activity loss (%)
None	1	12	8.03	
	31	23	4.19	48
	32[b]	16	6.02	25
0.1 mM pyridoxal phosphate	1	12	8.03	
	32	15	6.34	21
0.1 mM pyridoxal phosphate + 0.1 mM α-ketoglutaric acid	1	12	8.03	
	32	12.3	7.84	2.5

[a] This refers only to the number of days batch reactions were run; no reactions were run on weekends.

[b] This assay contained 0.1 mM pyridoxal phosphate and 0.1 mM α-ketoglutaric acid.

each run was only 12 h and this type of discontinuous operation does impart stability to some systems, the actual lifetime probably does not approach commercial viability [13].

4.5 POLYAZETIDINE-IMMOBILIZED CELLS

Each of the systems mentioned above has obvious drawbacks due to the immobilization matrix and the short half-lives. At Purification Engineering, Inc., we developed a continuous system for the production of L-alanine which overcomes the disadvantages seen in each of the above systems. Polyazetidine immobilization of *P. dacunhae* provides a catalyst which is quite stable and rigid, thus allowing long lifetimes, and which can withstand use in either a continuous stirred tank reactor or a fluidized bed reactor [19].

4.5.1 Immobilization Method

To immobilize *P. dacunhae* with polyazetidine, the cells were prepared by fermentation of *P. dacunhae* in the literature-optimized culture medium under aerobic conditions for 23 h. A mixture of 3.62 g (wet weight) of cell paste (80% moisture) and 3.6 g of aqueous polyazetidine solution (Hercules Polycup™ 172) was stirred to homogeneity at 25 °C by hand mixing with a wooden stick. This mixture was dispersed on 3.6 g of Amberlite IRA 938 ion exchange beads (Rohm & Haas) which had been previously air dried at 25–35 °C for 60 hours to remove more than 85% of the moisture associated with the beads. These beads, as received from the manufacturer, are quite friable. The thin film of the mixture on the beads was allowed to air dry at 25 °C for 24 h. The resulting free-flowing beads occupied a volume of 20 ml. At this point the beads are rigid and quite strong. Whereas the IRA 938 beads may be easily crushed between the fingers, the polyazetidine-coated beads can withstand pressures of over 170 kPa. These beads were washed with 5 volumes of normal saline and an 18-ml aliquot was placed in a flask containing 0.5 l of 1.5 M ammonium aspartate solution, pH 5.5 (adjusted with H_2SO_4), which was held at 37 °C. The entire mixture was stirred for a period of 48 h while the pH was maintained by the addition of sulfuric acid as needed. Analysis of the liquid at the end of this period on HPLC (μNH_2 Bondapak column, Waters Assoc.; eluant CH_3CN, 25%; 5 mM $KHPO_4$ pH 4.4, 75%) showed that L-aspartic acid had been decarboxylated to form L-alanine in 98% yield.

4.5.2 Reactors for the Immobilized Cells

Attempting to place these beads in a column is quite futile as the build-up of gas is immediate and the occlusion of the bead surface due to the high production of CO_2 immediately begins to propel the beads out of the column. We were unable to pack a column in substrate solution. On the other hand, in a gently stirred batch reactor the CO_2 is built up on the beads so that the beads gently mix by rising until the bubble size is such that the bead and bubble separate. The bead gently drifts downward until sufficient gas has once again built up for it to rise again, thus aiding the mixing.

For pilot plant experiments, a continuous flow stirred tank reactor was filled with polyazetidine-immobilized *P. dacunhae* with an activity of 4,000 μmol/g cell wet wt/h. The catalyst contained 200 g *P. dacunhae* cells wet wt/l beads and was fed 1.5 M ammonium aspartate, pH 5.5, continuously. The effluent contained 1.5 M L-alanine (> 99% conversion). The pH was controlled by addition of 10 N sulfuric acid to maintain the optimal pH of 6.8. In this reactor, the evolution of CO_2 is not a problem since the produced CO_2 rises to the top of the continuous stirred tank reactor and can be drawn off. As noted above, using either the carrageenan, polyacrylamide, or polyurethane systems, a stirred tank reactor was not suitable due to disruption of the foam/gel particles and subsequent destruction of catalytic activity. The half-life, measured by alanine produced over a 30-day period of continuous operation (Figure 4.9), was projected to be 133 days. This compares quite favorably with the projected half-life of the carrageenan system.

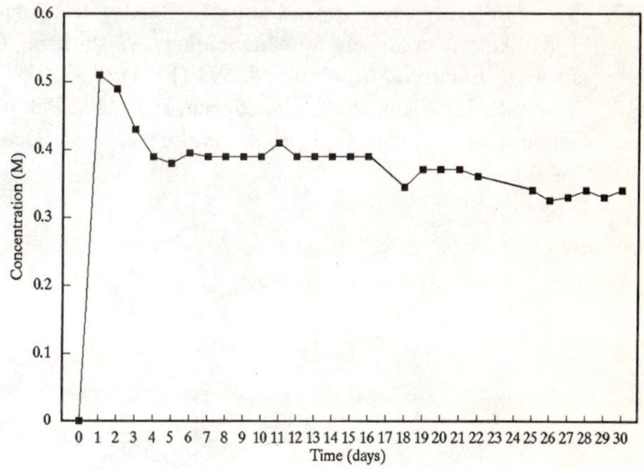

Figure 4.9. L-Alanine Production with Polyazetidine-Immobilized *P. dacunhae* in a Batch Reactor.

The large-scale manufacture of L-alanine by either the pressurized column or by continuous stirred tank reactor are commercial realities which will not be significantly improved upon economically without resorting to an alternative feedstock.

4.6 REFERENCES

1. Calton, G. J. In *ACS Symposium Series 334, Agricultural Biotechnology in Chemistry*; LeBaron, H. *et al.*, Eds.; American Chemical Society, Washington, DC, 1987; pp 181–189.

2. Moser, H., Ribs, G., Sauter, H., Z. *Naturforsch. 87b*, 451 (1982).
3. Novogrodsky, A., Meister, A., *J. Biol. Chem. 239*, 879 (1964).
4. Meister, A., Sober, H. A., Tice, S. V., *J. Biol. Chem. 189*, 577 (1951).
5. Meister, A., Sober, H. A., Tice, S. V., *J. Biol. Chem. 189*, 591 (1951).
6. Tate, S. S., Meister, A., *Biochemistry 8*, 1660 (1969).
7. Chibata, I., Kakimoto, T., Kato, J., *J. Appl. Microbiol. 13*, 638 (1965).
8. Chibata, I., Kakimoto, T., Kato, J., U.S. Pat. 3,458,400 (1969).
9. Shibatani, T., Kakimoto, T., Chibata, I., *Appl. Environ. Microbiol. 38*, 359 (1979).
10. Chibata, I., Tosa, T., Sato, T., Yamamoto, K., U.S. Pat. 3,898,128 (1975).
11. Yamamoto, K., Tosa, T., Chibata, I., *Biotechnol. Bioeng. 22*, 2045 (1980).
12. Takamatsu, S., Yamamoto, K., Tosa, T., Chibata, I., *J. Ferment. Technol. 59*, 489 (1981).
13. Furui, M., Yamashita, K., *J. Ferment. Technol. 61*, 587 (1983).
14. Takamatsu, S., Umenura, I., Yamamoto, K., Sato, T., Tosa, T., Chibata, I., *Eur. J. Appl. Microbiol. Biotechnol. 15*, 147 (1982).
15. Jandel, A.-S., Hustedt, H., Wandrey, C., *Eur. J. Appl. Microbiol. Biotechnol. 15*, 59 (1982).
16. Kula, M. R. In *Applied Biochemistry and Bioengineering*; Wingard, L. B., Jr., Katchalski-Katzir, E., Goldstein, L., Eds.; Academic Press, New York, 1979; Vol. 2, pp 71–95.
17. Swann, W. E., Fusee, M. C., Calton, G. J., Abstracts Annual Meeting Am. Soc. Microbiol. 1981, American Society for Microbiology, Washington, DC, #041; Fusee, M. C., Webber, J., *Appl. Environ. Microbiol. 48*, 694 (1984).
18. Wood, L. L., Hahn, P. A., Hartedegen, F. J., U.S. Pat. 4,312,946 (1976).
19. Wood, L. L., Calton, G. J., *Biotechnology 2*, 1081 (Dec. 1984).

MICROBIAL AND ENZYMATIC PRODUCTION OF D-AMINO ACIDS FROM DL-5-MONOSUBSTITUTED HYDANTOINS

Christoph Syldatk, Ralf Müller, Martin Siemann, Karsten Krohn, and Fritz Wagner

CONTENTS

5.1 INTRODUCTION

The term *hydantoin* was invented when Baeyer reacted the naturally occurring compound allantoin, which he had previously isolated from plant cells (the systematic name today is 5-ureidohydantoin), with HI and isolated sweet-tasting, white crystals (Figure 5.1) [1]. Baeyer named the newly isolated product after the reaction type (hydrogenation) and the precursor (allantoin): "*hyd-antoin*". This name, as well as the names *imidazolidine-2,4-dione* or *2,4-diketotetrahydroimidazole* can be found in the literature. Today "hydantoin" is not only used as the systematic name for the compound shown in Figure 5.1, but also as the group name for a family of different hydantoin derivatives. The most important members of this family are the 5-monosubstituted hydantoins, which are used as precursors for the chemical synthesis of DL-amino acids (see Section 5.2). Other compounds of interest are 5,5-disubstituted hydantoins, which are of pharmacological interest as hypnotic and narcotic agents (for a detailed review see [2].).

Figure 5.1. Formation of Hydantoin from Allantoin by Chemical Reaction with Hydrogen Iodide.

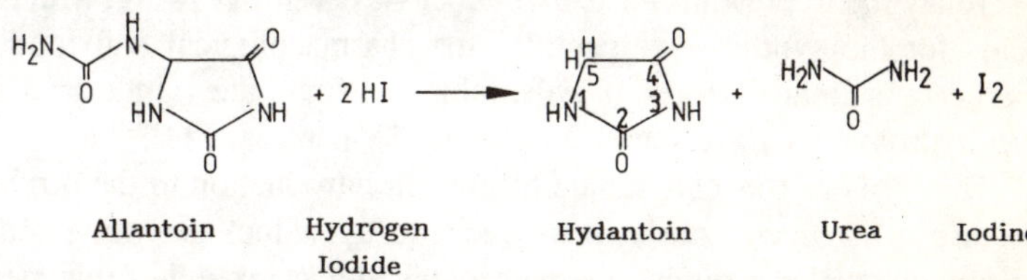

There are reports in the literature from as far back as the 1930s that hydantoin derivatives can be substrates for microbial and enzymatic cleavage. In 1926, Gaebler and Keltch showed that some hydantoin was excreted as hydantoic acid (also called *N*-carbamoylglycine) after injection into dogs [3]. In 1932, Sobotka [4] reported on the enzymatic cleavage of the levorotatory form of 5-phenyl-5-ethylhydantoin by growing cells of an *Aspergillus* sp. In 1934, Wada measured the production of urea when hydantoin was incubated with milk or other tissue suspensions [5]. In 1946, Bernheim and Bernheim [6] detected the ability of liver suspensions of omnivorous animals to rapidly hydrolyze hydantoin, while extracts of other organs or those of herbivores were not active. In 1947, they were able to detect this activity in various plant seeds [7]. In the 1950s, Grisolia *et al.* studied the role of 5,6-dihydropyrimidinase and a so-called *N*-carbamoyl-β-alanine decarbamoylase in the metabolism of pyrimidines [8, 9]. Other investigations on the microbial cleavage of various DL-5-monosubstituted hydantoins, which will be discussed in detail in Section 5.3, showed that the ability to hydrolyze these compounds was widespread in nature, and that D- as as well L-selective enzymes existed.

A biotechnological application of these hydantoin-cleaving reactions became interesting only in the 1970s, along with the growing interest of industry in producing the D-amino acids D-phenylglycine, D-*p*-hydroxyphenylglycine and D-thienylglycine as side chains for semisynthetic penicillins and cephalosporins. Starting from investigations of Dudley *et al.* [10, 11] on the metabolism of racemic forms of *N*-substituted 5-phenylhydantoins, which was postulated to be D-selective, Cecere *et al.* [12] found in 1975 that dihydropyrimidinase (EC 3.5.2.2) from calf liver could be used for the production of several *N*-carbamoyl-D-amino acids from the corresponding DL-5-monosubstituted hydantoins. In 1978, Dinelli *et al.* [13] successfully tried to use immobilized calf liver dihydropyrimidinase for the production of *N*-carbamoyl-D-phenylglycine and its *p*-chloro and *p*-hydroxy derivatives from the corresponding racemic hydantoin derivatives. At the same time Yamada and co-workers published on using microbial cells for this purpose [14, 15, 16, 17, 18]. In the late 1970s and 1980s various groups in Italy [19, 20, 21], Japan [22, 23, 24], France [25], Korea [26, 27] and Germany [28, 29, 30, 31, 32, 33] investigated the D-specific hydantoin cleavage in detail using microorganisms, microbial enzymes, or immobilized biocatalysts.

Today the interest in D-selective processes is still growing, with the use of D-amino acids for the synthesis of peptides for pharmacological and food uses. A current market estimate for D-hydantoinase for the industrial production of D-*p*-hydroxyphenylglycine is 2 million US$ per year [34].

The goals of this chapter are to give an introduction to the production of D-amino acids by D-selective cleavage of racemic hydantoin derivatives, and to review the historical development and the physiological background of this method as well as of industrial processes from an economic viewpoint.

5.2 CHEMICAL SYNTHESES AND PROPERTIES OF *C*-5-MONOSUBSTITUTED HYDANTOIN DERIVATIVES

The history, natural occurrence, and pharmacological importance [35, 36] of hydantoins is briefly mentioned in Section 5.1. Hydantoin derivatives display a broad array of biological activities. Most important is the anticonvulsant activity in the treatment of epilepsy [36, 37, 38, 39, 40]. In addition, antimicrobial [41], virucidal [42], tuberculostatic [43], as well as herbicidal [42, 44] and fungicidal [45] properties are known. The high nitrogen content of these compounds initiated investigations on their biomineralization and use as fertilizer [46]. More information on these topics and hydantoin synthesis can be found in reviews [2, 47, 48, 49, 50]. The purpose of this short review is to present existing methodology in the light of modern retrosynthetic analysis. Emphasis is given to 5-substituted hydantoins as possible substrates for microbial conversion to D- and L-amino acids.

Hydantoins are much more stable and can thus be handled more comfortably than the α-aminonitriles that are the intermediates of the Strecker amino acid synthesis from aldehydes. Bucherer and Steiner [51] in their classical work clearly recognized the importance of the hydantoins in amino acid synthesis: "hat man also eine Methode, um auf einfachem Wege Hydantoine herzustellen, so ist es nicht schwer, daraus die entsprechenden α-Aminosäuren zu erhalten". However, the great stability of hydantoins requires quite drastic conditions for their hydrolysis to α-amino acids. The authors themselves used 60% sulfuric acid at 120–130 °C to effect this transformation. 5-(Phenylmethyl)hydantoin is hydrolyzed with hydrochloric acid in a modern process by Degussa to manufacture DL-phenylalanine [52]. Barium hydroxide is often employed in the alkaline hydrolysis [53, 54; see also 2]. Evidently, both conditions lead to racemic amino acids. The full potential of hydantoins as precursors for amino acids could only be exploited with the recent discovery of stereoselective enzymatic or microbial cleavage to optically active amino acids (see Section 5.3).

5.2.1 Retrosynthetic Analysis and the Cyanate or Isocyanate Method

In a *retrosynthetic analysis*, a target molecule is reduced by breaking bonds to give simpler fragments (*synthons*) in a direction opposite to the implied synthesis. This method of analysis was largely influenced by Corey [55]. If a bond is broken heterolytically the formal charge can be distributed oppositely in the resulting synthons, as is shown in Scheme 1. Breaking the bond between N-3 and C-4 results in the two synthons **I** and **II**.

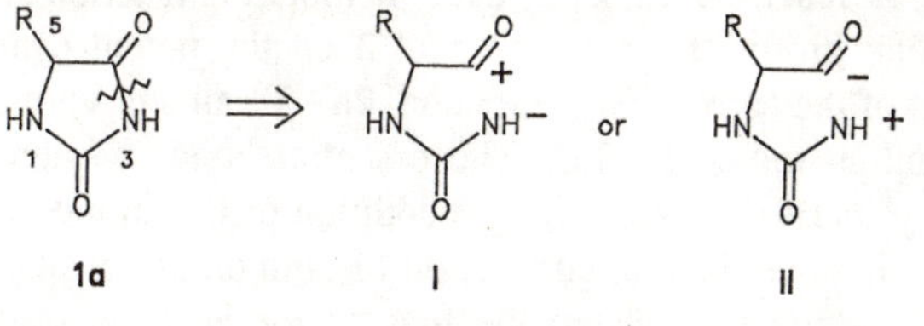

(S1)

The corresponding building block for synthon **I** in terms of the usual polarization of the functional groups could be hydantoic acid, **2** (Scheme 2). The synthesis of a synthetic equivalent for the polarization represented in **II** could be effected by "umpolung" methods but would require much more effort. In fact, **2** is an intermediate in the very efficient synthesis of hydantoins such as **1**, starting from amino acids or esters and alkali metal cyanates [56, 57]. The cyclization is slow under neutral conditions [58] and acid is needed to accelerate the cyclization step, as has been shown by recent kinetic investigations [11, 58, 59, 60]. Most probably the attack of the ureido group at the protonated carbonyl group leads to a transition state like **III**, and elimination of water then yields the hydantoin **1** [58, 59], as shown in Scheme 2. The cyclization can also be induced under mild neutral conditions by addition of dicyclohexylcarbodiimide (DCC) or by activation as the *p*-nitrophenyl ester [61].

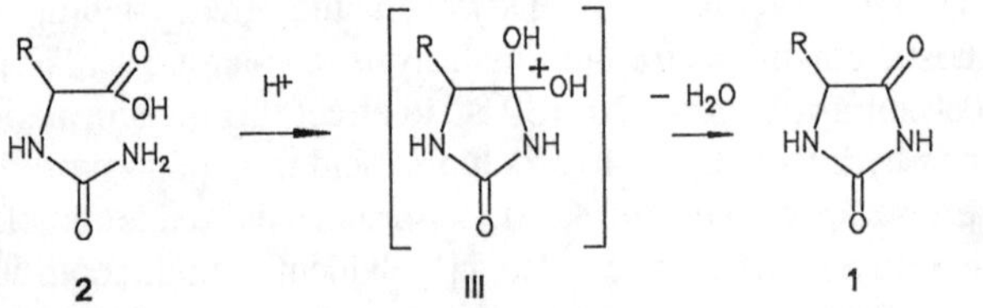

(S2)

It is important to recognize that the equilibrium in acidic medium is shifted towards the cyclic form, **1**; the order of magnitude for *K* is from 10^3 to 10^6 [58]. However, the hydantoin ring can easily be opened under alkaline conditions to afford initially the hydantoic acid, **2**. This is the disadvantage of the alkaline conditions in the cyclization of α-ureido esters [62].

This type of cyclization to the thermodynamically stable product is found very often in hydantoin synthesis. In addition to acids [56, 63], the corresponding esters [57, 64, 65], amides [66], or even nitriles [67, 68, 69] can be cyclized. The reaction usually proceeds in high yield and has even been used to prepare biologically active heterocyclic prostaglandin analogs by Caldwell *et al.* [70]. Tetrabutylammonium fluoride is a mild reagent for the cyclization of carbamates under virtually neutral conditions [71].

The thermodynamically favored cyclization is quite trivial from a chemical point of view. The crucial step in a retrosynthetic analysis can be the formation of the first bond formed in an intermolecular reaction. In the cyanate method this reaction is the attack of the nucleophilic amino group of the amino acid **3** on the potent cyanate electrophile **4** to afford the salt of the carbamoyl compound, **2a**. Treatment with acid then leads to the target hydantoin molecule, **1a**. The essential bond scission for retrosynthetic analysis is shown in **1b** (Scheme 3). This addition of the amino group to cyanate ion is related to the famous thermal conversion of ammonium cyanate to urea by Wöhler [72] in 1828, which represented the first synthesis of an organic compound from "inorganic" material.

The intermediate hydantoic acid, **2**, can be isolated by careful neutralization of the cooled reaction medium [2, 47, 73]. Usually this α-ureido acid precipitates and can be collected by filtration. Interestingly, the reverse process of opening of the hydantoin, **1**, to the carbamoyl compound, **2**, by microbial hydantoinase is the first step in the biochemical transformation of hydantoins to amino acids. The hydantoic

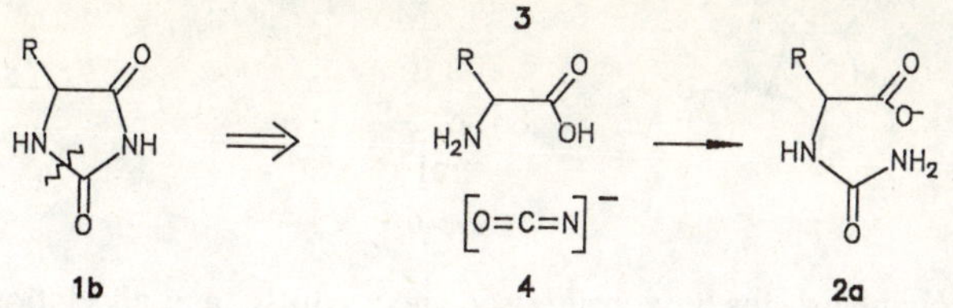

(S3)

acids, **2a**, obtained as intermediates in the reaction of cyanate with amino acids can thus be used for analytic purposes in the analysis of biochemical conversions.

Practical procedures for the conversion of amino acids to hydantoins by treatment with cyanate were developed by Dakin [65] and Read [68] and have found numerous applications (for reviews see [2, 47]). The poor solubility of some amino acids and the corresponding intermediate hydantoic acids (*e.g.*, tryptophan or tyrosine) can be overcome in the modification of Suzuki *et al.* [74] by the addition of about 30% pyridine or acetic acid to the aqueous reaction mixture.

With the exception of glycine, all amino acids yield 5-substituted hydantoin derivatives. An important question in the conversion of optically active amino acids to hydantoins is the integrity of chirality located at C-5. Careful investigations by Dudley [11] and also by Suzuki *et al.* [74] (ORD measurements) showed that the chirality is almost entirely preserved in the *cyclization* step, in contrast to the chemical ring *cleavage* procedures in alkaline or strongly acidic conditions. (For racemization kinetics under alkaline conditions see [75]).

The substituent at C-5 is derived from the amino acid used as the starting material in the reaction with cyanate to hydantoins. *N*-1–substituted hydantoins can similarly be prepared from *N*-alkyl amino acids. Thus, sarcosine (*N*-methylglycine), **5**, was converted to the corresponding *N*-1-methylhydantoin, **6** [76, 77] (Scheme 4). This is in contrast to the Bucherer-Bergs reaction, which yields *N*-unsubstituted hydantoins (see below).

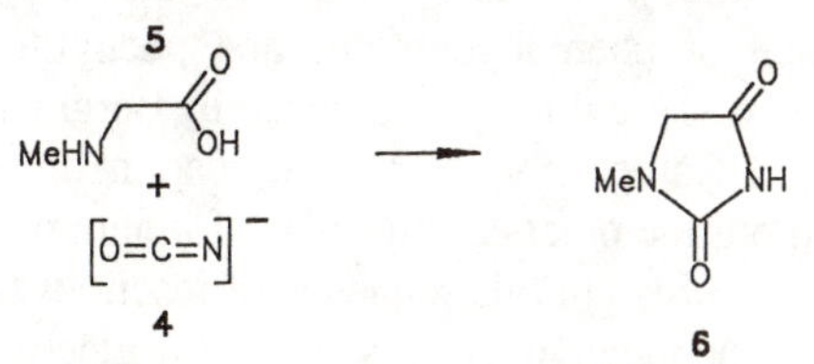

(S4)

Alkylation under basic conditions first occurs at the more acidic position at N-3 (around pH 9 [78]). A number of hydantions of the general formula **1** were methylated at N-3 to the methylhydantoins **7** [79, 80, 81, 82]. The same result was obtained by treatment of **1** with diazomethane [81, 83] (Scheme 5). In addition to

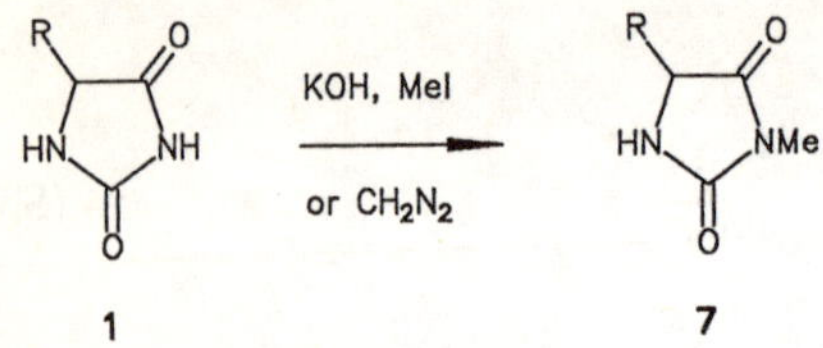 (S5)

aqueous alkali, basic ion exchange resins have been used successfully in *N*-alkylations of hydantoins [84]. Alkylation of 4-thiohydantoins under alkaline conditions leads to *S*-alkylation products [85] (for a review see [50]). *N*-1,*N*-3–Disubstituted hydantoins are obtained in strongly alkaline or phase transfer conditions [86].

The nitrogen at N-3 is also reative in aminoalkylation reactions [83, 87, 88]. The aminomethyl group can be cleaved under mild acidic or alkaline conditions. This was exploited by Orazi and Corral [88] to effect selective alkylation at N-1, affording **9** via the transient *N*-3–aminoalkylation product **8**, as shown in Scheme 6.

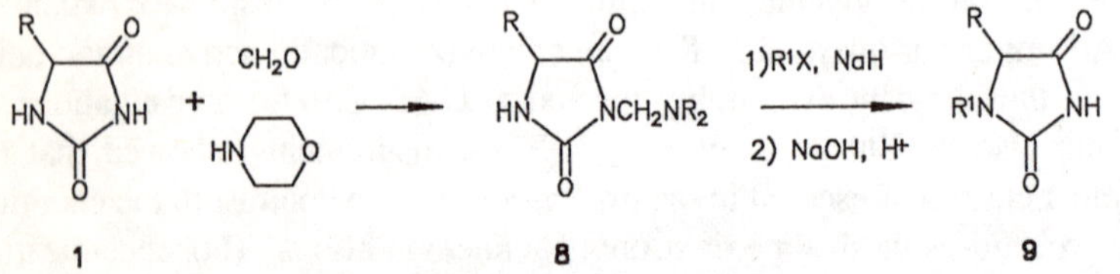 (S6)

N-3–Alkylated or *N*-3–arylated hydantoins can easily be prepared from amino acids [89, 90] or amino nitriles [91, 92, 93, 94, 95] and alkyl or aryl isocyanates. Similarly, the corresponding *N*-3–substituted 2-thiohydantoins are obtained from amino acids and isothiocyanate [64, 96, 97, 98, 99, 100, 101]. The related 2-selenohydantoins have been synthesized in a similar manner.

Occasionally chlorosulfonyl isocyanate [102] or diphenylcarbonate and a secondary amine [103] have also been used in the condensation reaction.

The reaction of peptides with phenyl isothiocyanate is the basis of the famous Edman degradation. Although the potential of phenylhydantoins and phenylthiohydantoins in amino acid and peptide analysis was previously recognized by Bergmann and Miekeley [104] as well as Schlack and Kampf [99] (for identification of the C-terminal end), it was Edman who worked out the practical procedure for automatic peptide analysis [98, 105] (see also [106, 107, 108]). In this sequence of reactions the free amino group of the N-terminal end of the peptide, **10**, attacks the highly electrophilic phenylisothiocyanate, **11**, to afford the thiourea intermediate, **12**. Under acidic conditions cyclization occurs to give *N*-3-phenyl-2-thiohydantoin, **13**, and the new peptide fragment is thereby shortened by one amino acid (Scheme 7). The amino acid incorporated into the thiohydantoin can be easily detected and characterized. For more detailed information on thiohydantoins see a review by Edward [50].

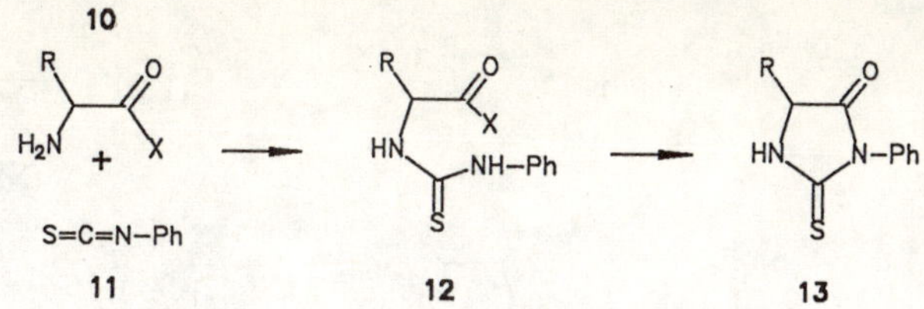

(S7)

5.2.2 The Bucherer-Bergs Reaction

The reaction of α-amino nitriles with cyanates [69, 109] or alkyl or aryl isocyantes [94, 95] resembles the Bucherer-Bergs reaction in which α-amino nitriles such as **16** (Scheme 8) are also early intermediates. The intermolecular N-C bond is formed as shown in **1b** (Scheme 3). The reaction was described in 1929 by Bergs [110] and was investigated later more carefully by Bucherer and co-workers [2, 51, 111]. The process continues to be of great importance due to its flexibility, operational simplicity, and the availability of the reagents. In fact, the procedure is so reliable that ^{13}C-labeled derivatives for pharmaceutical research have been prepared this way [112]. In this multicomponent reaction, aldehydes or ketones are treated in aqueous alcoholic solution at 40–60 °C with cyanide in the presence of excess ammonium carbonate. 5,5-Disubstituted hydantoins are obtained from ketones, and this process was used for analytical purposes in the characterization of ketones [113]. Spiro compounds result from the reaction of cyclic ketones [114, 115, 116, 117]. The unsubstituted hydantoin is an important starting material for many derivatives and is manufactured in a Bucherer-Bergs reaction from glycinonitrile, carbon dioxide, and a tertiary amine, as described in a patent [118, 119]. The 5-monosubstituted hydantoins are valuable intermediates for D-amino acid synthesis and are prepared from aldehydes, cyanide, and ammonium carbonate.

Mechanistic schemes for this transformation were proposed by Bucherer *et al.* [51, 111], but recently more details have been elucidated [114; see also 47]. As in the Strecker amino acid synthesis, an equilibrium is established in the presence of cyanide and ammonia between the aldehyde, **14**, the corresponding imine, **15**, and the α-aminonitrile, **16** [120]. The aminonitrile is carboxylated to **17** in the presence of excess ammonium carbonate, and this product is then cyclized to the oxazolidine intermediate, **18**. A mechanistically interesting step is the subsequent rearrangement of **18** to the thermodynamically stable hydantoin, **20**. It could be shown that an isocyanate such as **19** must be an intermediate in the rearrangement process. Aminonitriles of primary amines fail to undergo the rearrangement because they cannot form the isocyanate intermediate, **19**. In the presence of ammonia the unsubstituted hydantoins are formed by replacement of the secondary amino group of the aminonitrile [121] (Scheme 8). The same mechanistic considerations are supported

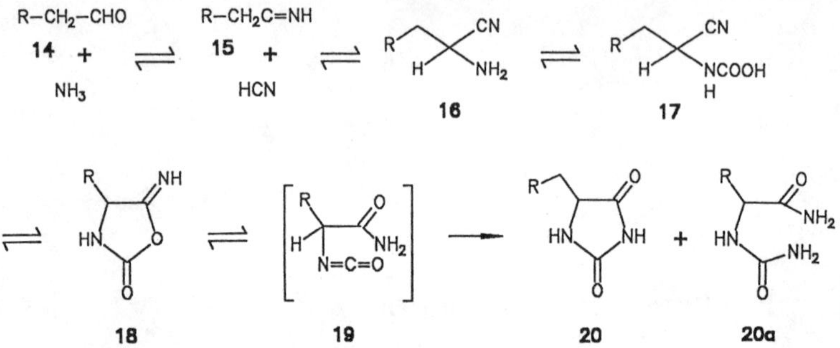

(S8)

by similar observations in the thiohydantoin series [122, 120]. *Thus, in contrast to the cyanate method, the Bucherer-Bergs reaction always yields N-unsubstituted hydantoins.* The second major difference results from the equilibrium in the initial α-aminonitrile formation. Thus, two different diastereoisomeric hydantoins result from the Bucherer-Bergs reaction of 4-*tert*-butylcyclohexanone [114].

A similar isocyanate intermediate, **22**, is expected in the Hofmann-type degradation of malonodiamides or cyanoacetamides such as **21** with sodium hypobromide, affording the 5,5-disubstituted hydantoins **23** [68, 123, 124] (Scheme 9). Isocyanates are also postulated as intermediates in a Schmidt-type degradation of α-cyanocarboxylic acid azides [125].

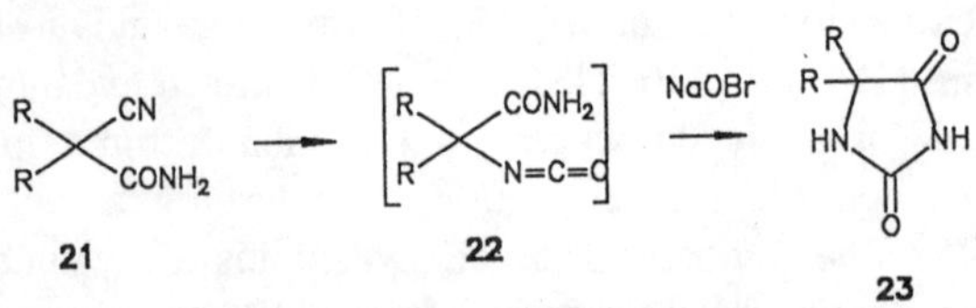

(S9)

A side product in the Bucherer-Bergs reaction may result from addition of ammonia to the isocyanate intermediate **22** to form the hydantoic acid amide **20a**. In fact, some patents describe treatment of the crude Bucherer-Bergs products with acid to complete cyclization to the hydantoin **20** [119, 126]. Careful NMR investigation of the crude products from Bucherer-Bergs reactions revealed traces of hydantoic acid amides [127].

The yields of the Bucherer-Bergs reaction vary from about 20% to 90%, depending on the conditions. In addition to the traditional procedure a number of variations have been introduced. To prevent undesired condensation reactions of the aldehyde, Bucherer himself [111] and others [128] successfully used the bisulfite adducts (*e.g.*, of phenylacetaldehyde). Other derivatives such as oximes, hydrazones, semicarbazones, azomethines, etc., served the same purpose and generally gave good yields [129]. Usually an aqueous alcoholic solution is used to provide sufficient solubility for the cyanide as well as the aldehyde. Detergents have been employed to increase the dispersion of the aldehyde in aqueous medium [130]. In some cases an autoclave was used as the reaction vessel to prevent the evaporation of volatile reagents [36, 131]. Remarkable progress has been made by systematic optimization studies using the simplex evolutionary method. Seven [132] or six [131] parameters, such as concentration of reactants, temperature, time, and solvent composition, have been varied, and yields could be increased to 88%.

5.2.3 Hydantoins from Urea and Derivatives

Urea is a fragment of bond cleavage at **1** and **2** in the hydantoin **1c** (Scheme 10). In fact, in one of the early syntheses of hydantoin, A. von Bayer converted α-bromoacetylurea, **24**, to hydantoin, **1**, by treatment with ammonia. Later, the synthesis of *N*-1–alkylated hydantoins from α-halogen acids, primary amines, and urea (or guanidine) found many other applications [133, 134, 135, 136]. However, in all these syntheses replacement of the halogen by the amine possibly precedes the ring closure by formation of bond **3** as shown in **1c** (Scheme 10). The amino acids, **25**, initially resulting from that replacement can also be used directly as a starting material in the reaction with urea to form hydantoins, **1**. Elimination of ammonia could lead to the intermediate isocyanate, **26**, in a fashion similar to that postulated for the Bucherer-Bergs reaction, but the actual existence of **26** has not yet been confirmed.

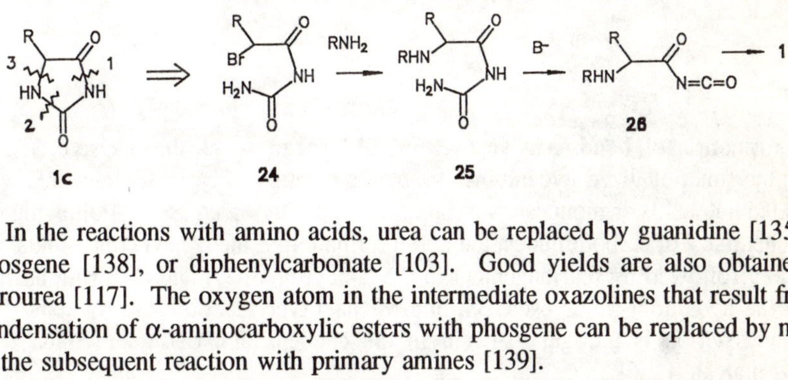

1c **24** **25** (S10)

In the reactions with amino acids, urea can be replaced by guanidine [135, 137], phosgene [138], or diphenylcarbonate [103]. Good yields are also obtained with nitrourea [117]. The oxygen atom in the intermediate oxazolines that result from the condensation of α-aminocarboxylic esters with phosgene can be replaced by nitrogen in the subsequent reaction with primary amines [139].

The reaction of α-oxocarboxylic acids such as pyruvic acid, **27**, with urea, **28**, was studied by several researchers [138, 140, 141, 142] and gives a number of products such as homoallantoic acid, homoallantoin, etc., in addition to the primary cyclization product 5-hydroxy-5-methylhydantoin, **29**. The 5-hydroxy compound, **29**, can easily be dehydrated in glacial acid and a small amount of hydrochloric acid to 5-methylhydantoin [87] (Scheme 11).

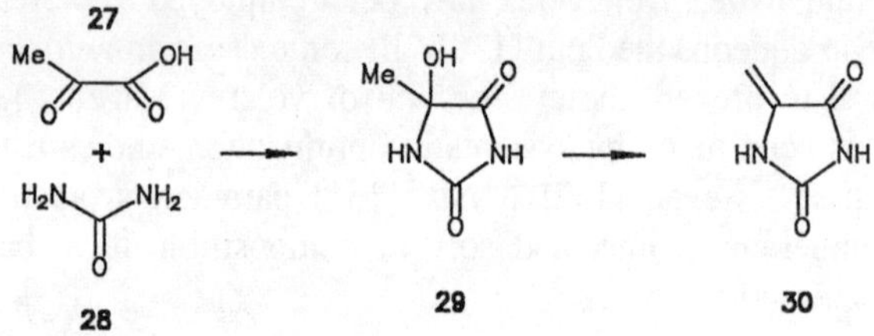

(S11)

5-Methylhydantoin rapidly polymerizes in pure form but can be generated *in situ* by the reaction of indole, formaldehyde and unsubstituted hydantoin, **1**, as exploited in the production of 5-indolylmethylenehydantoin [143] (see Chapter 6 also).

An interesting reaction was discovered by Biltz in the alkaline treatment of 1,2-dicarbonyl compounds such as benzil, **31**, with urea. The intermediate diol, **32**, underwent a pinacol-pinacolone rearrangement affording the 5,5-disubstituted hydantoin, **33** (Scheme 12). Other 1,2-dicarbonyl compounds and substituted ureas react similarly (see [2]). The preparation of unsubstituted hydantoin from glyoxal and urea by a similar mechanism is described by Pauly and Sauter [144] and in a recent patent [145].

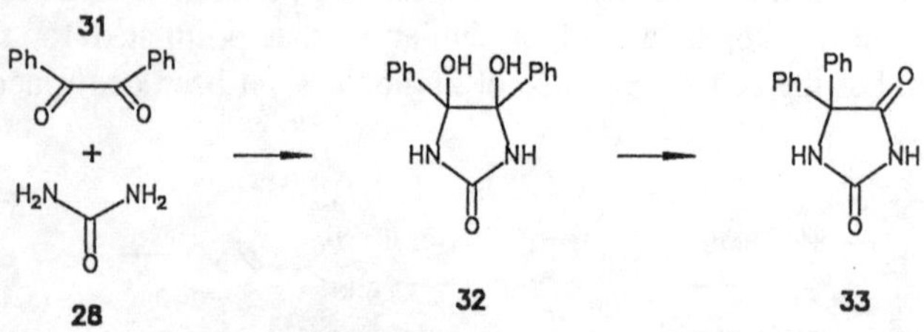

(S12)

The formation of bond 1 in **1c** (Scheme 10) seems to be the decisive step in forming the functionalized hydantoins, **38**, by the reaction of carbodiimides, **25**, with unsaturated ketoacids or monoesters of unsaturated diesters such as **34**. Following the known chemistry of carbodiimides, the initial formation of the *C*-acyl compounds, **36**, is probably followed by rearrangement to the *N*-acyl urea, **37**, and then ring closure to yield the hydantoin acetic ester, **38**, in a Michael-type reaction [60] (Scheme 13). A similar reaction is thought to occur in the condensation of dimethylurea with dimethyl maleate [146].

As in the reaction with carbodiimides, a C_4-carboxylic acid such as maleic acid can also be incorporated in a product, **42**, that results from methoxide-mediated rearrangement of methyl-(*S*)-dihydroorotate, **39** [147] (Scheme 14). In an attempted benzylation of dihydroorotate, **39**, with methoxide in methanol, the benzylated hydantoin, **42**, was isolated. The authors proposed a mechanism in which the dihydroorotate is attacked by methoxide to **40** and opened to **41**, as shown in Scheme 14. Ring closure

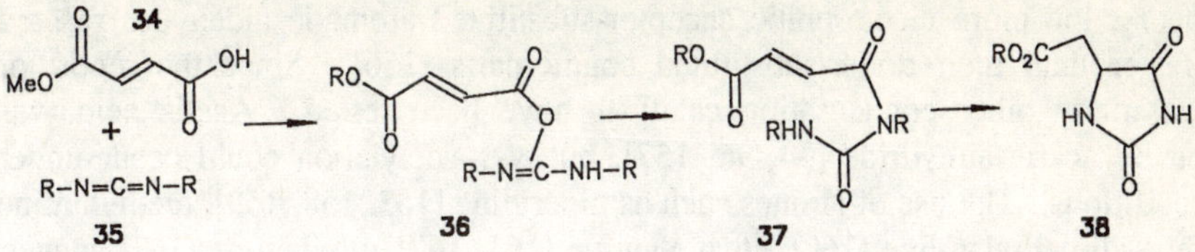

(S13)

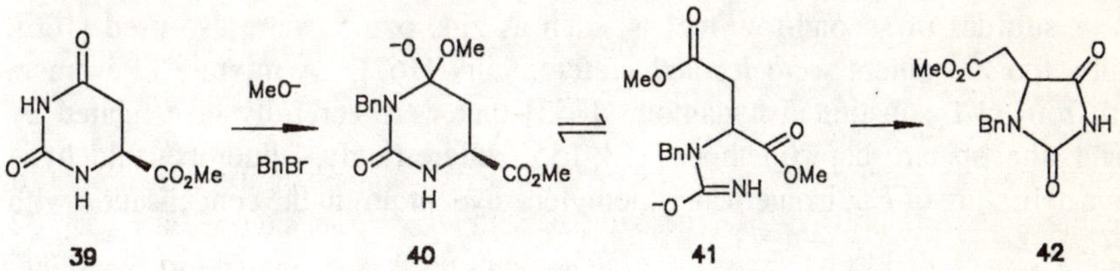

(S14)

preferentially leads to the hydantoin, **42**, once again demonstrating the thermodynamic stability of that ring system.

For related ring contractions of other pyrimidine derivatives see [47]. Ring extension of 3-dimethylamino-2*H*-azirines and dipolar cycloadditions with heterocumulenes (compare [47]) as well as condensation of imines with isonitriles and thiocyanic acid [148] are of minor importance for D-amino acid synthesis.

5.2.4 5-Substituted Hydantoins by Condensation of Hydantoin with Aldehydes

One of the most versatile routes to 5-substituted hydantoins such as **46** proceeds via the condensation of aldehydes, **44**, with unsubstituted hydantoin, **43**, whereas the Michael-type addition of activated nitriles [149] is not significant. This method always yields 5-alkyl or 5-arylmethyl hydantoins (Scheme 15).

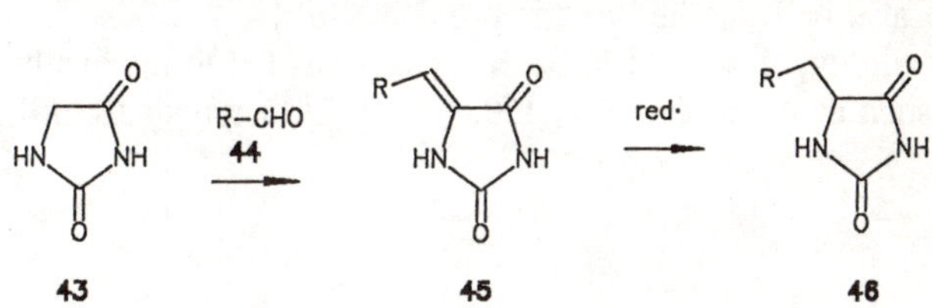

(S15)

The Perkin-type condensation of aromatic aldehydes was first investigated by Wheeler and Hoffman [150], who used sodium acetate in acetic acid as base. This method was later applied to numerous substituted aromatic and heteroaromatic aldehydes (for some examples see [2, 151, 152, 153, 154, 155]). In perfect agreement with theory, the more electrophilic acceptor-substituted aromatic aldehydes reacted much faster than their donor-substituted counterparts [156]. Since the report of Wheeler, many other condensation catalysts have been tested. Acetic acid was replaced by acetic anhydride [54, 96, 157], but *N*-1–acetylation could occur under these conditions. The use of amines such as piperidine [153, 158, 159], diethylamine [151], *N,N*-dimethylaniline [160], ethanolamine [161, 162], ethylamine [163] as well as ammonium acetate [164] or ammonium phosphate [165] has also been described. Oxides or sulfides of second-row metals, such as zinc oxide, were also used [166]. Generally, the *Z* isomers were formed preferentially [167]. A mixture of isomers resulted from *N*-1–substituted hydantoins [153] that were carefully investigated by NMR and other spectroscopic methods [153, 155]. Interestingly, 4-fluorobenzaldehyde also gave a mixture of *E*/*Z* isomeric arylmethylene hydantoins in the condensation with **43** [127].

With enolizable aliphatic aldehydes, side reactions predominated in the condensation reaction [2]. Nicolet [168] condensed the phenylhydrazone of acetaldehyde with the more reactive 2-thiohydantoins [97] to the corresponding ethylene compound. However, two recent patents have described the condensation of ketones [169] and aliphatic aldehydes [170] with hydantoin, **43**, in dilute alkali (pH 9 to 10) in the presence of amino acids such as glycine.

A direct hydrolysis of the olefinic condensation products leads to α-keto acids [171]. For the purpose of amino acid synthesis the reduction to **46** is required. There are a number of methods to reduce the unsaturated hydantoins, **45**. Concentrated hydroiodic acid [151, 172], as well as hydroiodic acid and red phosphorus [150, 173, 174], sodium amalgam [54, 157, 175, 176, 177], aluminum amalgam [150], tin [172] or zinc [178] in hydrochloric acid, hydrogen sulfide [158], and ammonium sulfide [90, 96, 179] have been used. Under many of these rather drastic reaction conditions the hydantoin ring was also cleaved to yield the corresponding amino acid [see 2]. Currently, catalytic hydrogenation with palladium [180] or Raney nickel [181] is the preferred method. Elevated temperatures and hydrogen pressures are required to effect the hydrogenation of the poorly reactive conjugated double bond of **45**. Electrochemical reductions have also been applied to this problem [179].

In particular, the methodology described in Section 5.2.4 has led to the synthesis of valuable amino acids such as tryptophan [77, 159, 175, 182] or tyrosine [150].

5.3 DISTRIBUTION OF HYDANTOIN CLEAVAGE IN NATURE

5.3.1 Naturally Occurring Hydantoin Derivatives and Their Biological Degradation

A naturally occurring hydantoin derivative, which is widely distributed in nature and was briefly discussed in Section 5.1, is allantoin. This compound is an important metabolite in the degradation of purine nucleotides (see Figure 5.2). This catabolic pathway starts from guanosine or adenosine, and leads via hypoxanthine or xanthine to uric acid, which is the end product of the degradation of purines in humans. All other organisms degrade uric acid to allantoin as the next step. This compound is the substrate for the enzyme allantoinase (EC 3.5.2.5), also called 5-ureidohydantoinase. It is found in many organisms—including microorganisms, plants and animals—either in combination with an allantoicase (EC 3.5.3.4) or an allantoate amidohydrolase (EC 3.5.3.9). The latter hydrolyzes allantoin to urea and glyoxylic acid [183], which are the end products of purine degradation in fish. Microbial allantoinases were isolated and characterized by various authors: Lee and Roush [184], Okumura *et al.* [185], and Trijbels and Vogels [186] described this enzyme as inducible and (+)-specific. Besides allantoin, some other compounds possessing a free ureido group like *N*-carbamoyl-L-asparagine, *N*-carbamoyl-L-aspartate (the corresponding D-compounds were ineffective), hydantoate (*N*-carbamoylglycinate) and diureidomethane were found to enhance the synthesis of this enzyme [187]. Investigation of the substrate specificity of allantoinases for other hydantoin derivatives is limited, according to literature reports, but, as an example, 5-aminohydantoin was shown to be accepted as substrate, although poorly [183]. Nonstereoselective allantoin hydrolysis and association of the allantoinase with a cofactor-independent allantoin racemase (EC 5.1.99.3) has been reported [183, 188], enabling some microorganisms to use (–)-allantoin as substrate, too.

Besides allantoin, only two other naturally occurring hydantoin derivatives are described in the literature: carboxymethylenehydantoin and carboxyethylenehydantoin.

L-5-Carboxymethylenehydantoin, which corresponds to the amino acid L-aspartic acid, was isolated by Lieberman and Kornberg after they incubated orotic acid with crude cell extracts of the anaerobic bacterium *Clostridium oroticum* [189, 190]. This compound was postulated to be a by-product in the L-selective degradation of dihydroorotic acid, a six-membered cyclic amide (see Figure 5.3).

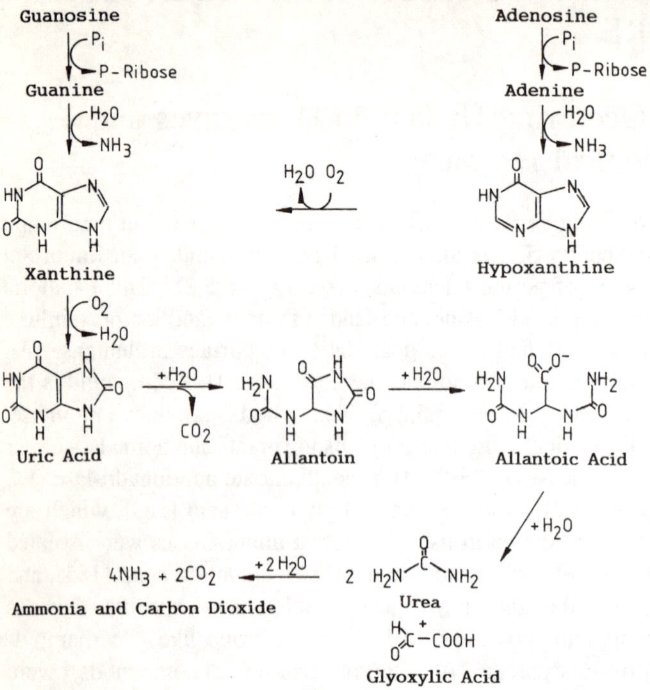

Figure 5.2. Metabolism of Purine Nucleotides and Degradation of Allantoin.

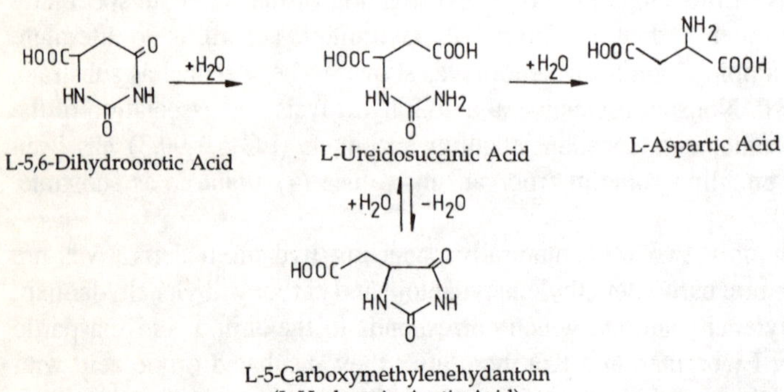

Figure 5.3. Metabolism of L-5,6-Dihydroorotic Acid and Formation of L-5-Carboxymethylenehydantoin.

L-5-Carboxyethylenehydantoin is closely related to the compound mentioned above. It was isolated first by Brown [191] from the urine of rats, monkeys, and humans after feeding ^{14}C-histidine and postulated to be a by-product (see Figure 5.4).

Both carboxymethylenehydantoin and carboxyethylenehydantoin can be used as substrates for L-selective enzymatic cleavages by microorganisms, resulting in the amino acids L-aspartic and L-glutamic acid, respectively. Akamatsu [192] proved

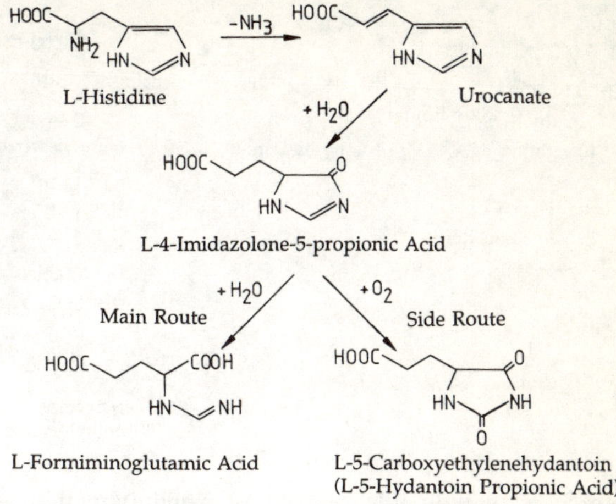

Figure 5.4. Metabolism of L-Histidine and Formation of L-5-Carboxyethylenehydantoin.

by induction experiments that the L-carboxyethylenehydantoinase described also by Tsugawa *et al.* [193] and Hassall and Greenberg [194] for the production of L-glutamic acid from the corresponding hydantoin, is not identical to the enzyme able to hydrolyze L-carboxymethylenehydantoin.

Besides the enzymes discussed above, other enzymes which have six-membered cyclic amides as natural substrates, but are able to cleave chemically synthesized DL-5-monosubstituted hydantoin derivatives too can be found in nature. Thus, 5,6-dihydrouracil is the natural substrate of the so-called "D-hydantoinase", which is discussed in detail in Section 5.3.2.

The dihydroorotase (EC 3.5.2.3) is able to hydrolyze the hydantoin corresponding to L-aspartic acid [192]. It is related to ureidosuccinase (EC 3.5.1.7) [195]. The natural function of these two enzymes was postulated to be a side branch in the biosynthesis of pyrimidine nucleotides [190].

Three other hydantoinases which have not yet been classed in the *Enzyme Nomenclature* are described in the literature [196]. Siedel *et al.* [197] and also Yamada *et al.* [198, 199] found a new ATP-dependent 1-methylhydantoinase in different bacteria. This inducible enzyme, which was also shown to act on unsubstituted hydantoin and 5-methylhydantoin [197], is involved in the degradation of the cyclic amide creatinine after its deimination in the 2- position to 1-methylhydantoin to result in sarcosine (*N*-methylglycine) via *N*-carbamoylsarcosine (*N*-carbamoyl-*N*-methylglycine) [198, 199] (see Figure 5.5). It is associated with a so-called D-*N*-carbamoylsarcosine hydrolase [199] (see also Section 5.4).

Nishida *et al.* [200], Syldatk *et al.* [32, 201, 202], Yamashiro *et al.* [203], and Yokozeki *et al.* [204, 205, 206] found a new hydantoinase and *N*-carbamoyl-L-amino acid amidohydrolase involved in the L-selective cleavage of arylalkylenehydantoins,

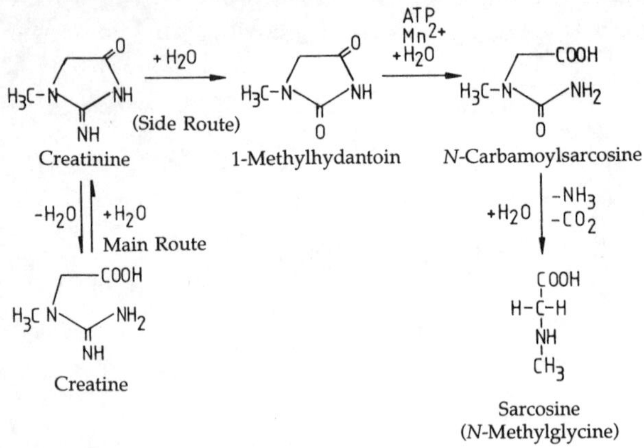

Figure 5.5. Degradation of Creatinine and Formation of 1-Methylhydantoin.

which were inducible best by DL-5-indolylmethylenehydantoin out of a variety of 5-monosubstituted hydantoins and naturally occurring cyclic amides [200, 202, 205]. The natural function of these enzymes is not yet known. This enzymatic pathway is discussed in detail in Chapter 6 [208].

5.3.2 Distribution of D-Hydantoin Cleavage in Nature

In the 1950s and early 1960s there were some publications on the enzyme dihydropyrimidinase (EC 3.5.2.2) and its role in pyrimidine metabolism [6, 7, 8, 9, 192, 209, 210]. In 1970 and 1973 Dudley *et al.* [10, 11] published on the D-selective cleavage of 5-phenylhydantoin to the corresponding *N*-carbamoyl-D-amino acid and on the spontaneous *in vivo* racemization of the residual *S*-isomer. In 1975 Cecere *et al.* [12] first published on the production of other *N*-carbamoyl-D-amino acids using a partially purified fraction of the dihydropyrimidinase from calf liver and stressed that this enzyme might find application for the preparation of optically active D-amino acids. In 1978 the same group [13] published on the production of various *N*-carbamoyl-D-amino acids by an immobilized dihydropyrimidinase preparation (see Section 5.7.3). At the same time Yamada *et al.* were able to show that the reason for the widespread ability in microorganisms to hydrolyze DL-5-monosubstituted hydantoin derivatives D-selectively was this enzyme, too [14, 15].

The natural substrates of dihydropyrimidinase (EC 3.5.2.2) are the six-membered rings 5,6-dihydrouracil and 5,6-dihydrothymine, which are comparable with 5,6-dihydroorotate (see Section 5.3.1). Both 5,6-dihydrouracil and 5,6-dihydrothymine are important intermediates in the degradation of pyrimidine nucleotides. The widely distributed, inducible (see Section 5.4.2) catabolic enzyme dihydropyrimidinase or

D-hydantoinase [196] interestingly is strictly D-selective (see Section 5.6.1) in contrast to the L-selective dihydroorotase (EC 3.5.2.3) which is involved in the opposite anabolic pathway (see Section 5.3.1). In some cases, the dihydropyrimidinase was shown to be associated with a D-specific D-*N*-carbamoylamino acid amidohydrolase, called β-ureidopropionase (EC 3.5.1.6) after its natural substrate *N*-carbamoyl-β-alanine [19, 24, 33, 209] (see Section 5.6.2), or a hydantoin racemase [21, 25]. The mechanism of the enzymatic degradation of dihydropyrimidine in comparison with the enzymatic cleavage of DL-5-monosubstituted hydantoin derivatives is shown in Figure 5.6.

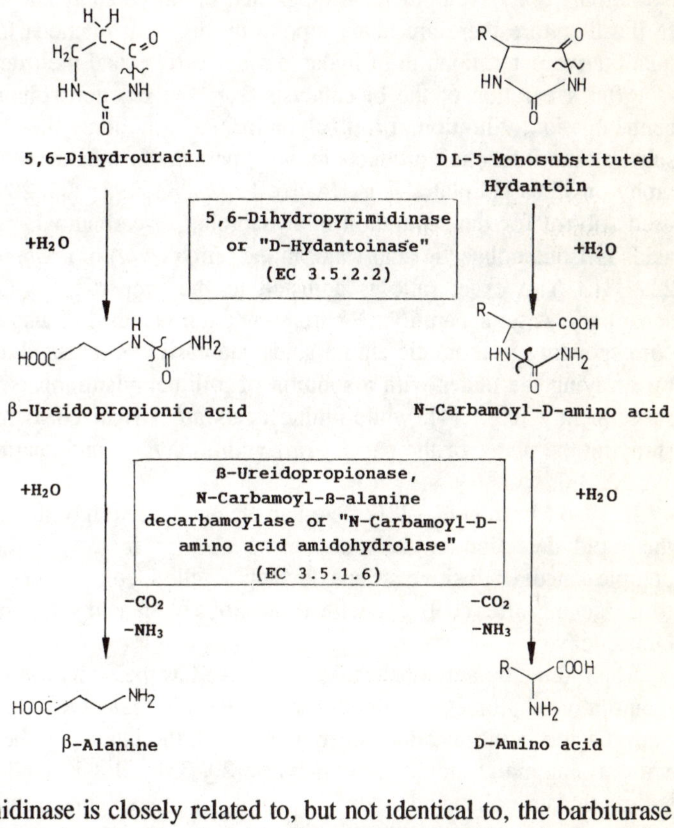

Figure 5.6. Degradation of 5,6-Dihydrouracil and DL-5-Monosubstituted Hydantoins by 5,6-Dihydropyrimidinase and β-Ureidopropionase.

The dihydropyrimidinase is closely related to, but not identical to, the barbiturase (EC 3.5.2.1) able to hydrolyze barbituric acid and methylbarbituric acid [8]. These substrates differ from uracil and thymine in the presence of a keto group instead of a methyl or a hydrogen group in the 6- position of the ring (see Figure 5.6). Barbiturase was detected first by Hayashi and Kornberg [210] in bacteria of the genera *Mycobacterium* and *Corynebacterium* and postulated to catalyze a side branch in the degradation of pyrimidines. Unfortunately there are no further data in the literature on substrate specificity and stereoselectivity of this enzyme, which would allow comparison with the D-hydantoinase.

5.4 SCREENING METHODS

5.4.1 Analytical Detection of Hydantoin Cleavage

Before starting screening for biocatalysts able to hydrolyze DL-5-monosubstituted hydantoins, the development of quick and effective analytical methods is necessary. In the literature there are many reports on different methods for the qualitative and quantitative determination of hydantoin derivatives and their degradation products.

After separation of the biocatalysts from the reaction solutions (for example, by centrifugation, filtration, or precipitation) a commonly used method to separate substrates and reaction products in the supernatants is paper or thin layer chromatography on silica gel plates [*e.g.*, 14, 16, 19, 22, 33, 198, 205, 209, 211]. A frequently used solvent for the separation of hydantoins, *N*-carbamoylamino acids, and amino acids is 1-butanol/acetic acid/water in the ratio (v/v/v) of about 4:1:1 [*e.g.*, 14, 16, 19, 22, 33]. The exact ratio is adjusted to the properties of the compounds to be separated. After separation, *N*-carbamoylamino acids and also hydantoin derivatives corresponding to aromatic amino acids can easily be detected as yellow or red spots by spraying the plates with a solution of *p*-dimethylaminobenzaldehyde and heating subsequently [*e.g.*, 14], while amino acids are visualized as red or violet spots by spraying the plates or the paper strips with ninhydrin and heating subsequently [*e.g.*, 19, 22, 33].

In 1986 Morin *et al.* [28] offered an elegant variation of this qualitative method for the rapid detection of microorganisms able to hydrolyze dihydrouracil on solid complex media. Positive strains develop a yellow color around the colonies within 5 to 10 seconds after challenge with 10-μl drops of an acid solution of *p*-dimethylaminobenzaldehyde.

The *p*-dimethylaminobenzaldehyde as well as the ninhydrin method can be used as quantitative photometric methods for the determination of *N*-carbamoylamino acids and amino acids in the reaction supernatants, too; the latter may be done in combination with an automatic amino acid analyzer [20, 33]. But it has to be stressed that the *p*-dimethylaminobenzaldehyde method cannot be applied to aromatic *N*-carbamoylamino acids, because the corresponding hydantoins also react.

The activity of enriched or purified *N*-carbamoylamino acid amidohydrolases can be studied by manometric methods based on CO_2 production [7] or by measuring the NH_4^+ released during the reaction [192, 199, 210].

Since the 1980s in most studies HPLC methods on reversed phase columns are used for the parallel analysis of hydantoins, *N*-carbamoylamino acids, and amino acids.

Using HPLC, aromatic compounds can easily be quantified photometrically at 220 nm or 280 nm [200, 202, 205, 212, 213], whereas for the detection of 5-alkylhydantoins and their degradation products pre- or post-column derivatization methods, such as with ninhydrin or *o*-phthaldialdehyde [214] have to be applied.

For screening experiments it is also necessary to assess the enantiomeric purities of the products. A classical method for quantification of enantiomeric purities is to isolate the products and to measure their optical rotation in comparison with the optically pure D- and L-compounds. But in the case of hydantoin hydrolysis this implies the separation and purification of the optically active hydantoin, the *N*-carbamoylamino acid, and in some cases of the amino acid as well.

Another possible method for detecting enantiomeric purities of amino acids is the use of the commercially available enzymes D- and L-amino acid oxidases. These can be applied to a wide range of amino acids in combination with the photometric ninhydrin method (see above) [31, 33]. To detect the configuration of the *N*-carbamoylamino acids produced, decarbamoylation is necessary. This can be done with nitrite under mild reaction conditions without a change of their optical rotation [*e.g.*, 16, 33].

Recently, with the development of chiral columns, HPLC has become an important tool for the detection of enantiomeric purities. In most cases a preceding separation of substrate and reaction products is not necessary. By using chiral supports (for example, copper-II–D-amino acid complexes), it is possible to determine not only the enantiomeric purity of *N*-carbamoylamino acids and amino acids, but also that of 5-monosubstituted hydantoins [24, 206, 215, 216]. Also, chiral thin layer chromatography with comparable supports may be used as a screening method [31].

5.4.2 Growth and Induction of Microorganisms with D-Hydantoinase and *N*-Carbamoyl-D-Amino Acid Amidohydrolase Activity

In 1978 Yamada *et al.* [14] investigated the distribution of hydantoin hydrolyzing activity in 391 microorganisms. They found high activity for forming *N*-carbamoyl-D-methionine from DL-5-(2'-methylthioethylene)hydantoin in various bacteria, yeasts, and molds using DL-5-(2'-methylthioethylene)hydantoin as inducer. After using unsubstituted hydantoin and DL-5-methylhydantoin as inducers of hydantoinase activity in *Pseudomonas striata*, which was the best strain in the screening, even higher activity was observed as when using DL-5-(2'-methylthioethylene)hydantoin as inducer [14]. Allantoin (see Section 5.3.1) caused no induction of D-hydantoinase activity. Meat extract in combination with glycerol was shown to be advantageous for D-hydantoinase formation, which was maximal at the beginning of the stationary growth phase [14].

Olivieri *et al.* used *Agrobacterium radiobacter* NRRL B 11291 for the complete conversion of racemic hydantoins to D-amino acids [19, 20]. This microorganism had been previously isolated using an enrichment mineral medium containing 0.5% glucose as the carbon source and 0.2% *N*-carbamoyl-D-phenylglycine as the sole source of nitrogen. The cultivation of this microorganism in a medium containing 0.2% uracil as the sole nitrogen source allowed the production of both D-hydantoinase and *N*-carbamoyl-D-amino acid amidohydrolase.

Wanru [26, 27] showed *Pseudomonas putida* to have the highest activity of dihydropyrimidinase among 275 bacterial strains belonging to 19 genera. Unsubstituted hydantoin, 5-methylhydantoin, 5-phenylhydantoin, uracil, adenine, and orotic acid were effective inducers for enzyme formation. Glycerol or lactic acid as C-sources in combination with 1–2% of yeast extract gave the best results in the synthesis of dihydropyrimidinase, while beef extract and polypeptone gave no enzyme activity.

Yokozeki *et al.* [22] used DL-5-cyanoethylenehydantoin and DL-5-(2'-methylthioethylene)hydantoin as inducers for screening experiments. Their goal was to isolate microorganisms able to form directly D-*p*-hydroxyphenylglycine from the corresponding DL-5-monosubstituted hydantoin. Only 2 of 430 microbial strains from stock cultures tested were able to perform this reaction. In a second screening starting with 2000 strains from natural sources, only 2 were able to form D-*p*-hydroxyphenylglycine. The best of these strains was identified as *Pseudomonas* sp. AJ 11220. This bacterium showed the highest activity after induction with DL-5-(2'-methylthioethylene)hydantoin, DL-5-cyanoethylenehydantoin, and DL-5-*p*-hydroxyphenylhydantoin [23]. Hydantoin, 5-methylhydantoin, uracil, and dihydrouracil caused only slight induction, while uric acid (see Section 5.3.1) gave 49% induction in comparison with DL-5-(2'-methylthioethylene)hydantoin. DL-5-Cyanoethylenehydantoin strongly inhibited the growth of *Pseudomonas* sp. AJ 11220 at concentrations of more than 3 g/l and therefore was added stepwise to the growth medium.

Möller *et al.* [33] used DL-5-hydroxymethylhydantoin as sole C- and N-source in a mineral salts medium for screening experiments. They were able to isolate 88 strains with D-hydantoinase activity, among them 12 able to perform the enzymatic decarbamoylation of the *N*-carbamoyl-D-serine produced. Experiments with the best strain, *Arthrobacter crystallopoietes* AM2, showed that unsubstituted hydantoin and 5-methylhydantoin were better inducers than the natural substrate dihydrouracil. While cells from the middle of the exponential growth phase only poorly formed free D-amino acids, maximal conversion was obtained with cells harvested at the end of the exponential growth phase. Brans *et al.* [214] were able to show that the compound used as inducer not only had a significant influence on total hydantoin cleaving activity, but on the ratio of free D-amino acids to *N*-carbamoyl-D-amino acid (see Table 5.1).

In screening experiments using crude cell extracts, Morin *et al.* [31] could induce D-hydantoinase activity with a variety of compounds, but obtained the highest hydantoin-hydrolyzing activities with dihydrouracil as the substrate for screening.

TABLE 5.1 Influence of Hydantoin Derivatives and Related Compounds on DL-5-Methylhydantoin Cleaving Activity of *Arthrobacter crystallopoietes* AM2 [214]

Inducer	*N*-Carb-D-Alanine (mmol/l)	D-Alanine (mmol/l)	Total (mmol/l)
None	9.3	3.8	13.1
1-Methylhydantoin	13.2	–	13.2
Barbituric Acid	10.0	6.7	16.7
5-Dimethylhydantoin	8.7	9.1	17.8
DL-5-Indolylmethyl-3-*N*-methylhydantoin	19.7	–	19.7
DL-5-Methylthioethyl-hydantoin	20.0	–	20.0
DL-5-Phenylhydantoin	20.3	–	20.3
Allantoin	25.6	7.0	32.6
DL-5-(2-Thienyl)-hydantoin	28.4	5.6	34.0
DL-5-Hydroxymethylhydantoin	21.5	43.5	65.0
Hydantoin	16.9	51.2	68.1

(Cells were grown with 4.4 mmol/l of each compound as inducer; the standard assay was incubated with 10 g/l DL-5-methylhydantoin for 2.5 h at 50 °C with shaking; for further details see [214].)

5.5 D-SELECTIVE MICROBIAL BIOCONVERSION OF 5-MONOSUBSTITUTED HYDANTOIN DERIVATIVES WITH WHOLE MICROBIAL CELLS

5.5.1 Microbial Production of *N*-Carbamoyl-D-Amino Acids

All microorganisms described in this section were able to form only *N*-carbamoyl-D-amino acids from the corresponding hydantoins. No *N*-carbamoyl-D-amino acid amidohydrolase activity was detected in these microorganisms. For industrial applications of these microorganisms, the isolation of the intermediary products as well as a second chemical reaction step are necessary to obtain the free D-amino acids [18, 25] (see Sections 5.7.1 and 5.7.2).

All microorganisms mentioned in this section have in common:

(i) that they can hydrolyze D-selectively a wide range of DL-5-monosubstituted hydantoins,

(ii) that they can be used as resting cells for this reaction up to temperatures of about 50 °C, and

(iii) that the pH optimum for the cleavage of the different hydantoin derivatives is about pH 8.0–9.0.

Pseudomonas striata was shown by Takahashi *et al.* [15] to have a wide substrate specificity even for producing *N*-carbamoyl derivatives of nonproteinogenic amino acids as *p*-chloro derivatives of DL-5-phenylhydantoin. Comparable results were obtained by the same authors with intact resting cells of *Aerobacter cloacae*, *Corynebacterium sepedonicum*, and *Streptomyces griseus* [16].

Guivarch *et al.* [52] used a *Pseudomonas* sp. for the production of the *N*-carbamoyl derivatives of D-methionine, D-phenylalanine, D-cyanopropyleneglycine, D-leucine, and D-valine. In contrast to others, they were able to prove enzymatic racemization of L-5-(2'-methylthioethylene)hydantoin.

Wanru [27] found the highest activity in intact resting cells of *Pseudomonas putida* for 5-phenylhydantoin of all substrates tested. He reported that the dihydropyrimidinase activity was increased 2–3 times when the cells were preincubated at 40 °C under weak acidic or basic conditions by enhancing their permeability.

The substrate specificity of dihydropyrimidinases from different sources is discussed in detail in Section 5.6.1 and can be seen in Table 5.5.

Other publications deal with a D-specific microbial hydantoin cleavage leading to the free, optically pure D-amino acids in two enzymatic reactions and are discussed in Section 5.5.2.

5.5.2 Microbial Production of D-Amino Acids

When producing D-amino acids from DL-5-monosubstituted hydantoins with microorganisms, at least two enzymatic reactions—in some publications the existence of a hydantoin racemase as third enzyme is postulated [21, 25]—are necessary (see Figure 5.6).

Comparable to the use of only the D-hydantoinase reaction (see Section 5.5.1) it would be possible to use only the *N*-carbamoyl-D-amino acid amidohydrolase reaction for the production of free D-amino acids from the corresponding *N*-carbamoyl-DL-amino acids, too. This was investigated by Olivieri *et al.* [19]. The enzyme from *Agrobacterium radiobacter* had a substrate specificity comparable to the D-hydantoinase with high activity for the production of D-phenylglycine (see Section 5.6.2 and Table 5.8), but this strategy had some disadvantages. The *N*-carbamoyl-D-amino acid amidohydrolase reaction was not suitable for a process starting from racemic substrates, because the L-enantiomers caused a distinct inhibition. Further *N*-carbamoylamino acids were so poorly taken up by intact cells that permeabilization with toluene was necessary. For these reasons Olivieri *et al.* [19] postulated that it would be better to start from the DL-5-monosubstituted hydantoins as substrates and to employ the *N*-carbamoyl-D-amino acid amidohydrolase in combination with the D-hydantoinase.

But for optimal performance of this process problems have to be solved, too. So compromises have to be found, not only for the induction and syntheses of both enzymes (see Section 5.4.2), but also between the optimal reaction conditions for the D-hydantoinase and the *N*-carbamoyl-D-amino acid amidohydrolase reaction.

While intact cells of *Agrobacterium radiobacter* NRRL B 11291 showed a maximal conversion of the DL-5-phenylhydantoin at 60 °C, maximal D-amino acid formation was obtained at 50 °C [20]. The *N*-carbamoyl-D-amino acid amidohydrolase began to lose activity at pH values higher than pH 9.0, when reaction conditions for the D-hydantoinase were optimal. At pH values of pH 7.0–7.5—optimal for the second reaction—the D-hydantoinase reaction became rate-limiting. A repetitive use of *Agrobacterium radiobacter* cells for the formation of D-*p*-hydroxyphenylglycine was possible, but the stability of the D-hydantoinase seemed to be higher [20]. The substrate specificity of the *N*-carbamoyl-D-amino acid amidohydrolase can be seen in Table 5.8.

Results similiar to those obtained with *Agrobacterium radiobacter* were obtained by using resting cells of *Arthrobacter crystallopoietes* AM2 [32, 33]. Optimal conditions for intact cells were found to be 55 °C and pH 9.2 [33].

Yokozeki *et al.* [22, 23, 24] described the optimal reaction conditions for the production of D-*p*-hydroxyphenylglycine from the corresponding racemic hydantoin by resting cells of *Pseudomonas* sp. AJ 11220 to be pH 8.0 and 43 °C. In contrast to Olivieri *et al.* [19, 20] and Möller *et al.* [32, 33], no *N*-carbamoyl-D-amino acids were observed as intermediary products. In contrast to other publications [12, 13, 15, 16, 32, 33], *Pseudomonas* sp. AJ 11220 was able to form D-glutamic acid from the corresponding hydantoin. Purification of the enzymes of *Pseudomonas* sp. AJ 11220 resulted in a D-hydantoinase, an *N*-carbamoyl-L-, and an *N*-carbamoyl-D-amino acid amidohydrolase [24], and is discussed in Sections 5.6.1, 5.6.2, and Chapter 6.

Battilotti and Barberini [21] used resting cells of *Agrobacterium radiobacter* for the production of D-valine from the corresponding hydantoin derivative. When comparing the reaction rate for the conversion of DL-5-isopropylenehydantoin with the rate for chemical racemization at different pH values at 40 °C they postulated the existence of a hydantoin racemase.

5.5.3 Bioconversions with Immobilized Microbial Cells

In the late 1970s Italian and Japanese groups [13, 18] reported first on the production of optically pure D-amino acids using immobilized biocatalysts. Only a few years later the Japanese company Kanegafuchi in Singapore [217] took advantage of using immobilized microbial cells when they began industrial production of D-(–)-4-hydroxyphenylglycine (see also Section 5.7.2.).

The prior investigation using immobilized biocatalysts for the development of biocatalytical processes for the production of optically pure amino acids has enormous advantages. Some of these advantages are:

(i) favorable processing conditions by means of biocatalyst support systems or cell recycle systems [218, 219];
(ii) enhanced operating stability [18]; and
(iii) reuse of biocatalysts [13, 217].

It was mentioned in Section 5.3.2 that in 1978, Dinelli *et al.* [13] successfully tried to use immobilized calf-liver dihydropyrimidinase for the production of *N*-carbamoyl-D-phenylglycine and its *p*-chloro- and *p*-hydroxyphenyl derivatives from the corresponding racemic hydantoin derivatives. Nearly 10 years later, the Beecham group published a patent [111] describing enzyme preparations of a hydantoinase-producing organism (*Bacillus brevis*) for the D-enantioselective hydrolysis of DL-5-hydroxyphenylhydantoin to D-*N*-carbamoylhydroxyphenylglycine. Nevertheless, in order to create favorable process conditions, immobilization of whole microbial cells seemed to be more successful. A detailed overview of the suitability of microorganisms was given especially by Yamada *et al.* [18]. Both the bacteria *Pseudomonas striata* and an alkalophilic *Bacillus* sp. gave a nearly complete conversion of phenyl-, *p*-hydroxyphenyl-, or thienylhydantoin to the corresponding D-*N*-carbamoylamino acids, when the cells were entrapped into a matrix of polyacrylamide. Immobilization of cells of either *Bacillus* sp. or *Pseudomonas striata* did not cause any significant changes in their activities and substrate specificities.

Based on literature published in recent years, the most important application of immobilized dihydropyrimidinase (for example, D-hydantoinase) seemed to be the formation of D-phenylglycine–related amino acids (see Table 5.2).

In respect to the formation of D-phenylglycine–related amino acids, Yamada *et al.* [18] investigated the D-enantioselective hydrolysis of different racemic 5-substituted hydantoins (for example, *p*-chlorophenylhydantoin). In addition Dinelli *et al.* [13] described the D-enantioselective hydrolysis of amino acid hydantoins, for example, of leucine, α-alanine, methionine, glycine, or valine (Table 5.2). Chevalier *et al.* [218, 219] took advantage of the wide substrate specificity of the dihydropyrimidinase reaction when using immobilized active cells of *Pseudomonas putida* for the conversion of dihydrouracil to *N*-carbamoyl-β-alanine, the precursor for D-alanine (Table 5.2).

In order to create optimal process conditions for the production of optically active amino acids, much work has been done on the establishment of classical (and mostly proved) immobilization methods (Table 5.2).

Naturally the immobilization method is strictly guided by the character of the biocatalyst used (for example, whole cell or enzyme preparation). As mentioned above, Dinelli *et al.* [13] entrapped calf-liver dihydropyrimidinase into cellulose triacetate fibers (see also Section 5.5.4.). In 1987, the Beecham group has tested the ability of different commercially available anion exchange resins (*e.g.*, Duolite, Purolit, Amberlite, Lewatit, and Diaion) for the D-enantioselective hydrolysis of racemic *p*-hydroxyphenylhydantoin to D-*N*-carbamoylhydroxyphenylglycine by a partially purified enzyme solution of *Bacillus brevis* [111]. In the same work they also mentioned that cross-linking with glutaraldehyde enhanced the working stability of the enzyme preparation.

The most ubiquitous method for the immobilization of whole cells is the entrapment of microorganisms into a calcium alginate gel matrix [111, 217, 218, 219]. Only Yamada *et al.* [18] used polyacrylamide gels for the entrapment of microorganisms (for example, *Bacillus* and *Pseudomonas*). In 1988, an alternative method of relevance was described by Chevalier *et al.* The hydantoinase activity of *Pseudomomas putida* was evaluated using cells immobilized in a microporous membrane system or a hollow fiber bioreactor [218, 219]. In addition, Chevalier *et al.* compared these cell recycled systems with systems involving calcium alginate as carrier. Process efficiency (percent conversion rate, productivity) was distinctly reduced with the membrane, involving problems like the so-called "membrane fouling" or insufficient cell support. Therefore the hydrolysis of dihydrouracil to *N*-carbamoyl-β-alanine seemed to be more effective when using packed bed columns of calcium alginate biocatalysts or a continuous fermentation process involving calcium alginate biocatalysts in a stirred tank reactor. Although the alginate and membrane processes were carried out in a continuous mode, no data were given concerning the reuse of the biocatalyst systems for further processes. Only Yamada *et al.* [18] examined the

TABLE 5.2 Survey of Biocatalytic Systems Used (*e.g.*, dihydropyrimidinase enzyme preparation or dihydropyrimidinase-active microbial cells), Immobilization Systems, Substrates Tested (*e.g.*, hydantoin derivatives or dihydrouracil) and the Corresponding Products Obtained, as Published in the Literature.

Biocatalytic System	Immobilization Method	Substrate	Product	Reference
Calf-liver enzyme preparation	Cellulose triacetate fiber	Amino acid hydantoins Phenylglycine derivatives	*N*-Carbamoyl-D-amino acids *N*-Carbamoyl-D-phenylglycine	[13]
Bacillus brevis enzyme preparation	Diverse anion exchange matrices Glutardialdehyde cross-linking	DL-5-Hydroxyphenylhydantoin	*N*-Carbamoyl-D-hydroxyphenylglycine	[111]
Pseudomonas striata *Bacillus brevis*	Polyacryamide gel	Different 5-substituted phenylhydantoins	Corresponding *N*-carbamoyl-D-amino acid	[18]
Corynebacterium sepedonicum *Aerobacter rhizogenes*		Phenyl-, *p*-hydroxyphenyl-, thienylhydantoin	-"-	

Pseudomonas putida	— Ca alginate beads in packed bed columns or submerged culture — microporous hollow fiber bioreactor — free cell recycle system using microporous membranes	Dihydrouracil	*N*-Carbamoyl-β-alanine	[218, 219]
Agrobacterium radiobacter	Cellulose triacetate fibers	DL-5-Isopropylhydantoin	D-Valine	[21]
-"-	-"-	Diverse DL-5–substituted hydantoins	Data not shown	[19]
Bacillus brevis	Ca alginate beads in a fed-batch process	DL-OH-phenylhydantoin	Biotechnol. conversion to *N*-Carbamoyl-D-phenylglycine; chemical conversion to D-Phenylglycine	[in 217]

process stability of polyacryamide-entrapped cells of an alkalophilic *Bacillus* strain for the production of *N*-carbamoyl-D-amino acids. Compared with free cells the hydrolyzing activity of the alkalophilic *Bacillus* strain was maintained after 10 reaction cycles. The possibility of reusing biocatalyst systems results in realizing optimal (cost-effective) process conditions, which was demonstrated by Tramper and Luyben [217]. As mentioned at the beginning of this section, the Japanese company Kanegafuchi started the biotechnological production of D-4-hydroxyphenylglycine (see also Section 5.7.2). The corresponding D-*N*-carbamoyl-4-hydroxyphenylglycine was produced by a calcium alginate-immobilized *Bacillus* strain, whereas fed-batch conditions were found to be optimal. The same biocatalyst batch charge could be used up to 15 times (see also Section 5.7.2.). In this connection it should be mentioned that the half-life of one batch charge was estimated to be 28 days [217].

All the processes mentioned above included only the stereoselective conversion of racemic hydantoin derivatives or dihydrouracil to their corresponding D-*N*-carbamoylamino acid. If performed, further hydrolysis of the D-*N*-carbamoylamino acid to the D-amino acid was realized by chemical methods (see also Section 5.7.2.). Battilotti *et al.* [21] described a biocatalytic one-step conversion of DL-isopropylhydantoin to D-valine. The first requirement for the development of such a one-step conversion process was an *N*-carbamoyl-D-amino acid amidohydrolase-active microorganism of good activity. Both Olivieri *et al.* [19] and, nearly 10 years later, Battilotti *et al.* [21] took advantage of the Gram-negative bacterium *Agrobacterium radiobacter*, which was both dihydropyrimidinase- and *N*-carbamoyl-D-amino acid amidohydrolase–active. Referring to this, Olivieri *et al.* studied the fully enzymatic hydrolysis of DL-5-phenylhydantoin with cellulose triacetate-immobilized *Agrobacterium radiobacter*. These investigations indicate that it is feasible to prepare D-amino acids in a simple and economical way if immobilizing both enzyme systems in a suitable matrix.

5.6 ISOLATION AND CHARACTERIZATION OF THE ENZYMES

5.6.1 Isolation and Characterization of D-Hydantoinase from Various Sources

D-Hydantoinase, which is identical with the enzyme dihydropyrimidine amidohydrolase (dihydropyrimidinase, EC 3.5.2.2) was isolated from animal cells as well as from bacteria.

5.6.1.1 D-Hydantoinase from Animal Cells

Three papers, by Wallach *et al.* [8], Cecere *et al.* [220] and Brooks *et al.* [221], reported on the isolation and characterization of D-hydantoinase from bovine liver. Table 5.3 gives a short overview of the purification procedures and characteristic properties of this enzyme. The bovine liver dihydropyrimidinase consists of four subunits and it could be demonstrated that every active enzyme contains four Zn(II) cations [221] which are tightly bound ($K_s > 1.33 \times 10^9$ M^{-1}).

TABLE 5.3 Purification and Characteristic Properties of D-Hydantoinase from Animal Cells

Source	acetone powder from beef liver	catalase fraction from beef liver
Reference	[8]	[221]
Purification Steps	acid and heat treatment, $(NH_4)_2SO_4$ and acetone precipitation	hydrophobic chromatography or preparative electrophoresis
Yield	25%	13%
Purification Factor	200	24.2
Purity	80%	homogeneous
Optimal pH	8.2	
Metal Ion Requirements	Mn^{2+} and Mg^{2+} only when dihydrouracil is the substrate!	Zn^{2+} and Co^{2+}
Molecular Weight		226 kd
Subunits		4 × 56.5 kd

The substrate specificity is listed in Table 5.5. At pH values around 8.5 the cyclic amides are hydrolyzed. The optimal pH of the reverse reaction (ring closure) is on the alkaline side but the cleavage of hydantoin is irreversible [8]. Cecere *et al.* [220] demonstrated that only substrates having the *R* configuration were accepted by the enzyme.

5.6.1.2 Microbial D-Hydantoinases

Several publications describe D-selective hydantoinases isolated from various microorganisms, such as *Arthrobacter crystallopoietes* AM2 [33], *Pseudomonas fluorescens* DSM 84 [30], *Pseudomonas striata* [15], and *Pseudomonas* sp. AJ 11220 [24]. A survey of the isolation and some characteristic data are given in Table 5.4. The substrate specificities of the D-hydantoinases are presented in Table 5.5. It can be seen that all these D-hydantoinases are rather similar to the dihydropyrimidinase from animal tissue.

Another 1-methylhydantoinase was isolated, by ammonium sulfate and polyethyleneimine precipitations and several chromatographic methods [197], from *Arthrobacter* sp. DSM 2563, DSM 2564, *Brevibacterium* sp. DSM 2843, *Micrococcus* sp. DSM 2565, and *Moraxella* sp. DSM 2562. This ATP-dependent enzyme accepts a broad spectrum of substrates, but there are no reports on the stereospecificity.

5.6.2 Isolation and Characterization of *N*-Carbamoyl-D-Amino Acid Amidohydrolases from Various Sources

D-Specific *N*-carbamoylamino acid amidohydrolases (D-*N*-carbamoylases) were purified from microbial cells and from rat liver. There are great differences in the properties of this group of enzymes.

5.6.2.1 D-*N*-Carbamoylase from Rat Liver

The rat liver D-*N*-carbamoylase was isolated by Caravaca and Grisolia [9] as described in Table 5.6. It was found also in the supernatants of liver homogenates of dog, pigeon, and rabbit. The substrate specificity of the enriched enzyme is listed in Table 5.8. This D-*N*-carbamoylase is involved in the pyrimidine degradation pathway.

5.6.2.2 Microbial D-*N*-Carbamoylases

Catalytic activity of D-*N*-carbamoylase was detectable in microorganisms like *Agrobacterium radiobacter* [20], *Clostridium uracilicum* [209], *Pseudomonas putida* 77 [222] and *Pseudomonas* sp. AJ 11220 [24]. The enzymes were induced either by *N*-carbamoylamino acids or by hydantoin derivatives and purified as shown in Table 5.7.

TABLE 5.4 Purification and Characteristic Properties of Microbial D-Hydantoinases

Source	*Arthrobacter crystallopoietes* AM2	*Pseudomonas fluorescens*	*Pseudomonas striata*	*Pseudomonas* sp. AJ 11220
Reference	[33]	[30]	[15]	[24]
Inducer	dihydrouracil, hydantoin and various DL-5-monosubstituted hydantoins		hydantoin	5-cyanoethylene-hydantoin
Purification Steps	protamine sulfate and $(NH_4)_2SO_4$ precipitation, ion exchange chromatography (DEAE-Cellulose)	hydrophobic chromatography (Phenyl-Sepharose), size exclusion chromatography (Sephacryl S-400) and preparative electrophoresis	protamine sulfate and $(NH_4)_2SO_4$ precipitation, ion exchange chromatography (DEAE-Cellulose), hydroxylapatite and size exclusion chromatography (Sephadex G-200), crystallization	ion exchange chromatography (DEAE-Toyopearl)
Yield		1%	3%	63%
Purification Factor		0.53	300	27
Purity	crude enzyme	homogeneous	homogeneous	crude enzyme
Optimal Temperature	50–60 °C	55 °C	45–55 °C	55 °C
Temperature Stability	< 50 °C	< 40 °C	< 60 °C	
Optimal pH	8.2–9.2	9.0	8.0–9.0	8.0
pH Stability	6.5	5.5–8.5	6.0–7.0	
Metal Ion Requirements	none	Fe^{2+}	Fe^{2+}, Co^{2+}	
Molecular Weight		230 kd	190 kd	
Subunits		4 × 60 kd		

TABLE 5.5 Substrate Specificities of D-Hydantoinase Enzymes

Enzyme	Bovine Liver Dihydropyrimidinase		Microbial D-Hydantoinases		
Source	acetone powder from beef liver	catalase fraction from beef liver	*P. fluorescens*	*P. striata*	*P.* sp. AJ 11220
References	[8, 220]	[221]	[30]	[15]	[24]
Ring Cleavage:					
Dihydrothymine	+	+	o	o	o
Dihydrouracil	+	+	+	+	+
5-Br-Dihydrouracil	o	+	o	o	o
5-I-Dihydrouracil	o	+	o	o	o
Barbituric acid	–	o	o	o	o
Orotic acid	–	o	o	o	o
Urocanic acid	–	o	o	o	o
Hydantoin	+	+	+	+	+
2-Thiohydantoin	–	o	o	o	o
1-Methylhydantoin	o	+	o	o	o
5-Methylhydantoin	o	+	o	+	+
5-*i*-Propylhydantoin	+	o	+	+	+
5-*i*-Butylhydantoin	+	o	o	+	+
5-Hydroxymethylenehydantoin	o	o	o	o	+
5-Methoxymethylenehydantoin	o	o	o	o	+
5-(2-Methylthioethylene)hydantoin	+	o	+	+	+
5-(4-Aminobutylene)hydantoin	o	o	o	o	+
5-Carboxymethylenehydantoin	–	o	o	o	+
5-Carboxyethylenehydantoin	–	o	o	o	+
5-(Carboxymethylidine)hydantoin	–	o	o	o	o

Enzyme	Bovine Liver Dihydropyrimidinase		Microbial D-Hydantoinases		
Source	acetone powder from beef liver	catalase fraction from beef liver	*P. fluorescens*	*P. striata*	*P.* sp. AJ 11220
5-Phenylhydantoin	+	o	o	+	+
5-*p*-Hydroxyphenylhydantoin	+	o	o	+	+
5-*m*-Hydroxyphenylhydantoin	o	o	o	+	+
5-*o*-Hydroxyphenylhydantoin	o	o	o	o	+
5-*p*-Methoxyphenylhydantoin	o	o	o	+	o
5-*p*-Chlorophenylhydantoin	o	o	o	+	o
5-*m*-Chlorophenylhydantoin	o	o	o	+	o
5-(2,4-Dichlorophenyl)hydantoin	o	o	o	+	o
5-Benzylhydantoin	+	o	+	o	+
5-*p*-Hydroxybenzylhydantoin	o	o	o	o	+
5-(3,4-Dihydroxybenzyl)hydantoin	o	o	o	o	+
5-(3,4-Dimethoxybenzyl)hydantoin	o	o	o	o	+
5-Benzyloxymethylenehydantoin	o	o	o	o	+
5-(3,4-Methylenedioxybenzyl)-hydantoin	o	o	o	o	+
5-Indolylmethylenehydantoin	–	o	o	o	+
5,5-Dimethylhydantoin	o	o	o	–	o
5,5-Diphenylhydantoin	–	o	o	–	o
Imidazole 4,5-dicarboxylic acid	–	o	o	o	o
4,5-Aminoimidazolecarboximide	–	o	o	o	o
Reverse Reaction:					
N-Carbamoyl-β-alanine	+	+	o	o	o
N-Carbamoyl-β-alanine amide	–	o	o	o	o
N-Carbamoyl-β-aminoisobutyrate	+	+	o	o	o
N-Carbamoylglycine	–	o	o	o	o

+ = conversion; – = no conversion; o = not determined

TABLE 5.6 Purification and Characteristic Properties of D-*N*-Carbamoylase from Animal Tissues

Source	rat liver
Reference	[9]
Purification Steps	acetone precipitation and acid treatment
Yield	12%
Purification Factor	20.4
Purity	crude enzyme
Optimal pH	6.75–7.0
Metal Ion Requirements	none

Like the enzyme isolated from rat liver, the inducible *Clostridium* D-*N*-carbamoylase is involved in the degradation of pyrimidines. Only a few compounds were tested as substrates for this enzyme (see Table 5.8), but it is obvious that the other three hydrolases are different. The *N*-carbamoylsarcosine amidohydrolase from *Pseudomonas putida* 77 has its biological function in the metabolic pathway of creatinine.

The two enzymes of *Agrobacterium radiobacter* and *Pseudomonas* sp. AJ 11220 are likely to be identical. They hydrolyze aliphatic as well as aromatic compounds and accept only the D-enantiomers.

5.6.3 Genetic Engineering

There is general agreement that literature methods involving genetic engineering are of great relevance for the future development of biotechnological processes. Nevertheless, little has been published in recent years concerning gene manipulation methods related to D-hydantoinase–active organisms. Only the German company BASF published a patent involving gene manipulation methods in order to create D-hydantoinase-active microorganisms [223].

TABLE 5.7 Purification and Characteristic Properties of Microbial D-*N*-Carbamoylases

Source	*Agrobacterium radiobacter*	*Clostridium uracilicum*	*Pseudomonas putida* 77	*Pseudomonas* sp. AJ 11220
Reference	[20]	[209]	[222]	[24]
Inducer	*N*-carbamoyl-D-phenylglycine	*N*-carbamoyl-β-alanine	1-methyl-hydantoin	5-cyanoethylene-hydantoin
Purification Steps	none	$MnCl_2$, $(NH_4)_2SO_4$ and acetone precipitation, hydroxylapatite chromatography	$(NH_4)_2SO_4$ precipitation, ion exchange chromatography (DEAE-Cellulose) crystallization	ion exchange chromatography (DEAE-Toyopearl)
Yield		18%	63.2%	36%
Purification Factor		98.7	27.4	17
Purity	disrupted cells	crude enzyme	homogeneous	crude enzyme
Optimal Temperature	60 °C	30–35 °C	37 °C	55 °C
Temperature Stability	< 40 °C	< 45 °C	< 40 °C	
Optimal pH	7.0	7.4–7.8	7.0–8.0	7.0
pH Stability	7.0–9.0		6.0–7.0	
Metal Ion Requirements	none	none		
Molecular Weight			102 kd	
Subunits			4 × 27 kd	

TABLE 5.8 Substrate Specificity of the D-*N*-Carbamoylases

Source	Rat liver	*Agrobacterium radiobacter*	*Clostridium uracilicum*	*Pseudomonas putida* 77	*Pseudomonas* sp. AJ 11220
Reference	[9]	[20]	[209]	[222]	[24]
N-Carbamoylglycine	–	o	–	+	+
N-Carbamoylalanine	–	+	–	+	+
N-Carbamoyl-β-alanine	+	o	+	o	+
N-Carbamoyl-α-amino-*n*-butyric acid	o	o	o	+	o
N-Carbamoyl-α-amino-isobutyric acid	o	o	o	–	o
N-Carbamoyl-β-amino-isobutyric acid	+	o	–	o	o
N-Carbamoyl-γ-amino-isobutyric acid	–	o	o	o	o
N-Carbamoylvaline	o	+	o	+	+
N-Carbamoylleucine	o	o	o	+	+
N-Carbamoylisoleucine	o	o	o	–	o
N-Carbamoylnorleucine	o	o	o	+	o
N-Carbamoylsarcosine	o	o	o	+	o
N-Methyl-*N*-carbamoyl-alanine	o	o	o	+	o
N-Carbamoylproline	–	o	o	o	o
N-Carbamoylserine	o	o	o	–	o
N-Carbamoylthreonine	o	–	o	+	o
N-Carbamoylmethionine	o	o	o	–	+
N-Carbamoylglutamic acid	–	+	o	–	+
N-Carbamoylaspartic acid	–	o	–	–	o
N-Carbamoylasparagine	o	o	o	–	+
N-Carbamoyl-phenylglycine	o	+	o	+	+
N-Carbamoyl-*p*-hydroxy-phenylglycine	o	+	o	+	+

Source	Rat liver	*Agrobacterium radiobacter*	*Clostridium uracilicum*	*Pseudomonas putida* 77	*Pseudomonas* sp. AJ 11220
N-Carbamoyl-*p*-chlorophenylglycine	o	+	o	o	o
N-Carbamoyl-*p*-methoxy-phenylglycine	o	+	o	o	o
N-Carbamoyl-*p*-methyl-phenylglycine	o	o	o	+	o
N-Carbamoyl-phenylalanine	o	+	o	+	+
N-Carbamoyl-3,4-dihydroxyphenylalanine	o	o	o	o	+
N-Carbamoyltyrosine	o	o	o	–	+
N-Carbamoyltryptophan	o	o	o	+	+
N-Carbamoylhistidine	o	o	o	–	o
N-Carbamoylthienyl-glycine	o	+	o	o	o
N-Acetylalanine	o	–	o	o	o
N-Acetylphenylalanine	o	–	o	o	o
L-Citrulline	–	o	o	o	o

+ = conversion; – = no conversion; o = not determined

The use of genetic engineering methods is mainly brought about by two different scenarios:

(i) well-directed and relevant increases in the production abilities (for example, fermentation yield, specific enzyme productivity) of common wild-type microorganisms; and

(ii) specific changes to the catalytic abilities of enzymes.

A notable increase of a desired enzyme inside a single cell (that often leads to a desired increase in production abilities) could be realized in two different ways:

(i) directed changes to flanking regions of the structural gene, such as the promoter, operator, or terminator sequences; or

(ii) amplification of the gene sequence encoding the enzyme within an entire gene section (for example, the bacterial chromosome).

The strategy of gene manipulation methods, such as finding dihydropyrimidinase-encoding gene sequences in relevant microorganisms and developing dihydropyrimidinase-overproducing microorganisms, is discussed in this section.

Gene cloning investigations of D-hydantoinase by the BASF company [223] serve as an example and are presented in simplified terms.

First of all there must be a microorganism of sufficient D-hydantoin–hydrolyzing activity (along with other microorganisms, the investigators used a thermophilic *Bacillus* sp., internal designation LU 1220).

In the next step, the chromosomal DNA of the thermophilic *Bacillus* was isolated and partially digested into well-defined fragments (DNA lengths > 20 kilobases) using the DNA-restriction endonuclease SAU 3A. Now, after these DNA fragments were separated from DNA fragments of insufficient lengths, the fragments were combined with a so-called "cloning vector". The cosmid vector pHC79 was used for this purpose, because of its high cloning capacity (the foreign "donating" fragment DNA could be about 50 kilobase pairs in length). Basically, the ligation product (vector DNA + insert DNA) forms the starting point for the construction of a so-called "cosmid gene library". After *in vivo* packaging of the recombinant vector DNA into the bacteriophage *lambda*, the recombinant cosmid was amplified exploiting the mesophilic host strain *E. coli* HB 101 ("shotgun" cloning procedure). The identification of D-hydantoinase–positive clones within the cosmid gene library was performed using an immunological assay (polyclonal antibody). This immunological assay is strictly dependent on a sufficient expression rate of the D-hydantoinase gene, so that it is of enormous importance that the cloned gene fragment comprises gene signal sequences which could serve the host strain as transcription promoter, terminator, or ribosome binding site.

Unfortunately, and probably due to the enhanced cloning capacity of cosmid vectors, the genetic stability of clones often is reduced. In addition to this, one cannot avoid separating D-hydantoinase structural gene sequences from signal sequences or even from sequences which do not accompany the D-hydantoinase sequences. This was realized by subcloning D-hydantoinase–active cosmid clones into derivatives of the plasmid pBR327. Recombinant plasmid-containing cells of the host strain *E. coli* were tested employing serologic assays. In addition, serologically-positive clones were also tested using an enzymatic assay (conversion of DL-methylthioethylhydantoin to *N*-carbamoyl-D-methionine—both compounds detected by HPLC chromatography). Only these biologically active recombinants were used for the subsequent gene sequencing. At this point, it should be mentioned that gene sequencing requires additional subcloning in plasmids (method of "overlapping deletion plasmids"). The total D-hydantoinase gene sequence, including diverse associated gene regions, derived from the thermophilic *Bacillus* sp. LU 1220 is presented in Figure 5.7 [221]. The D-hydantoinase–encoding sector was published starting with nucleodide 391 and finishing at nucleotide position 1746. Both the codon which initiates the translation (start codon) and the codon which terminates the translation (stop codon) are marked in Figure 5.7.

The stated purpose for using genetic engineering methods is an attempt to drastically increase the production parameters, such as the overproduction of a

Nucleotide Position	Sequence
1 - 50	CGGTTATGGA TATTGTTGGG AAGTCAGTTC ATAAATATTC CAATTGGGGG
51 - 100	GACCTCGACC TTCATTATAA AGTGGTTGTT CGCATTAACA AAGATGTTTG
101 - 150	CATTAATTGC AATAAATGTT ATATCTCCTG TGAAGATGCT TCTCATCAAT
151 - 200	GCATTGATCG TTTAACGGAT GAAAATGGAA AAGAGTATTT AAAAGTGCGC
201 - 250	GAAGAAGATT GCGTAGGGTG TAATTTATGT TCGATCGTCT GTCCGGTGGA
251 - 300	TGGTGCGATT GACATGGTCG AAATGCCAAG CGACAATCTG CCGATGACAT
301 - 350	GGAATGAACG CCAAGCGGCC ATTAGCGGGC TGAGCAGCTG TAGCGTTGAT
351 - 400	GTGAAATAAA ACGAAATTTC CAGCGGAGGA GGATGTTGAA **ATG**ACAAAAA
401 - 450	TAATAAAAAA TGGAACGATT GTTACCGCAA CCGATACGTA TGAAGCGGAC
451 - 500	TTGCTCATTA AAGACGGAAA AATTGCCATG ATAGGCCAAC ATTTAGAAGA
501 - 550	AAAAGGCGCT GAAGTGATTG ATGCCAAAGG CTGTTACGTA TTTCCAGGCG
551 - 600	GTATTGATTC GCACACGCAT TTAGATATGC CGTTTGGCGG CACGGTGACA
601 - 650	AAGGATGATT TCGAATCTGG AACGATTGCG GCGGCATTTG GCGGAACAAC
651 - 700	GACCATCATC GACTTTTGTT TAACGAATAA AGGGGAGCCA TTAAAAAAAG
701 - 750	CGATTGAAAC TTGGCACAAC AAAGCGAAGG GAAAAGCGGT TATTGATTAT
751 - 800	GGCTTCCATT TAATGATTAG CGAAATTACG GATGACGTAT TAGAAGAGCT
801 - 850	GCCAAAAGTC ATTGCCGAAG AAGGGATAAC ATCCTTTAAA GTGTTTATGG
851 - 900	CGTATAAAAA CGTATTTCAG GCAGATGATG GAACGTTATA CCGCACGCTA
901 - 950	GTGGCTGCCA AAGAACTTGG CGCGCTTGTC ATGGTTACTG CGGAAAATGG
951 - 1000	GGATGTGATT GATTACTTAA CGAAAAAAGC GCTTGCGGAA GGGAATACGG
1001 - 1050	AGCCGATTTA CCATGCTTTA ACGCGGCCTC CAGAAGTAGA AGGAGAAGCG
1051 - 1100	ACCGGGCGCG CCTGTCAATT GACAGAGCTT GCCGGTTCAC AACTTTACGT
1101 - 1150	TGTTCACGTG ACATGTGCGC AAGCGGTGGA AAAAATTGCA CAAGCGCGCA
1151 - 1200	ATAAAGGGTT GGATGTGTGG GGAGAAACGT GTCCGTAATA TCTTGTTCTC
1201 - 1250	GACCAATCGT ATTTAGAAAA GCCTGATTTT GAAGGCGCGA AATATGTTTG
1251 - 1300	GTCCCCTCCG CTTCGTGAAA AATGGCATCA AGAAGTATTG TGGAATGCGC
1301 - 1350	TGAAAAACGG CCAGCTGCAA ACGCTTGGAT CGGACCAATG TTCATTTGAC
1351 - 1400	TTTAAAGGCC AAAAAGAACT TGGCAGAGGA GATTTTACTA AAATTCCAAA
1401 - 1450	CGGCGGGCCG ATGGTCGAGG ATCGGGTCAG CATTCTTTTC AGTGAAGGGG
1451 - 1500	TTAAAAAAGG AAGAATCACG TTAAATCAAT TTGTCGATAT TATGTCGACA
1501 - 1550	AGAATTGCCA AATTGTTCGG GTTATTCCCG AGAAAAGGAA CGATCGCGGT
1551 - 1600	AGGTTCAGAC GCAGACTTAG TCATTTTTGA CCCGGATATC GAACGGGTGA
1601 - 1650	TTTCGGCGGA AACACACCAT ATGGCCGTCG ACTATAATGC ATTTGAAGGA
1651 - 1700	ATGAAAGTAA CGGGTGAACC GGTATCGGTT CGTGCAGAGG CGAATTTGTT
1701 - 1750	GTCCGTGATA AACAATTTGT CGGAAAACCA GGGTACGGCC AATATT**TAA**A
1751 - 1800	ACGGGCAAAA TACGGAACCT CAAAGATTTC CAAGCAGAAC GAGAAATTAA
1801 - 1850	CCATTTAAAA GAATAACAAC CTACTCTTGC CCCTTAAAAT GCCAATAAAA
1851 - 1900	TGCAACACTT AGCTTTATTC CCGTTCTAAG CAG

Figure 5.7. The total D-hydantoinase gene sequence, including diverse associated gene regions, derived from the thermophilic *Bacillus* sp. LU 1220. The D-hydantoinase–encoding sector was proclaimed starting with nucleotide 391 and finished at nucleotide position 1746. Both the start codon and the stop codon are marked in bold type and underlined.

desired enzyme by a suitable microorganism. This could be achieved employing highly competitive "over-expression" plasmid vectors. BASF took advantage of using such over-expression plasmid vectors, when combining the isolated D-hydantoinase structural gene with derivatives of the plasmid pEX31 and pCL547. The efficacy of over-expression is based on strong and efficient gene signal sequence systems, such as the pL-promoter region of the *lambda* phage, which is combined with the MS2-polymerase region. These D-hydantoinase combined vector systems had been transfected into the host strain *E. coli* N4830; the induction of gene expression was performed by heat-shocking. After selection of D-hydantoinase–positive clones (serological assay as well as enzymatic assay) the production abilities of relevant clones were tested in a fermentation process. The specific enzyme activity was found to be approximately 4-fold higher than the wild-type thermophilic *Bacillus* LU 1220 strain. Other cloning systems for other D-hydantoinase structural genes from various thermophilic *Bacillus* strains gave a 4- to 40-fold higher enzyme activity.

5.7 INDUSTRIAL APPLICATION

5.7.1 Survey

Many well-established methods are described in the literature for the production of optically pure amino acids:

(i) by chemical synthesis in combination with diastereomeric salt crystallization [52],
(ii) by biosynthetic pathways starting from low-priced C-sources and complex nutrients [224], and
(iii) by bioconversions starting from racemic precursors [225].

So an industrial application of the hydantoin method will only be of interest if there are decisive advantages in comparison with the processes reviewed in the literature mentioned above.

The production of D-amino acids from DL-5-monosubstituted hydantoin derivatives is of industrial interest because of three main reasons:

(i) besides the hydantoin method in the literature, only a few other methods are described for the direct production of optically pure D-amino acids [226, 227, 228, 229, 230, 231], of which the hydantoin method has the widest substrate specificity (see Sections 5.5 and 5.6);
(ii) most of the DL-5-monosubstituted hydantoins can be easily synthesized from low-priced chemical precursors (see Section 5.2); and
(iii) in contrast to other bioconversion methods using esterases, amidases, or acylases as biocatalysts [226, 228, 229, 230, 231; for a review see 225],

the racemic hydantoin derivatives can be totally converted to 100% optically pure products. The reasons are either the chemical properties of the hydantoin derivatives (see Section 5.2) or the occurrence of hydantoin racemases (see Section 5.5).

In spite of many patents, which reveal the industrial interest in this process [*e.g.*, 220, 222, 232, 233, 234, 235, 236, 237], D-*p*-hydroxyphenylglycine is the only compound produced by this method at the moment. This D-amino acid is used for the synthesis of the semisynthetic antibiotics amoxycillin, cefadroxiol and cefatrizine.

Alternative industrial processes have been developed for the production of D-*p*-hydroxyphenylglycine by the Japanese companies Kanegafuchi, using immobilized cells of *Bacillus brevis*, Ajinomoto, using immobilized cells of *Pseudomonas* sp. (see Sections 5.5.3 and 5.7.2), and by the Italian company Snamprogetti, using immobilized dihydropyrimidinase from calf liver (see Sections 5.6.1 and 5.7.3). A survey of the different strategies of these processes, which will be discussed in Sections 5.7.2 and 5.7.3, is given in Figure 5.8. A detailed comparison of the feasibility of these enzymatic methods for the production of D-*p*-hydroxyphenylglycine in comparison with other enzymatic methods was given by Tramper and van Luyben in 1982 and 1984 [217, 238].

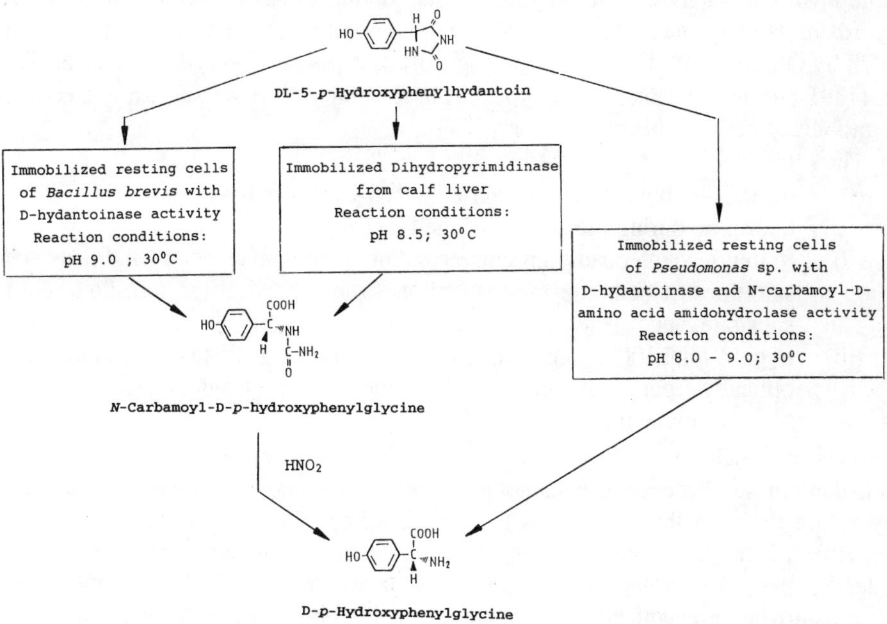

Figure 5.8. Industrial Processes for the Biotechnological Production of D-*p*-Hydroxyphenylglycine from DL-5-*p*-Hydroxyphenylhydantoin.

5.7.2 Process Development for the Production of D-*p*-Hydroxyphenylglycine with Microorganisms

Two strategies can be found in the literature for the production of D-*p*-hydroxyphenylglycine using resting intact cells of microorganisms (see Figure 5.8): Yamada *et al.* and the Japanese company Kanegafuchi developed a process using only the D-hydantoinase activity of immobilized cells of *Bacillus brevis* [14, 15, 16, 18, 239, 240, 241]. After immobilization in polyacrylamide (see Section 5.5.3), the cells were used for the repetitive production of various *N*-carbamoyl-D-amino acids of industrial interest [18]. Because of the quick chemical racemization of the substrates under reaction conditions (pH 8.5 and temperature of 33 °C), 10 g/l of the respective substrates are totally converted to the *N*-carbamoyl-D-amino acid within 14–18 h. After separation of the immobilized biocatalyst, which is used again in a repeat of the reaction, a second chemical reaction has to be performed—decarbamoylation of *N*-carbamoyl-D-hydroxyphenylglycine with nitrite—to get the free optically pure D-amino acids. This method can also be applied to the conversion of, for example, DL-5-phenylhydantoin, its chloro derivatives, and DL-5-thienylhydantoin.

The second strategy is to use both D-hydantoinase and *N*-carbamoyl-D-amino acid amidohydrolase activities of microbial cells for the direct conversion of DL-5-*p*-hydroxyphenylglycine to the corresponding D-amino acid. This method was postulated 1978 by Olivieri *et al.* [19, 20] for resting cells of *Agrobacterium radiobacter* NRRL B 11291 and at the same time by the Japanese company Ajinomoto using free and immobilized cells of different bacteria [234].

The problems in using this method are:

(i) to induce both D-hydantoinase and *N*-carbamoyl-D-amino acid amidohydrolase during the growth of the biocatalyst,

(ii) to find reaction conditions corresponding to optimal reaction conditions and stabilities of both enzymes as well as to the chemical racemization of the substrate, and

(iii) to get a total 100% conversion of the DL-5-monosubstituted hydantoin to the optically pure D-amino acid without any residual substrate or *N*-carbamoylamino acid.

This is much more difficult in the case of two enzymes than of only the D-hydantoinase, because D-hydantoinase and *N*-carbamoyl-D-amino acid amidohydrolase differ in their properties (see Section 5.6). Because of the low enzyme activities of the Ajinomoto biocatalysts [234], Tramper and van Luyben [217, 238] selected the first strategy—using only the D-hydantoinase and performing the decarbamoylation chemically—as the more feasible one of these two methods. The following assumptions were made [238]: annual (300 workdays) production capacity of 300 t, overall yield of 90%, bioconversion productivity of 5.4×10^{-2} mol·h^{-1}·kg^{-1} immobilized biocatalyst (= 8.95 g·h^{-1}·kg^{-1} biocatalyst for D-*p*-hydroxyphenylglycine).

In 1987 Yokozeki *et al.* of Ajinomoto published on the isolation and characterization of a new biocatalyst for the second process, *Pseudomonas* sp. AJ 11220 [22, 23, 24]. With this microorganism they observed high enzyme activities and no intermediary product at pH 8.0 and 43 °C after induction with the newly synthesized compound DL-5-cyanoethylenehydantoin [23]. So it would be of interest to compare the feasibilities of the Kanegafuchi process and the Ajinomoto process—using immobilized cells of this newly isolated biocatalyst—again.

5.7.3 Process Development for the Production of D-Phenylglycine and D-*p*-Hydroxyphenylglycine with Immobilized Calf Liver Dihydropyrimidinase

Dinelli *et al.* [13] reported in 1978 on a new method for the production of D-amino acids using immobilized D-hydantoinase from calf liver. They applied the fiber-entrapment method using cellulose triacetate for the immobilization of the dihydropyrimidinase with a recovery of 80%. The process included three steps:

(i) chemical synthesis of the hydantoin derivative;
(ii) bioconversion to the D-configured *N*-carbamoylamino acids by the immobilized biocatalyst; and
(iii) chemical decarbamoylation.

The production of D-phenylglycine and D-*p*-hydroxyphenylglycine at a scale of about 3 kg per day has been investigated in a stirred-tank reactor at 30 °C and a pH of 8.5 which was kept constant by the addition of NaOH. Because of the very rapid racemization of the corresponding hydantoin derivatives under reaction conditions, a nearly total conversion of the DL-substrates to the *N*-carbamoyl-D-amino acids was achieved within five hours. The half-life of the entrapped D-hydantoinase was about 20 days. The *N*-carbamoylamino acids were hydrolyzed by adding calcium hydroxide at 100 °C.

5.7.4 Future Outlook

Optically pure D-amino acids are important as biomedically active compounds, as chiral directing auxiliaries, and chiral synthons in organic synthesis. At the moment the most important product of the D-hydantoinase process is D-*p*-hydroxyphenylglycine, which is used for the synthesis of semisynthetic penicillins and cephalosporins (see Section 5.7.1). But recent papers and patents report on process developments for the production of other D-amino acids, for example, of D-naphthylglycine for the synthesis of pharmaceuticals [242] or of D-valine for the synthesis of pesticides [21]. In future an important application of D-amino acids will be the use as synthons for biologically active peptides [243].

For the production of D-amino acids of interest in these applications by the hydantoin process, increasing the productivities of the biocatalysts and enhancing the enzyme activities will be the goals of biotechnology in the next few years. Different strategies are possible:

(i) using chemically modified inducers during the growth of the biocatalysts to get higher enzyme activities (see Chapter 6);

(ii) using special continuous processes with cell recycle in the case of product inhibitions [218, 219, 244]; and

(iii) altering enzyme syntheses and properties by genetic engineering (see Section 5.6.3).

Wada [5] reports on another application of the D-hydantoinase. He describes the production of optically active hydantoins used as chiral building blocks from the corresponding *N*-carbamoyl-DL-amino acids by this enzyme. And Siedel *et al.* of the German company Boehringer [197] report on the use of a 1-methylhydantoinase and *N*-carbamoyl-D-sarcosine amidohydrolase for diagnostic uses in combination with other enzymes.

5.8 CONCLUSIONS

The enzymatic production of optically pure D-amino acids from DL-5-monosubstituted hydantoin derivatives has become an important method in biotechnology in the last 10 years. The great advantages of this method are the wide applicability to a broad range of various substrates, the high enantioselectivity and stability of the biocatalysts, and the complete conversion of 100% of racemates of chemically synthesized hydantoin derivatives to the corresponding *N*-carbamoyl-D-amino acids or D-amino acids. As a result of the desire for D-amino acids, for example, for the synthesis of peptides for pharmacological and food uses as well as for pesticides, industrial interest in this process is still growing. Successful experiments with immobilized microorganisms and enzymes, as well as in genetic engineering, are a good starting point for further developments. The use of the enzymatic hydantoin racemization in combination with D-hydantoinase and *N*-carbamoyl-D-amino acid amidohydrolase will be of particular interest in future.

5.9 REFERENCES

1. von Baeyer, A., *Liebigs Ann. Chem. 117*, 178 (1861).
2. Ware, E., *Chem. Rev. 46*, 403 (1950).
3. Gaebler, O. H., Keltch, A. K., *J. Biol. Chem. 70*, 763 (1926).
4. Sobotka, H., U.S. Pat. 1,861,458 (1932).
5. Wada, M., *Proc. Imp. Acad. Japan 10*, 17 (1934).

6. Bernheim, F., Bernheim, M. L. C., *J. Biol. Chem. 163*, 683 (1946).
7. Eadie, G. S., Bernheim, F., Bernheim, M. L. C., *J. Biol. Chem. 181*, 449 (1949).
8. Wallach, D. P., Grisolia, S., *J. Biol. Chem. 226*, 277 (1957).
9. Caravaca, J., Grisolia, S., *J. Biol. Chem. 231*, 357 (1958).
10. Dudley, K. H., Bius, D. L., Butler, T. C., *J. Pharmacol. Exp. Ther. 175*, 27 (1970).
11. Dudley, K. H., Bius, D. L., *J. Heterocycl. Chem. 10*, 173 (1973).
12. Cecere, F., Galli, G., Morisi, F., *FEBS Lett. 57*, 192 (1975).
13. Weetall, H. H. In *Enzyme Engineering 3*; Dinelli, D., Marconi, W., Cecere, F., Galli, G., Morisi, F., Pye, E. K., Eds.; Plenum, New York, 1978; pp 477–481.
14. Yamada, H., Takahashi, S., Kii, Y., Kumagai, H., *J. Ferment. Technol. 56*, 484 (1978).
15. Takahaski, S., Kii, Y., Kumagai, H., Yamada, H., *J. Ferment. Technol. 56*, 492 (1978).
16. Takahashi, S., Ohashi, T., Kii, Y., Kumagai, H., Yamada, H., *J. Ferment. Technol. 57*, 328 (1979).
17. Shimizu, S., Takahashi, H., Shimada, S., Ohashi, T., Tani, Y., Yamada, H., *Agric. Biol. Chem. 44*, 2233 (1980).
18. Yamada, H., Shimizu, S., Shimada, H., Tani, Y., Takahashi, S., Ohashi, T., *Biochimie 62*, 395 (1980).
19. Olivieri, R., Fascetti, E., Angelini, L., Degen, L., *Enzyme Microb. Technol. 1*, 201 (1979).
20. Olivieri, R., Fascetti, E., Angelini, L., Degen, L., *Biotechnol. Bioeng. 23*, 2173 (1981).
21. Battilotti, M., Barberini, U., *J. Mol. Cat. 43*, 343 (1988).
22. Yokozeki, K., Nakamori, S., Eguchi, C., Yamada, K., Mitsugi, K., *Agric. Biol. Chem. 51*, 355 (1987).
23. Yokozeki, K., Nakamori, S., Yamanaka, S., Eguchi, C., Mitsugi, K., Yoshinaga, F., *Agric. Biol. Chem. 51*, 715 (1987).
24. Yokozeki, K., Kubota, K., *Agric. Biol. Chem. 51*, 721 (1987).
25. Guivarch, M., Gillonier, C., Brunie, J.-C., *Bull. Soc. Chim. Fr. 1-2*, 91 (1980).
26. Wanru, S., *Acta Microbiologica Sinica 23*, 133 (1983).
27. Wanru, S., *Acta Microbiologica Sinica 23*, 257 (1983).
28. Morin, A., Hummel, W., Kula, M.-R., *Biotechnol. Lett. 8*, 573 (1986).
29. Morin, A., Hummel, W., Kula, M.-R., *Appl. Microbiol. Biotechnol. 25*, 91 (1986).
30. Morin, A., Hummel, W., Schütte, H., Kula, M.-R., *Biotechnol. Appl. Biochem. 8*, 564 (1986).
31. Morin, A., Hummel, W., Kula, M.-R., *J. Gen. Microbiol. 133*, 1201 (1987).
32. Syldatk, C., Cotoras, D., Möller, A., Wagner, F., *BTF-Biotechforum 3*, 9 (1986).
33. Möller, A., Syldatk, C., Schulze, M., Wagner, F., *Enzyme Microb. Technol. 10*, 618 (1988).
34. Polastro, E. T., Walker, A., Teeuwen, H. W. A., *Bio/Technology 7*, 1238 (1989).
35. Triadan, H. *Über die parodontale Nebenwirkung der chronischen Hydantoinbehandlung*; Hüthig, Heidelberg, 1968.
36. Spurlock, J. J., U.S. Pat. 2,366,221 (1945). [*Chem Abstr. 39*, 1736 (1945)].
37. Mehta, N. B., Diuguid, C. A. R., Soroko, F. E., *J. Med. Chem. 24*, 465 (1981).
38. Spinks, A., Waring, W. S., *Prog. Med. Chem. 3*, 313 (1963).

39. Bogoch, S., Dreyfus, J. *The Broad Range of Use of Diphenylhydantoin*; Dreyfus Medical Foundation, New York, 1970.
40. Perry, J. K., Newmark, M. E., *Ann. Intern. Med. 89*, 207 (1979).
41. Tahara, S., Miura, Y., Mizutani, J., *Agric. Biol. Chem. 43*, 919 (1979).
42. Thibault, T. D., St. Clair, R. L., U.S. Pat. 4,345,936 (1982). [*Chem. Abstr. 98*, 16687p (1983)].
43. Tielemann, H., *Sci. Pharm. 39*, 8 (1971).
44. Ohta, H., Jikihara, T., Wakabayashi, K., Fujita, T., *Pestic. Biochem. Physiol. 14*, 153 (1980).
45. Mappes, C. J., Pommer, E.-H., Rentzea, C., Zeeh, B., U.S. Pat. 4,198,423 (1980). [*Chem. Abstr. 93*, 71784 (1980)].
46. Shimizu, T., *Soil Sci. Plant. Nutr. (Tokyo) 32*, 373 (1986).
47. López, C. A., Trigo, G. G., *Adv. Heterocycl. Chem. 38*, 177 (1985).
48. Bateman, J. H. In *Kirk Othmer's Encycl. Chem. Technol.*; 3rd ed.; Wiley-Interscience, New York, 1980; Vol. 5, p 573; Vol. 11, pp 141–164; Vol. 12, pp 692–711; Vol. 13, pp 135ff.
49. Schipper, E. S., Day, A. R. In *Heterocyclic Compounds*; Elderfield, R. C., Ed.; John Wiley & Sons, New York, 1957; Vol. 5, p 254.
50. Edward, J. T., *Chem. Org. Sulfur Comp. 2*, 287 (1966).
51. Bucherer, H. T., Steiner, W., *J. Prakt. Chem. 140*, 291 (1934).
52. Hoppe, B., Martens, J., *Chemie in unserer Zeit 18*, 73 (1984).
53. Elks, J., Hems, B. A., Ryman, B. E., *J. Chem. Soc. 1948*, 1386.
54. Deulofen, V., *Ber. Dtsch. Chem. Ges. 69*, 2456 (1936).
55. Corey, E. J., Cheng, X.-M. *The Logic of Chemical Synthesis*; John Wiley & Sons, New York, 1989.
56. Urech, F., *Liebigs Ann. Chem. 165*, 99 (1872).
57. Harries, C., Weiss, M., *Ber. Dtsch. Chem. Ges. 33*, 3418 (1900).
58. Güler, F., Moodie, R. B., *J. Chem. Soc., Perkin Trans. 2 1980*, 1752.
59. Stella, V., Huguchi, T., *J. Org. Chem. 38*, 1527 (1973).
60. Schulte, B., Jakob, W., Dünwald, W., Meyer, K.-H., Zecher, W., Eur. Pat. 0080 079 (1986).
61. Nagasawa, H. T., Fraser, P. S., Elberling, J. A., *J. Org. Chem. 37*, 516 (1972).
62. Bailey, J. R., *Am. Chem. J. 28*, 386 (1902).
63. Boyd, W. J., *Biochem. J. 27*, 1838 (1933).
64. Thielemann, H., *Z. Chemie 18*, 174 (1978).
65. Dakin, H. D., *Biochem. J. 13*, 398 (1919).
66. Johnston, T. B., Nicolet, B. H., *J. Am. Chem. Soc. 36*, 355 (1914).
67. Rigler, N. E., Henze, H. R., *J. Am. Chem. Soc. 58*, 474 (1936).
68. Read, W. T., *J. Am. Chem. Soc. 44*, 1746 (1922).
69. Heintz, W., *Liebigs Ann. Chem. 169*, 120 (1873).
70. Caldwell, A. G., Harris, C. J., Stepney, R., Whittaker, N., *J. Chem. Soc., Perkin Trans. 1 1980*, 495.
71. Pless, J., *J. Org. Chem. 39*, 2644 (1974).
72. Wöhler, F., *Pogg. Ann. 12*, 253 (1828).
73. Hassall, H., Greenberg, D. M., *J. Biol. Chem. 238*, 3325 (1963).
74. Suzuki, T., Igarashi, K., Hase, K., Tuzimura, K., *Agric. Biol. Chem. 37*, 411 (1973).

75. Bovarnick, M., Clarker, H. T., *J. Am. Chem. Soc. 60*, 2426 (1938).
76. Jones, R. C. F., Sumaria, S., *Tetrahedron Lett. 1978*, 3173.
77. Miller, E. J., Robson, W., *J. Chem. Soc. 1938*, 1910.
78. Pickett, L. W., McLean, M., *J. Am. Chem. Soc. 61*, 423 (1939).
79. Siemonsen, L., *Liebigs Ann. Chem. 333*, 101 (1904).
80. Franchimont, A. P. N., Klobbie, E. A., *Rec. Trav. Chim. Pay-Bas 8*, 283.
81. Biltz, H., Slotta, K., *J. Prakt. Chem. 113*, 233 (1926).
82. Novelli, A., Lugones, Z. M., Velasco, P., *Anales Asoc. Quim. Argentina 30*, 225 (1942).
83. Orazi, O. O., Corral, R. A., *Tetrahedron 15*, 93 (1961).
84. Natarajan, P. N., *Acta Pharm. Suecica 8*, 537 (1971).
85. Johnson, T. B., Nicolet, B. H., *J. Am. Chem. Soc. 34*, 1048 (1912).
86. Pedregal, C., Trigo, G. G., Espada, M., Elguero, J., *J. Heterocycl. Chem. 21*, 477 (1984).
87. Werner, W., *Arch. Pharm. 299*, 513 (1966).
88. Orazi, O. O., Corall, R. A., *Experientia 21*, 508 (1965).
89. Mouneyrat, A., *Ber. Dtsch. Chem. Ges. 33*, 2393 (1900).
90. Boyd, W. J., Robson, W., *Biochem. J. 29*, 546 (1935).
91. Koenigs, E., Mylo, B., *Ber. Dtsch. Chem. Ges. 41*, 4427 (1908).
92. Patton, T. L., *J. Org. Chem. 32*, 383 (1967).
93. Delépine, M., *Bull. Soc. Chim. (Fr.) 29* (3), 1190 (1903).
94. Oldfield, W., Cashin, C. H., *J. Med. Chem. 8*, 239 (1965).
95. Krueger, U., Zinner, G., *Arch. Pharm. (Weinheim.) 311*, 1978 (39).
96. Cook, A. H. *et al.*, *J. Chem. Soc. 1950*, 1892.
97. Wheeler, H. L., Brautlecht, C. A., *J. Am. Chem. Soc. 45*, 446 (1911).
98. Edman, P., *Acta Chem. Scand. 4*, 274 (1950).
99. Schlack, P., Kumpf, W., *Z. Physiol. Chem. 154*, 125 (1926).
100. Haring, K. M., Johnson, T. B., *J. Am. Chem. Soc. 55*, 395 (1933).
101. Fujiwara, H., Bose, A. K., Manhas, M. S., Vander Veen, J. M., *J. Chem. Soc., Perkin Trans. 2 1979*, 653.
102. Sarges, R., Howard, H. R., Kelbaugh, P. R., *J. Org. Chem. 47*, 4081 (1982).
103. Iwata, K., Hara, S., *J. Heterocycl. Chem. 15*, 1231 (1978).
104. Bergmann, M., Kann, E., Miekeley, A., *Liebigs Ann. Chem. 458*, 40 (1927).
105. Edmann, P., *Acta Chem. Scand. 4*, 283 (1950).
106. Bates, G. R., Smith, D. G., *J. Biol. Chem. 238*, 214 (1963).
107. Stark, G. R., *Biochem. 7*, 1796 (1968).
108. Suzuki, T., Tomioka, T., Tuzimura, K., *Can. J. Biochem. 55*, 512 (1977).
109. Urech, F., *Ber. Dtsch. Chem. Ges. 4*, 526 (1871).
110. Bergs, H., Ger. Pat. 566 094 (1929). [*Chem. Abstr. 27*, 1001 (1933)].
111. Bucherer, H. T., Lieb, V. A., *J. Prak. Chem. 141*, 5 (1934).
112. Poupaert, J. H., Winland, M., Dumont, P., *J. Labelled Comp. Radiopharm. 18*, 1827 (1981).
113. Henze, H. R., Speer, R. J., *J. Am. Chem. Soc. 64*, 522 (1942).
114. Edward, J. T., Jitrangsri, C., *Can. J. Chem. 53*, 3339 (1975).
115. Munday, L., *J. Chem. Soc. 1961*, 4372.
116. Maki, Y., Masugi, T., *Chem. Pharm. Bull. (Tokyo) 21*, 685 (1973).

117. Schipper, E., Chinery, E., *J. Org. Chem. 26*, 3597 (1961).
118. Shoga, H., Morita, T., Nakayasu, K., Kawasaki, K., Ger. Pat. DE 3 525 922 (1986).
119. Takeuchi, K., Sato, Y., Okamura, T., Ogata, H., Fuse, K., Jpn. Pat. 6183 164 (1986). [*Chem. Abstr. 105*, 172458h (1986)].
120. Taillades, J., Commeyras, A., *Tetrahedron 30*, 127; *30*, 2493; *30*, 3407 (1974).
121. Trigo, G. G., Avendano, C., *An. Quim. 73*, 426 (1977).
122. Chubb, F. L., Edward, J. T., *Can. J. Chem. 59*, 2724 (1981).
123. Rinkes, J. J., *Rec. Trav. Chim. 46*, 268 (1927).
124. Dvonch, W., Fletcher, H., Alburn, H. E., *J. Org. Chem. 29*, 2764 (1964).
125. Knabe, J., Wumm, W., *Arch. Pharm. (Weinheim) 313*, 538 (1980).
126. Tamura, F., Togo, K., Eguchi, N., Yokozeki, K., Jpn. Pat. 7837 664 (1978). [*Chem. Abstr. 89*, 90191z (1978)].
127. Krohn, K., Bilitewski, U., Eickhoff, A., unpublished results, 1990.
128. Wieland, T. In *Houben-Weyl Methoden der Organischen Chemie*; Müller, E., Ed.; Thieme Verlag, Stuttgart; Vol. XI/2, pp 305, 307.
129. Coker, J. N., Kohlhase, W. L., Martens, T. F., Rogeres, A. O., Allan, G. G., *J. Org. Chem. 27*, 3202 (1962).
130. [Ajinomoto Co., Inc.], Jpn. Pat. 59222 454. [*Chem. Abstr. 103*, 22920x (1985)].
131. Pedregal, C., Trigo, G. G., Espada, M., Mathieu, D., Lamarca, J., Elguero, J., *J. Heterocycl. Chem. 21*, 1527 (1984).
132. Chubb, F. L., Edward, J. T., Wong, S. C., *J. Org. Chem. 45*, 2315 (1980).
133. Icli, S., Colebrook, L. D., *J. Pure Appl. Sci. 9*, 39 (1976).
134. Behera, M., Dhal, P. N., Nayak, A., *J. Indian Chem. Soc. 52*, 680 (1973).
135. Wandelin, W. *et al.*, *Monatsh. 112*, 1091 (1981).
136. Kochkanyan, R. O., Israelyan, Y. A., Zaritovskii, A. N., *Kim. Geterotskl. Soedin. 1*, 87 (1978).
137. Lippich, F., *Ber. Dtsch. Chem. Ges. 39*, 2953 (1906).
138. Roeda, D., Westera, G., *Int. J. Appl. Radiat. Isot. 32*, 843 (1981).
139. [Ube Industries, Ltd.], Jpn. Pat. 8259 874 (1982). [*Chem. Abstr. 97*, 182420v (1982)].
140. Murahashi, S., Yuki, H., Kosai, K., Doura, F., *Bull. Chem. Soc. Jpn. 39*, 1559 (1966).
141. Davidson, D., *J. Am. Chem. Soc. 47*, 255 (1925).
142. Ben-Ishai, D., Altmann, J., Bernstein, Z., *Tetrahedron 33*, 1191 (1977).
143. Kyora, T., Jpn. Pat. 6 122 081 (1986). [*Chem. Abstr. 105*, 42800c (1986)].
144. Pauly, H., Sauter, H., *Ber. Dtsch. Chem. Ges. 63*, 2063 (1930).
145. [Ube Industries, Ltd.], Jpn. Pat. 8 259 873 (1982). [*Chem. Abstr. 97*, 182419b (1982)].
146. [Bayer AG], Ger. Pat. DE 2539730-A (1977).
147. Keys, L. D., III, Folting, K., Streib, W. S., Johnston, M., *J. Org. Chem. 51*, 4721 (1986).
148. Ugi, I., Risendahl, K., Bodesheim, F., *Liebigs Ann. Chem. 666*, 54 (1963).
149. Daboun, H. A. F., Abdou, S. E., Hussein, M. M., Elnagdi, M. H., *Synthesis 1982*, 502.

150. Wheeler, H. I., Hoffman, C., *J. Am. Chem. Soc. 45*, 568 (1911).
151. Phillips, A. P., *J. Am. Chem. Soc. 67*, 744 (1945).
152. Grassmann, W., Wünsch, E., Deufel, P., Zwick, A., *Chem. Ber. 91*, 539 (1958).
153. Tan, S.-F., Ang, K.-P., Fong, Y.-F., *J. Chem. Soc., Perkin Trans. 2 1986*, 1941.
154. Nadeau, G., Gaudry, R., *Can. J. Res. 27B*, 421 (1949).
155. Tan, S.-F., Peng, K.-P., Jayachandran, H., Fong, Y.-F., *J. Chem. Soc., Perkin Trans. 2 1987*, 1043.
156. Phillips, A. P., Murphy, J. G., *J. Org. Chem. 16*, 954 (1951).
157. Barger, G., Easson, A. P. T., *J. Chem. Soc. 1938*, 2100.
158. Boyd, W. J., Robson, W., *Biochem J. 29*, 542 (1935).
159. Shabica, A. C., Howe, E. E., Ziegler, J. B., Tishler, M., *J. Am. Chem. Soc. 68*, 1156 (1946).
160. [Abbott Labs.], U.S. Pat. 2,305,501 (1940).
161. Bharucha, A. R., Pavilanis, V., Ajdukovic, D., Shrenk, H. M., Ger. Pat. 2329 745 (1974). [*Chem. Abstr. 80*, 95948d (1974)].
162. Inomata, M., Itakura, M., Odaka, K., Takeuchi, K., Inagaki, H., Jpn. Pat. 6168 472 (1986). [*Chem. Abstr. 105*, 172457g (1986)].
163. Luessling, T., Scherberich, P., Ger. Pat. DE 3 023 349 (1980). [*Chem. Abstr. 96*, 104248k (1982)].
164. Kleemann, A., Luessling, T., Pfeifer, W., Scherberich, P., Ger. Pat. DE 3 013 647 (1981). [*Chem. Abstr. 96*, 20099z (1982)].
165. Imaki, S., Takuma, I., Jpn. Pat. 6137 775 (1986). [*Chem. Abstr. 105*, 42806j (1986)].
166. Imaki, S., Takuma, I., Jpn. Pat. 6137 774 (1986). [*Chem. Abstr. 195*, 42807k (1986)].
167. Johnson, B., Bates, J. S., *J. Am. Chem. Soc. 37*, 383 (1915).
168. Nicolet, B. H., *J. Am. Chem. Soc. 37*, 2753 (1915).
169. Tanaka, A., Nakayasu, K., Jpn. Pat. 6 127 970 (1986). [*Chem. Abstr. 105*, 42801d (1986)].
170. Tanaka, A., Nakayasu, K., Jpn. Pat. 6 127 971 (1986). [*Chem. Abstr. 105*, 6511a (1986)].
171. Gaudette, R. R., Belg. Pat. 859290 (1987). [*Chem. Abstr. 89*, 43415u (1978)].
172. Henze, H. R., Whitney, W. B., Eppright, M. A., *J. Am. Chem. Soc. 62*, 565 (1940).
173. Johnson, H. I., Brautlecht, C. A., *J. Biol. Chem. 12*, 175.
174. Connors, T. A., Ross, W. C. J., Wilson, J. G., *J. Chem. Soc. (London) 1960*, 2994.
175. Majima, R., Kotake, M., *Ber. Dtsch. Chem. Ges. 55*, 3859 (1922).
176. Gordon, W. G., Jackson, R. W., *J. Biol. Chem. 110*, 151 (1935).
177. Johnson, T. B., Bengis, R., *J. Am. Chem. Soc. 34*, 1054 (1940).
178. Johnson, B., Bates, J. B., *J. Am. Chem. Soc. 37*, 383 (1915).
179. Boyd, W. J., Robson, W., *Biochem. J. 29*, 2256 (1935).
180. Hahn, D. A., Gilman, E., *J. Am. Chem. Soc. 47*, 2941 (1925).
181. Elks, J., *J. Chem. Soc. 1944*, 629.
182. Moriya, T., Hagio, K., Yoneda, N., *Chem. Pharm. Bull. 28*, 1891 (1980).
183. Vogels, G. D., Trijbels, F., Uffink, A., *Biochim. Biophys. Acta 122*, 482 (1966).
184. Lee, K. W., Roush, A. H., *Arch. Biochem. Biophys. 108*, 460 (1964).
185. Okumura, I., Kondo, K., Miyake, Y., Itaya, K., Yamamoto, T., *J. Biochem. 79*, 1013 (1976).

186. Trijbels, F., Vogels, G. D., *Biochim. Biophys. Acta 113*, 292 (1966).
187. Rijnierse, V. F. M., van der Drift, C., Vogels, G. D., *Can. J. Microbiol. 23*, 633 (1977).
188. van der Drift, L., Vogels, G. D., van der Drift, C., *Biochim. Biophys. Acta 391*, 240 (1975).
189. Lieberman, I., Kornberg, A., *Biochim. Biophys. Acta 12*, 223 (1953).
190. Lieberman, I., Kornberg, A., *J. Biol. Chem. 207*, 911 (1954).
191. Brown, D. D., Kies, M. W., *J. Biol. Chem. 234*, 3182 (1959).
192. Akamatsu, N., *J. Biochem. 47*, 809 (1960).
193. Tsugawa, R., Okumura, S., Ito, T., Katsuga, N., *Agric. Biol. Chem. 30*, 27 (1966).
194. Hassall, H., Greenberg, D. M., *J. Biol. Chem. 238*, 3325 (1963).
195. Liebermann, I., Kornberg, A., *J. Biol. Chem. 212*, 909 (1955).
196. International Union of Biochemistry, *Enzyme Nomenclature*; Academic Press, San Diego, 1984.
197. Siedel, J., Deeg, R., Röder, A., Ziegenhorn, J., Möllering, H., Gauhl, H., Ger. Pat. DE 3 406 770 A1 (1985).
198. Yamada, H., Shimizu, S., Kim, J. M., Shinmen, Y., Sakai, T., *FEMS Microbiol. Lett. 30*, 337 (1985).
199. Kim, J. M., Shimizu, S., Yamada, H., *J. Biol. Chem. 261*, 11832 (1986).
200. Nishida, Y., Nakamichi, K., Nabe, K., Tosa, T., *Enzyme Microb. Technol. 9*, 721 (1987).
201. Syldatk, C., Cotoras, D., Dombach, G., Groß, C., Kallwaß, H., Andree, H., Gebinoga, M., Wagner, F., *Hoppe-Seyler's Z. Physiol. Chem. 367*, 340 (1986).
202. Syldatk, C., Cotoras, D., Dombach, G., Groß, C., Kallwaß, H., Wagner, F., *Biotechnol. Lett. 9*, 25 (1987).
203. Yamashiro, A., Yokozeki, K., Kano, H., Kubota, K., *Agric. Biol. Chem. 52*, 2851 (1988).
204. Yokozeki, K., Sano, K., Eguchi, C., Yamada, K., Mitsugi, K., *Agric. Biol. Chem. 51*, 819 (1987).
205. Yokozeki, K., Sano, K., Eguchi, C., Iwagami, K., Mitsugi, K., *Agric. Biol. Chem. 51*, 729 (1987).
206. Yokozeki, K., Hirose, Y., Kubota, K., *Agric. Biol. Chem. 51*, 737 (1987).
207. Syldatk, C., Müller, R., Dombach, G. *Natural Substrates of Hydantoinase and L-N-Carbamoylase Involved in the Production of L-Amino Acids from DL-5-Monosubstituted Hydantoins*; 8th Int. Biotechnol. Symposium, Paris, 1988; Abstracts, p 154.
208. Syldatk, C., Müller, R., Pietzsch, M., Wagner, F. *this book*, Chapter 6.
209. Campbell, L. L., *J. Biol. Chem. 235*, 2375 (1960).
210. Hayashi, O., Kornberg, A., *J. Biol. Chem. 197*, 717 (1952).
211. Sano, K., Yokozeki, K., Eguchi, C., Kagawa, T., Noda, I., Mitsugi, K., *Agric. Biol. Chem. 41*, 819 (1977).
212. Groß, C., Syldatk, C., Wagner, F., *Biotechnol. Techniques 1*, 85 (1987).
213. Syldatk, C., Mackowiak, V., Höke, H., Groß, C, Dombach, G., Wagner, F., *J. Biotechnol. 14*, 345 (1990).
214. Brans, A., Syldatk, C., Wagner, F., in preparation 1992.

215. Syldatk, C., Mackowiak, V., Dombach, G., Groß, C., Röhrmann, A., Wagner, F. In *Proc. 4th Eur. Congr. Biotechnol.*; Neijssel, O. M., van der Meer, R. R., Luyben, K. C. A. M., Eds.; Elsevier Science Publishers B.V., Amsterdam, 1987; Vol. 2, pp 245–248.
216. Syldatk, C., Wagner, F., *J. Food Biotechnol. 4*, 87 (1990).
217. Tramper, J., Luyben, K. C. A. M., *PT/Procestechniek 37*, 97 (1982).
218. Chevalier, P., Roy, D., Morin, A., *Appl. Microbiol. Biotechnol. 30*, 482 (1989).
219. Chevalier, P., Roy, D., Morin, A., *Biotechnol. Techniques 3*, 135 (1989).
220. Cecere, F., Marconui, W., Morisi, F., Rappuoli, B., Ger. Pat. DE 2 615 594 A1 (1978).
221. Brooks, K. P., Jones, E. A., Kim, B.-D., Sander, E. G., *Arch. Biochem. Biophys. 226*, 469 (1983).
222. Watanabe, W., Snell, E. E., *Proc. Nat. Acad. Sci. 69*, 1086 (1972).
223. Jacob, E., Henco, K., Marcinowski, S., Schenk, G., Ger. Pat. DE 3535 987 (1987).
224. Soda, K., Tanaka, H., Esaki, N. In *Biotechnology*; Rehm, H. J., Reed, G., Eds.; Verlag Chemie, Weinheim, 1983; Vol. 3, pp 480–530.
225. Schmidt-Kastner, G., Egerer, P. In *Biotransformations, Biotechnology*; Kieslich, K., Rehm, H. J., Reed, G., Eds.; Verlag Chemie, Weinheim, 1984; Vol. 6a, pp 387–419.
226. Meijer, E. M., Boesten, W. H. J., Schoemaker, H. E., van Balken, J. A. M. In *Biocatalysis in Organic Syntheses*; Tramper, J., van der Plas, H. C., Linko, P., Eds.; Elsevier Science Publishers, Amsterdam, 1985; pp 135–156.
227. van den Tweel, W. J. J., Ogg, R. L. H. P., de Bont, J. A. M., Dutch Pat. 8702 449 (1987).
228. Schutt, H., Schmidt-Kastner, G., Arens, A., Preiss, M., *Biotechnol. Bioeng. 27*, 420 (1985).
229. Moriguchu, M., Ideta, K., *Environ. Microbiol. 54*, 2767 (1988).
230. Sugie, M., Suzuki, H., *Agric. Biol. Chem. 42*, 107 (1978).
231. Sugie, M., Suzuki, H., *Agric. Biol. Chem. 44*, 1089 (1980).
232. Takahashi, H., Takahashi, S., Ohashi, T., Yoneda, K., Watanabe, K., Jpn. Pat. 6225 990 (1987).
233. Degen, L., Viglia, A., Fascetti, E., Perricone, E., Ger. Pat. DE 2631 048 (1976).
234. Nakamori, S., Yokozeki, K., Mitsugi, K., Eguchi, C., Iwagami, H., U.S. Pat. 4,211,840 (1980).
235. Lungershausen, R., Martin, C., Marcinowski, S., Siegel, H., Kuesters, W., Ger. Pat. DE 3031 151 A1 (1982).
236. Gillonier, C., Guivarch, M., Ger. Pat. DE 3018 584 A1 (1980).
237. Tawaki, S., Tarukawa, H., Aikawa, T., Takeichi, M., Jpn. Pat. 63112 990 (1988).
238. Tramper, J., Luyben, K. C. A. M., *PT/Procestechniek 39*, 61 (1984).
239. Yamada, H. In *Proc. 4th Eur. Congr. Biotechnol.*; Neijssel, O. M., van der Meer, R. R., Luyben, K. C. A. M., Eds.; Elsevier Science Publishers B.V., Amsterdam, 1987; Vol. 4, pp 689–692.
240. Yamada, H., Shimizu, S. In *Biocatalysts in Organic Syntheses*; Tramper, J., van der Plas, H. C., Linko, P., Eds.; Elsevier Science Publishers, B.V., 1985; pp 19–37.
241. [Kogyo, Kanegafuchi Kagaku], Eur. Pat. 0175 312 A2 (1985).

242. [Kanegafuchi], Jpn. Pat. 145 510 (1986).
243. Jakubke, D., Jeschkeit, H. *Aminosäuren, Peptide, Proteine*; Verlag Chemie, Weinheim, 1982.
244. [Mitsui-Toatsu], Jpn. Pat. 325 384 (1989).

MICROBIAL AND ENZYMATIC PRODUCTION OF L-AMINO ACIDS FROM DL-5-MONOSUBSTITUTED HYDANTOINS

Christoph Syldatk, Ralf Müller, Markus Pietzsch, and Fritz Wagner*

*Parts of this chapter are the subject of C. Syldatk's "Habilitationsarbeit", TU Braunschweig, 1990.

CONTENTS

6.1 INTRODUCTION

The enzymatic resolution of amino acid derivatives by strictly L-selective, hydrolytic enzymes has been described starting from racemic esters, amides and *N*-acyl derivatives of amino acids (for reviews see [1], [2], [3], and [4]). For economic reasons, it is necessary in all cases to racemize the unreacted substrates in a second chemical reaction and to recycle the racemates thus obtained back to the first enzymatic reaction. In contrast, DL-5-monosubstituted hydantoins can be totally converted to optically pure amino acids without any additional chemical reaction steps: most of the 5-monosubstituted hydantoins already racemize under reaction conditions suitable for enzymes.

Many papers report on the enzymatic production of optically pure *N*-carbamoyl-D-amino acids or D-amino acids from DL-5-monosubstituted hydantoin derivatives (see Chapter 5). On the other hand, the L-selective hydantoin cleavage has so far been found only in bacteria of the genera *Arthrobacter* [5, 6, 7], *Bacillus* [8], *Clostridium* [9] and *Flavobacterium* [10, 11]; the enzymes of *Bacillus* and *Clostridium* have substrate specificities distinctly different from those of *Arthrobacter* and *Flavobacterium* (see Section 6.2).

The natural function of the D-specific enzyme D-hydantoinase is identical with that of the well-known enzyme dihydropyrimidinase (EC 3.5.2.2), and the natural function of the D-*N*-carbamoylase with that of the enzyme β-ureidopropionase (EC 3.5.1.6) (see Chapter 5). In contrast, the natural function of the enzymes involved in the L-selective cleavage of DL-5-monosubstituted hydantoins in *Arthrobacter* and *Flavobacterium* is not yet known [12, 13].

A great advantage of both the D- and the L-specific processes for the production of optically pure amino acids compared with other enzymatic methods is their wide substrate specificity—even nonproteinogenic amino acids can be produced by both methods [14, 15].

The goals of this work are to review investigations performed from 1984 to 1989 at the Institute of Biochemistry and Biotechnology A of the Technical University of Braunschweig (FRG) on the production of L-tryptophan from DL-5-indolylmethylenehydantoin, and to compare the results with those obtained by other working groups. We survey process development, including analytical and screening methods [15, 16], selection of strains with high enzyme activities but low L-tryptophan degrading activity [17], optimization of growth and enzyme synthesis of the biocatalysts [17, 18], optimization of the bioconversion by resting cells [19, 20, 21], and isolation and characterization of the enzymes involved in this process [7, 13, 22, 23, 24, 25]. Figure 6.1 gives a survey of all steps of this process development. We describe the detection of a 5-indolylmethylenehydantoin racemase and its separation from the enzymes L-hydantoinase and *N*-carbamoyl-L-amino acid amidohydrolase [26]. At the end we discuss further experiments and industrial applications of this process.

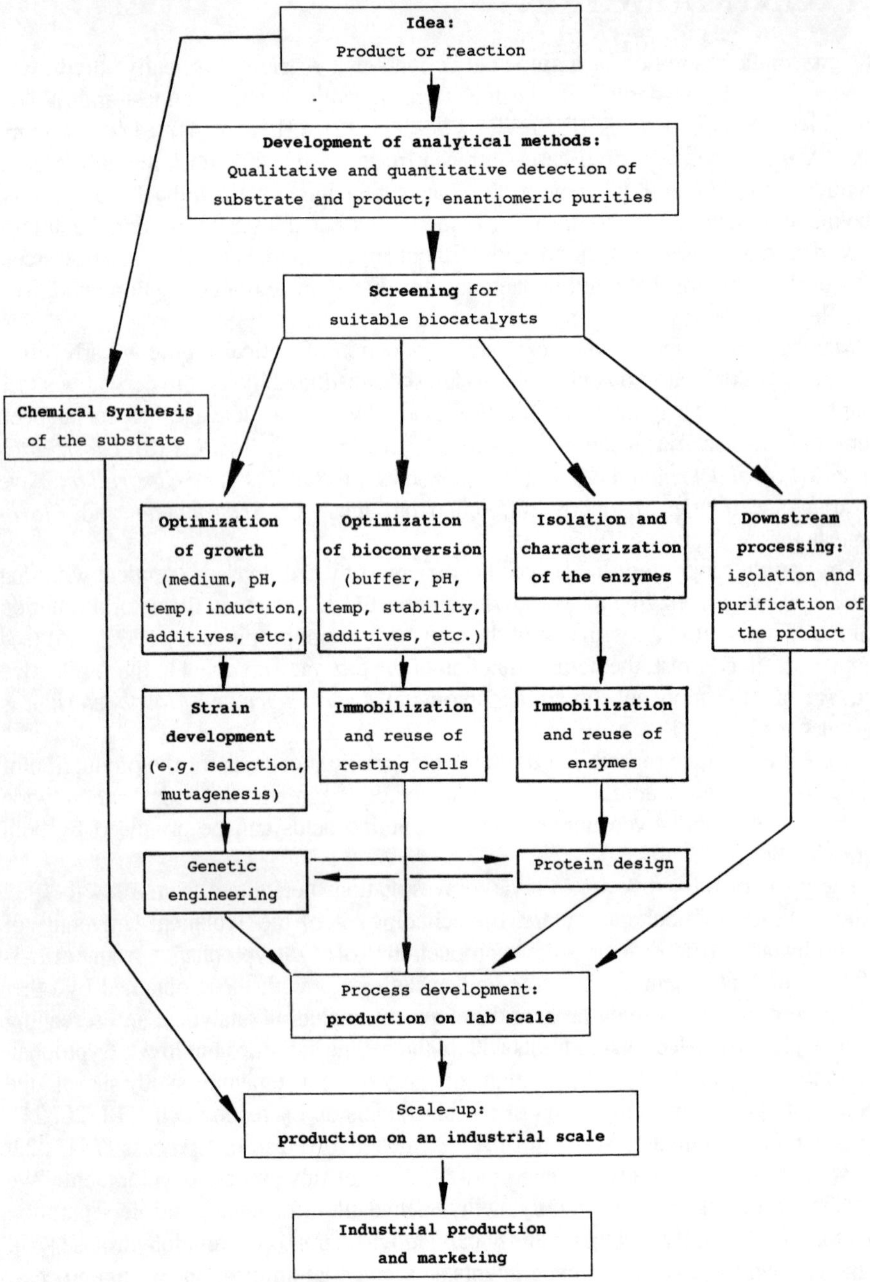

Figure 6.1. Development of an Industrial Bioconversion Method.

6.2 DISTRIBUTION OF L-SELECTIVE HYDANTOIN CLEAVAGE IN NATURE

Naturally occurring hydantoin derivatives and the distribution of hydantoin-cleaving enzymes in nature are discussed in detail in Chapter 5 and in [27]. While many papers report on the enzymatic production of optically pure *N*-carbamoyl-D-amino acids or D-amino acids from DL-5-monosubstituted hydantoin derivatives (see Chapter 5), L-selective hydantoin cleavage has so far only been studied in detail in bacteria of the genera *Arthrobacter* [5, 6, 7], *Bacillus* [8, 28, 29], *Clostridium* [9, 30, 31], *Flavobacterium* [10, 11, 32, 33, 34] and *Nocardia* [35].

Liebermann and Kornberg in the 1950s [9, 30, 31] were the first to describe L-selective cleaveage of a hydantoin derivative. They found that the enzymes from the anaerobic bacterium *Clostridium oroticum* involved in the orotic acid degradation pathway were able to produce L-aspartic acid from the corresponding L-5-carboxymethylenehydantoin. In the 1960s Hassall and Greenberg [36] and Tsugawa *et al.* [8] reported on the microbial L-selective cleavage of carboxyethylenehydantoin, which was shown by Akamatsu [37] by inducer experiments to be different from the former enzyme. In screening experiments, Tsugawa *et al.* [8] were able to isolate strains of the genera *Pseudomonas*, *Micrococcus*, *Aerobacter*, *Achromobacter*, and *Bacillus* able to produce L-glutamic acid from DL-5-hydantoin propionic acid (see Section 6.3.2, too). *Bacillus brevis* ATCC 8185 was the first microorganism used for the 90% biotechnological conversion of a racemic hydantoin derivative to an L-amino acid, L-glutamic acid [8]. But the enzymes of *Bacillus* and *Clostridium* have properties and substrate specificities distinctly different from those of *Arthrobacter* and *Flavobacterium* (for a detailed discussion see Sections 6.6.2 and 6.6.3 and Tables 6.10 and 6.14). Whether these enzymes are identical to those of *Nocardia* sp. DSM 3306 (described by Klages *et al.* [35]), or of *Bacillus brevis* AJ 12299 (described by Yamashiro *et al.* [28, 29] in 1988) has to be investigated.

So-called L-selective "arylalkylenehydantoinases" and "*N*-carbamoylarylalkyl-L-amino acid amidohydrolases" of comparable substrate specificities and properties were found by Guivarch *et al.* [5], Miyoshi *et al.* [6], and Syldatk *et al.* [7, 13, 17, 21, 24, 38, 39] in *Arthrobacter* sp. and by Yokozeki *et al.* [33, 34, 40] and Nishida *et al.* [10] in *Flavobacterium* sp., and will be discussed in detail in Sections 6.5 and 6.6.

6.3 SCREENING EXPERIMENTS

6.3.1 Development of Analytical Methods

The goal of these investigations was the development of a bioconversion process for the production of L-tryptophan (L-Trp) from DL-5-indolylmethylenehydantoin

(DL-5-IMH). First, analytical methods had to be developed for the qualitative and quantitative detection of DL-5-IMH, and the corresponding compounds *N*-carbamoyltryptophan (*N*-Carbtrp) and Trp as well as for the detection of enantiomeric purities of these compounds (see Figure 6.1). A detailed survey of various methods for the detection of hydantoin derivatives and degradation products is given in Chapter 5.

A qualitative analysis of 5-IMH, *N*-Carbtrp and Trp was possible after thin layer chromatography (TLC) of 10–20 μl of the reaction supernatants on silica gel plates in 1-butanol/acetic acid/H_2O (65:13:22, v/v/v), spraying with *p*-dimethylaminobenzaldehyde and subsequent heating of the plates [7, 39].

The concentrations of DL-5-IMH, *N*-Carbtrp, and Trp were determined quantitatively by HPLC on an RP-8 column (Serva, Heidelberg, FRG) using 50 mM KH_2PO_4/40% methanol as eluent and UV detection (280 nm). The flow rate was 1.0 ml/min; retention times were: Trp, 5.8 min; *N*-Carbtrp, 7.2 min; DL-5-IMH, 10.4 min (Figure 6.2) [26]. The specific activities of the enzymes hydantoinase and *N*-carbamoyl-L-amino acid amidohydrolase were defined as mmol product·g cell dry mass^{-1}·h^{-1}.

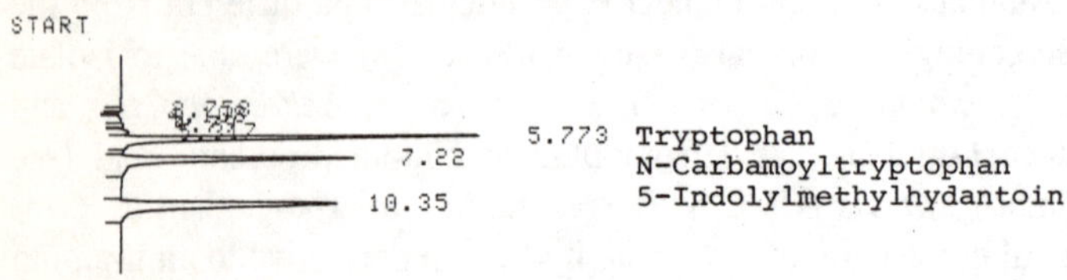

CHROMATOPAC C-R3A FILE 9
SAMPLE NO 0 METHOD 64
REPORT NO 28381 SAMPLE WT 100

PKNO	TIME	AREA	MK	IDNO	CONC	NAME
1	5.317	1087		1	0.0023	T
2	5.773	43573	V	1	0.0933	T
3	7.22	34997		2	0.0887	C
4	10.35	46049		3	0.099	I
	TOTAL	125707			0.2834	

Figure 6.2. Quantitative Analysis of DL-5-IMH, *N*-Carbtrp, and Trp by HPLC.

Indolylmethylenehydantoin racemase activity was investigated by incubating this enzyme with 0.2 g/l of optically pure D- or L-5-IMH in 0.1 M phosphate buffer solution for 20 min at 37 °C and pH 8.5. The reaction was stopped by adding methanol to the reaction solution. Analyses of the enantiomers of IMH as well as of *N*-Carbtrp and Trp were carried out by UV detection at 280 nm after separation by HPLC on a chiral D-proline-Cu-II column (Serva, Heidelberg, FRG) with 1 mM $CuSO_4$ as eluent [26]. Using this method, it was possible to detect in parallel the enantiomeric purities of 5-indolylmethylenehydantoin and *N*-carbamoyltryptophan (see Figure 6.3).

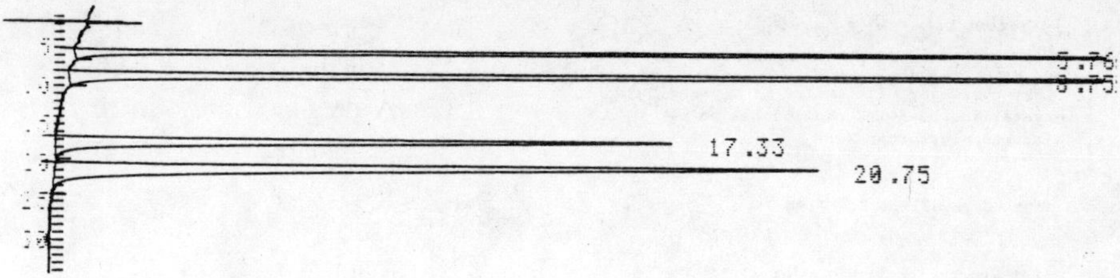

FILE: 1 CALC-METHOD: AREA% TABLE: 0 CONC: AREA

NO.	RT	AREA	CONC	NAME	HEIGHT
1	5.76	94339	28.938	D-5-IMH	6092
2	8.75	96371	29.561	L-5-IMH	3508
3	17.33	52104	15.982	D-Carbtrp	1225
4	20.75	83183	25.516	L-Carbtrp	1519
TOTAL					
		325997	100.000		12344

Figure 6.3. Separation of the Enantiomers of DL-IMH, L-5-IMH, *N*-Carb-D-Trp and *N*-Carb-L-Trp by HPLC [60].

Similiar methods for detecting the bioconversion products after performing the reaction with other microorganisms were developed by Sano *et al.* [32], Yokozeki *et al.* [33, 34, 40], and Nishida *et al.* [10]. Nishida *et al.* [10] used a Shimadzu zorbax SIL-column (Shimadzu, Japan) and 40% methanol for separation of DL-5-IMH, *N*-Carbtrp and Trp by HPLC; Yokozeki *et al.* [34] used a Hitachi gel #3020 resin and 0.1 M Na_2HPO_4/20% methanol, adjusted to pH 2.2 with phosphoric acid, as the mobile phase. Optical purities of the reaction products were determined by Nishida *et al.* [10] by measuring the optical rotation with a polarimeter. Yokozeki *et al.*[40] used a Chiralpak WH resin (Dicel Chemical Industries, Japan) and HPLC for this analysis.

For screening and selection experiments (see Sections 6.3.2 and 6.3.3) an overlay assay was developed by Groß *et al.* [16] for the quick detection on solid media of microorganisms able to form L-Trp from crystalline DL-5-IMH. The crystalline DL-5-IMH was in the bottom layer. After inoculation of a second medium with microorganisms, this was used as the overlay medium; the plates were incubated for 24 h at 27 °C. Hydantoinase-producing strains used the crystalline DL-5-IMH in the bottom layer as the substrate for bioconversion and formed clear areas around their colonies. A second overlay assay with an L-Trp-auxotrophic yeast allowed the detection of L-Trp-forming microorganisms without any other analytical methods necessary. For a scheme of this procedure see Figure 6.4.

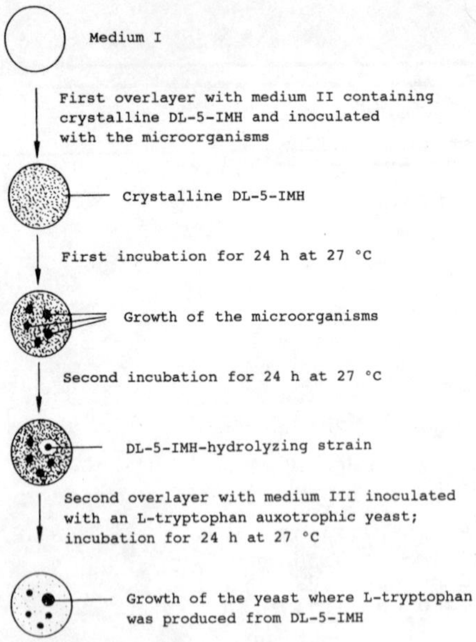

Figure 6.4.
Scheme for Screening by the Overlay Test of Microorganisms Able to Form L-Trp from DL-5-IMH [16].

6.3.2 Screening for DL-5-IMH-Cleaving Microorganisms

Starting from the analytical methods discussed above, a first screening was performed using a mineral salts medium containing DL-5-IMH as the sole carbon source [39]. Cotoras [41] isolated only two bacterial strains able to form L-Trp from DL-5-IMH, while many strains were able to form *N*-Carbtrp in different enantiomeric purities from this substrate. One of these two strains was identified as *Arthrobacter aurescens* BH 20 by the German collection of Microorganisms (DSM, Braunschweig, FRG) [7].

A second screening, using the optimized growth conditions of the strain BH 20, resulted in 11 bacterial strains. While 5 of these strains again showed only the formation of *N*-Carbtrp in different enantiomeric purities, the other 6 were able to produce optically pure L-Trp from DL-5-IMH. All 6 strains showed only slight differences in their morphological and physiological characteristics and belonged to the genus *Arthrobacter* [17], especially to the *Arthrobacter globiformis* group. They differed strongly in the ratio of *N*-Carbtrp to L-Trp formation. The best strain, *Arthrobacter* sp. DSM 3747, which had been isolated from compost, showed a ratio of 87% L-Trp to 13% *N*-Carbtrp under assay conditions and was chosen for further investigations.

Arthrobacter sp. DSM 3747 was able to utilize not only DL-5-IMH, but also natural cyclic amides such as uric acid, allantoin and creatinine (for physiological background see Chapter 5) as sole N-sources in a mineral salts medium. Interestingly, orotic acid, hydantoin propionic acid, and dihydrouracil, which are the natural substrates of dihydroorotase (EC 3.5.2.3), 5-carboxymethylenehydantoinamidohydrolase (EC 3.5.2.4), and dihydropyrimidinase (EC 3.5.2.2) (see Chapter 5), could not be utilized as N- or C-sources by this microorganism.

Using the same procedure, Sano *et al.* [32] were able to isolate 5 bacterial strains from soil, of which the best one was identified as *Flavobacterium* sp. DL-5-IMH was directly converted to L-Trp by this strain without production of *N*-Carbtrp. Nishida *et al.* [10] were able to isolate 20 L-Trp-forming strains using 1% DL-5-IMH as the sole C- and N-source. Only the best strain was used for further investigations; it was identified as *Flavobacterium* sp., too. When screening for microorganisms able to form L-valine from DL-5-isopropylenehydantoin, Yamashiro *et al.* [42, 28] found that of 2300 strains isolated from soil, only one was able to form the L-amino acid after growth on a medium containing 0.5% L-5-isopropylenehydantoin. This strain was identified as *Bacillus brevis*. L-5-IMH caused an induction of only 2% of the enzymatic activity in comparison with L-5-isopropylenehydantoin. Tsugawa *et al.* [8] were able to isolate bacteria of the genera *Pseudomonas*, *Aerobacter*, *Micrococcus*, *Flavobacterium*, and *Bacillus* able to form L-glutamic acid from DL-hydantoin-5-propionic acid after growth on a medium containing this hydantoin derivative as inducer.

6.3.3 Selection Experiments

The goal of these investigations was to isolate a strain blocked in the aerobic L-Trp degradation pathway. When performing bioconversions of DL-5-IMH under aerobic or microaerophilic conditions, this pathway (see Figure 6.5) caused serious problems. Using different methods to get genetically blocked mutants of the strain *Arthrobacter* sp. DSM 3747, including mutagenesis with *N*-methyl-*N'*-nitro-*N*-nitrosoguanidine (NG) [17] in combination with the screening methods described in Section 6.3.2, it was not possible to select a strain which was blocked in the first oxidative reaction of the L-Trp degradation pathway. Many strains were isolated showing higher DL-5-IMH-cleaving enzymatic activities than the wild strain. But at the same time these strains showed a worse ratio of L-Trp to *N*-Carbtrp. Therefore, a mutant of *Arthrobacter* sp. DSM 3747 able to form twice the amount of L-Trp as the wild strain, with a ratio of 85% L-Trp to 15% *N*-Carbtrp under assay conditions, was chosen for further experiments. For all further experiments the assay as well as the production of L-Trp were performed under a nitrogen atmosphere to prevent enzymatic degradation [17, 20]. Figure 6.6 shows the influence of aerobic, microaerophilic, and anaerobic conditions on the L-Trp degradation [43].

Yokozeki *et al.* [33] isolated a genetically blocked mutant of *Flavobacterium* sp., which had a relative growth rate in a medium containing L-Trp as the sole *N*-source that was only 10% of that of the wild strain, after mutagenesis with *N*-methyl-*N'*-nitro-*N*-nitrosoguanidine (NG).

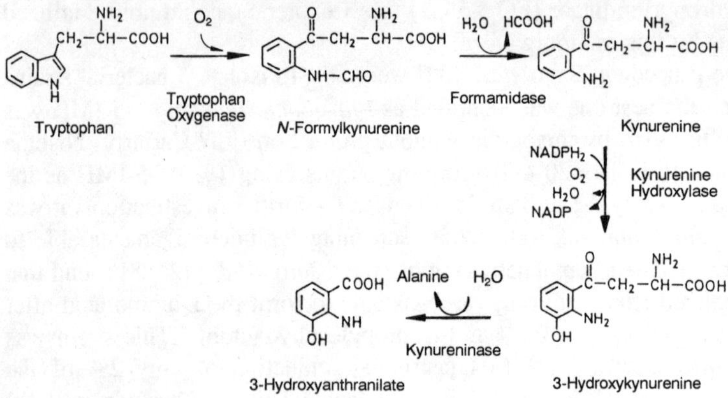

Figure 6.5. Aerobic Degradation of L-Tryptophan.

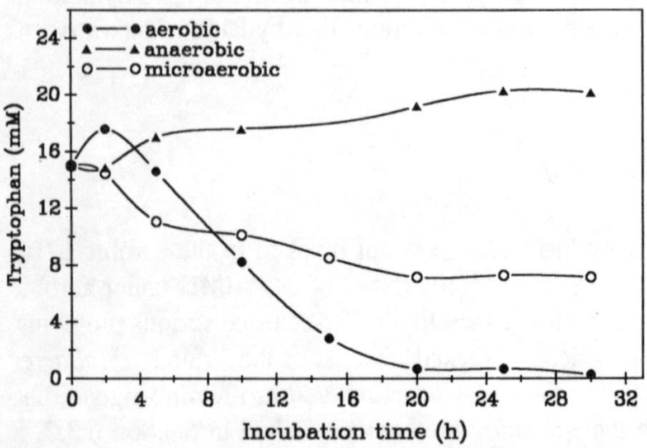

Figure 6.6. Degradation of L-Trp by Resting Cells of *Arthrobacter* sp. DSM 3747 under Various Reaction Conditions [43].

6.4 GROWTH AND INDUCTION OF MICROORGANISMS WITH L-HYDANTOIN-CLEAVING ACTIVITY

Starting from the mineral salts medium used for the screening, without any additional vitamins or complex nutrients, investigations were performed on growth and enzyme syntheses of the DL-5-IMH–cleaving enzymes of the mutant of *Arthrobacter* sp. DSM 3747 (see Section 6.3). The goal was to find a low-cost medium that allowed the production of 100–120 g/l of bacterial cell wet mass with high DL-5-IMH–cleaving activity within 24 h (for a detailed description of these experiments see [17]).

6.4.1 Induction of DL-5-IMH–Cleaving Activity

When mineral salts media were used without DL-5-IMH, only very low DL-5-IMH-cleaving activity could be detected in the different microorganisms isolated in the screening. Similiar results were obtained by Nishida *et al.* [10] for *Arthrobacter* sp. DK 200, an indication that the enzymes involved in this reaction were inducible.

Initially, experiments were done with *Arthrobacter aurescens* BH 20 on the induction of the enzyme syntheses by different hydantoin derivatives and related compounds. The results showed that of more than 30 compounds tested, DL-5-IMH was the most effective inducer of the L-Trp-forming enzyme activities [7, 39] (see Figure 6.7). Naturally occurring cyclic amides had no effect. These results were comparable to those obtained by Yokozeki *et al.* and Nishida *et al.* for different *Flavobacterium* species [10, 34]. Miyoshi *et al.* [6] reported on no inducer experiments for *Arthrobacter* sp. DK 200, but used either 1.0 g/l DL-5-IMH or 0.5 g/l *N*-Carbtrp as inducers for hydantoin-cleaving activity of this strain.

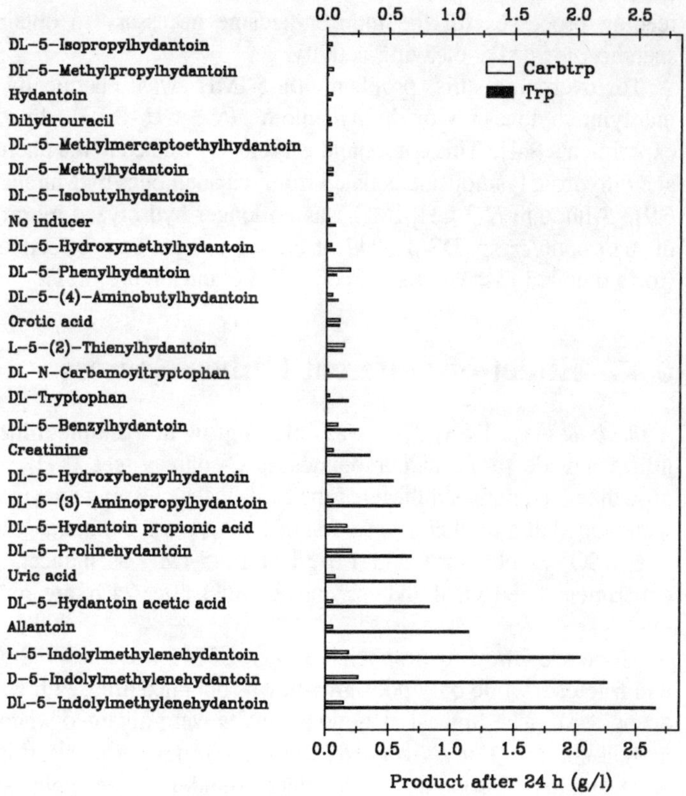

Figure 6.7. Induction of DL-5-IMH-cleaving Activity in *Arthrobacter aurescens* BH 20 [7].

The optimal concentration of DL-5-IMH for *Arthrobacter* sp. DSM 3747 was found to be 1.0 g/l for shake flask experiments. Higher concentrations had no further stimulating effect, but the cell yield was not negatively influenced by concentrations of up to 2.5 g/l DL-5-IMH. Yokozeki *et al.* [34] found that for *Flavobacterium* sp. AJ 3940 cells, 3.5–5.0 g/l of DL-5-IMH were necessary and that there was no difference in induction when using L-5-IMH or DL-5-IMH. Nishida *et al.* [10] observed that 5.0 g/l *N*-Carbtrp caused induction of DL-5-IMH-hydrolyzing enzymes in *Flavobacterium* sp. I-3, too. But they supposed that this effect was caused by L-5-IMH formed enzymatically from this compound (for further discussion see Section 6.6.2). DL-5-IMH was not an inducer of L-isopropylhydantoinase activity in *Bacillus brevis* AJ 12299 [42]. For this reaction only the bioconversion substrate DL-5-isopropylhydantoin, L-5-isobutylhydantoin, and L-5-*sec*-butylhydantoin were suitable inducers.

For *Arthrobacter* sp. DSM 3747 all following experiments in shake flasks were carried out with 1.0 g/l DL-5-IMH as inducer. During cell growth in a bioreactor the inducer DL-5-IMH was shown to be consumed by enzymatic cleavage in the exponential growth phase [17] (see Section 6.4.6 and Figure 6.8), so that special feeding processes for the inducer became necessary to obtain biomass with high specific DL-5-IMH–cleaving activity.

To overcome this problem, DL-5-IMH was chemically modified to DL-5-indolylmethylene-3-*N*-methylhydantoin (*N*-3-CH_3-IMH) by alkylation in later experiments [44]. This compound was closely related to the pheromones aplysinopsine and dihydroaplysinopsine isolated from sea anemones by Murata *et al.* [45] (see Figure 6.9). Although *N*-3-CH_3-IMH was no longer hydrolyzed by growing or resting cells of *Arthrobacter* sp. DSM 3747, it proved to be a very effective inducer [17, 26, 44] (for a detailed discusion see Section 6.4.6 and Figure 6.11).

6.4.2 Effect of Different Carbon Sources

Arthrobacter sp. DSM 3747 was able to grow in a simple mineral salts medium and utilize a wide range of compounds as C-sources (see [17]). To see the effect on growth and enzyme synthesis of the DL-5-IMH–cleaving enzymes, different C-sources were added at a final concentration of 20.0 g/l to the mineral salts medium containing $(NH_4)_2SO_4$ as N-source and 1.0 g/l of DL-5-IMH as inducer. The results of these experiments, cell yield and enzyme activities after 24 h, are summarized in Table 6.1 [17].

Good growth of *Arthrobacter* sp. DSM 3747 was observed with glycerol, glucose, and fructose, while only poor growth was obtained after 24 h with ethanol, citrate and acetic acid. The highest enzyme activities were obtained when glucose was used as carbon source. For *Flavobacterium* sp. I-3, dextrin and sorbitol as well as sucrose were suitable C-sources, while glucose and fructose gave only moderate activity

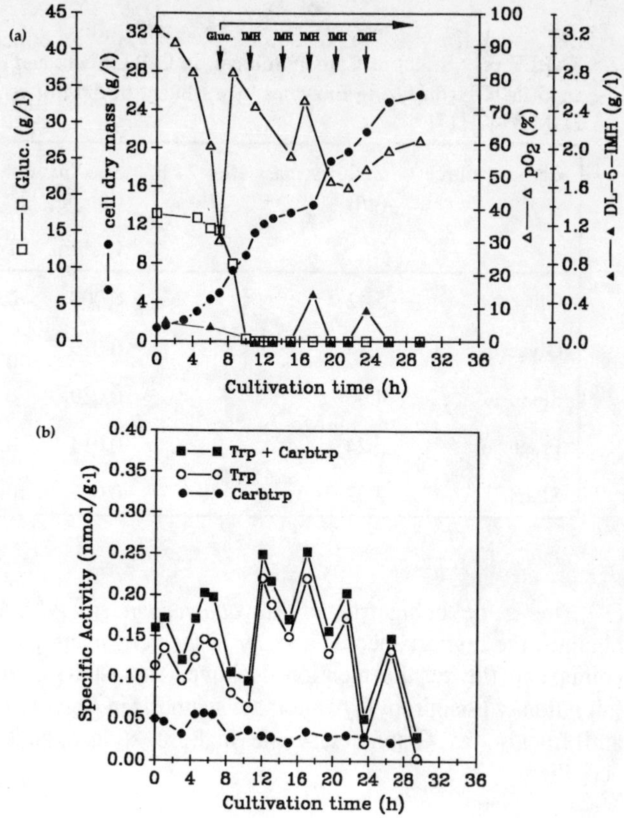

Figure 6.8. Growth of *Arthrobacter* sp. DSM 3747 in a 20-l Bioreactor with Feeding of Glucose and Stepwise Feeding of Inducer DL-5-IMH [17]; (a) growth; (b) enzymatic activities.

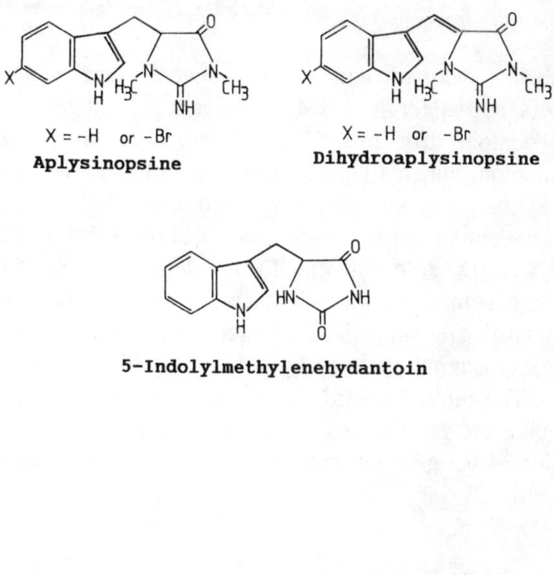

Figure 6.9. Structures of Aplysinopsine and Dihydroaplysinopsine in Comparison with 5-Indolylmethylenehydantoin [45].

TABLE 6.1 Effect of Carbon Sources on Cell Growth and Synthesis of the DL-5-IMH-Hydrolyzing Enzymes by a Mutant of *Arthrobacter* sp. DSM 3747 [17]

Carbon source	Cell dry mass after 24 h (g/l)	Specific enzyme activities ($mmol \cdot g^{-1} \cdot h^{-1}$)		
		Carbtrp	Trp	Carbtrp + Trp
Glucose	5.42	0.009	0.042	0.051
Glycerol	5.84	0.019	0.025	0.044
Sucrose	4.90	0.019	0.011	0.030
Fructose	5.24	0.014	0.010	0.024
Starch	3.99	0.007	0.024	0.031

[10], the same being true of *Flavobacterium* sp. AJ 3940 [34]: Yokozeki *et al.* obtained the highest specific activity of the cells using glutamic acid as C-source. In contrast to the results mentioned above, Yamashiro *et al.* [12] observed a strong inhibition of L-isopropylhydantoinase synthesis in *Bacillus brevis* AJ 12299 by glucose and fructose as C-sources, while maltose, xylose, and succinic acid gave high activities.

6.4.3 Effect of Different Nitrogen Sources

Various inorganic and organic N-sources were tested for growth and enzyme synthesis of *Arthrobacter* sp. DSM 3747, including natural cyclic amides closely related to the hydantoin structure. While *Arthrobacter* sp. DSM 3747 was able to grow with allantoin and creatinine as N-sources, dihydrouracil and dihydrothymine (the substrates of the enzyme dihydropyrimidinase [EC 3.5.2.2], which is identical with D-hydantoinase) as well as orotic acid (the substrate of the enzyme dihydroorotase [EC 3.5.2.3]) were not accepted (for a detailed discussion of the degradation pathways of these compounds see Chapter 5). Investigations of the purified hydantoinase of *Arthrobacter* sp. DSM 3747 later showed that neither the natural cyclic amide allantoin nor creatinine could be hydrolyzed by this enzyme (see Section 6.6.2).

The different nitrogen sources were added to the mineral salts medium, which contained glucose as C-source and 1.0 g/l DL-5-IMH as inducer, at a final concentration of 6.5 g/l. The results of these experiments are shown in Table 6.2 [17].

TABLE 6.2 Effect of Nitrogen Sources on Cell Growth and Synthesis of the DL-5-IMH-Hydrolyzing Enzymes by a Mutant of *Arthrobacter* sp. DSM 3747 [17]

Nitrogen source	Cell dry mass after 24 h (g/l)	Specific enzymatic activities ($mmol \cdot g^{-1} \cdot h^{-1}$)		
		Carbtrp	Trp	Carbtrp + Trp
$(NH_4)_2SO_4$	5.15	0.012	0.099	0.111
NH_4NO_3	5.15	0.011	0.063	0.074
$NaNO_3$	4.20	0.008	0.104	0.112
Urea	4.43	0.029	0.114	0.143

Cell growth of *Arthrobacter* sp. DSM 3747 after 24 h was optimal with urea- and ammonium-containing N-sources. High L-Trp-forming enzyme activities were observed with urea, sodium nitrate, and ammonium sulfate, the latter being chosen as N-source for further investigations.

Comparable results were obtained for growth and enzyme syntheses of the *Flavobacterium* species: the most favorable N-sources for *Flavobacterium* sp. I-3 and AJ 3940 were urea and $(NH_4)_2SO_4$ [10, 34].

6.4.4 Effect of Organic Nutrients

In many cases the addition of complex nutrients has a distinct influence on the synthesis of inducible enzymes. Therefore different complex compounds were added to the basal medium of *Arthrobacter* sp. DSM 3747 containing glucose as C-source, $(NH_4)_2SO_4$ as N-source and 1.0 g/l DL-5-IMH as inducer. Their effects on cell growth and enzyme synthesis are summarized in Table 6.3 [17]. While in some cases the addition of organic nutrients had a positive effect on the 24-h cell growth of *Arthrobacter* sp. DSM 3747, the L-Trp-forming enzyme activities were reduced as shown by the ratio of L-Trp to *N*-Carbtrp: the second enzymatic reaction leading to the free L-amino acid was negatively influenced. This effect was caused by all components except yeast extract. In further experiments 1.0 g/l yeast extract was found to be the optimal concentration for microbial growth and enzyme synthesis and therefore chosen as an additional nutrient for bioreactor experiments (see 6.4.6) [17].

For *Flavobacterium* sp. I-3 [10] and *Flavobacterium* sp. AJ 3940 [34], 0.1% yeast extract gave the most favorable results as regards cell growth and enzyme synthesis after addition of various complex nutrients to the mineral salts medium, too. Good effects were obtained as well by addition of a soybean hydrolysate [34].

TABLE 6.3 Effect of Organic Nutrients on Cell Growth and Synthesis of the DL-5-IMH-Hydrolyzing Enzymes by a Mutant of *Arthrobacter* sp. DSM 3747

Organic nutrient	Cell dry mass (g/l)	Specific enzymatic activities ($mmol \cdot g^{-1} \cdot h^{-1}$)		
		Carbtrp	Trp	Carbtrp + Trp
Control	5.42	0.009	0.042	0.051
Corn steep liquor	5.79	0.004	0.024	0.028
Malt extract	6.22	0.018	0.016	0.034
Meat extract	5.67	0.008	0.038	0.046
Peptone	5.28	0.009	0.030	0.039
Soy flour	5.56	0.007	0.031	0.038
Tryptone	6.15	0.007	0.039	0.046
Yeast extract	5.19	0.007	0.043	0.050

6.4.5 Effect of Metal Ions

In previous investigations on *Arthrobacter aurescens* BH20 [7, 39, 24], Mn^{2+} had been found to be essential for the synthesis and the activity of the enzyme hydantoinase (see Figure 6.10). That is why *Arthrobacter* sp. DSM 3747 was cultivated for several passages in mineral salts medium without Mn^{2+}, before shake flasks were inoculated containing the same medium and different concentrations of Mn^{2+} [17].

As in previous work [7, 39, 24], Mn^{2+} had no effect on cell growth, whereas it was found to be essential for the synthesis of the DL-5-IMH-hydrolyzing enzymes. A concentration of 2 mg $MnSO_4 \cdot 2H_2O$/g cell dry mass proved to be optimal for shake flask experiments [17].

Comparable results were obtained by Yokozeki *et al.* [34] for *Flavobacterium* sp. AJ 3940, where the addition of 2 mM $MnSO_4 \cdot 4H_2O$ or $CoCl_2 \cdot 6H_2O$ had a positive effect on the L-Trp-forming enzyme activities. Miyoshi *et al.* [6] reported on the same result for the DL-5-IMH–cleaving *Arthrobacter* sp. DK 200.

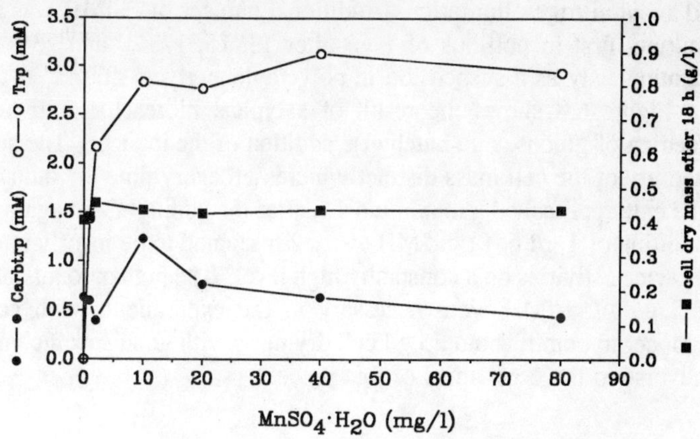

Figure 6.10. Influence of Mn^{2+} on Hydantoinase and *N*-Carbamoyl-L-amino Acid Amidohydrolase Activity in *Arthrobacter aurescens* BH 20 [15, 41].

6.4.6 Bioreactor Experiments

Starting from the investigations in the shake flasks (see above), bioreactor experiments were performed with *Arthrobacter* sp. DSM 3747 aiming at high yields of biomass with enhanced enzyme activities within short times for an application as biocatalyst in the production of L-Trp from DL-5-IMH (for a detailed discussion of these experiments see [17]).

At first, batch cultivation was performed under the optimal shake flask conditions at pH 6.8 and 30 °C to follow the time course of growth and enzyme synthesis of the DL-5-IMH-hydrolyzing enzymes. *Arthrobacter* sp. DSM 3747 was able to grow at temperatures up to 35 °C, but temperatures of 27–30 °C were found to be more favorable for enzyme synthesis than 35 °C. A comparable result was obtained by Nishida *et al.* [10] for *Flavobacterium* sp. I-3.

In the basal medium containing initial concentrations of 20.0 g/l glucose, 6.5 g/l $(NH_4)_2SO_4$ and 1.0 g/l DL-5-IMH, with *Arthrobacter* sp. DSM 3747, a biomass concentration of 10.5 g/l cell dry mass was obtained within 12 h (μ_{max} = 0.37; $y_{X/S}$ = 0.44). The highest specific enzyme activities were observed in the exponential growth phase, but decreased distinctly after 8 h. Further investigations showed that the reason was the consumption of the inducer DL-5-IMH, when it was added only once at the onset of cultivation.

Feeding experiments of DL-5-IMH and glucose were started, to avoid a decrease in the specific DL-5-IMH-hydrolyzing activity of the cells. The concentrations of Ca^{2+}, Fe^{2+}, Mg^{2+} and Mn^{2+} were doubled to avoid growth limitations from a lack of the bivalent cations. The pH value was regulated by addition of 10% aqueous ammonia to avoid nitrogen limitation. Additional inducer DL-5-IMH was added to the growing culture, first in portions of 1 g/l after 13, 15, 17, 20 and 23 h of growth, and later continuously as a suspension in polyethylene glycol 400.

Figure 6.8 shows the result of a typical bioreactor cultivation with continuous feeding of glucose and batchwise addition of the inducer. The specific L-Trp-forming activity of the cell mass distinctly increased every time the inducer had been added to the culture, reached a maximum 1 h after the addition, and again decreased. Even the addition of 1 g/l of DL-5-IMH every 2 h seemed to be insufficient to keep the specific enzyme activities on a constantly high level. The high amounts of inducer—more than 8.5 g/l DL-5-IMH were necessary in the experiments with continuous feeding of inducer to obtain about 25 g/l cell dry mass with good specific enzyme activity—were adverse to the economics of this process.

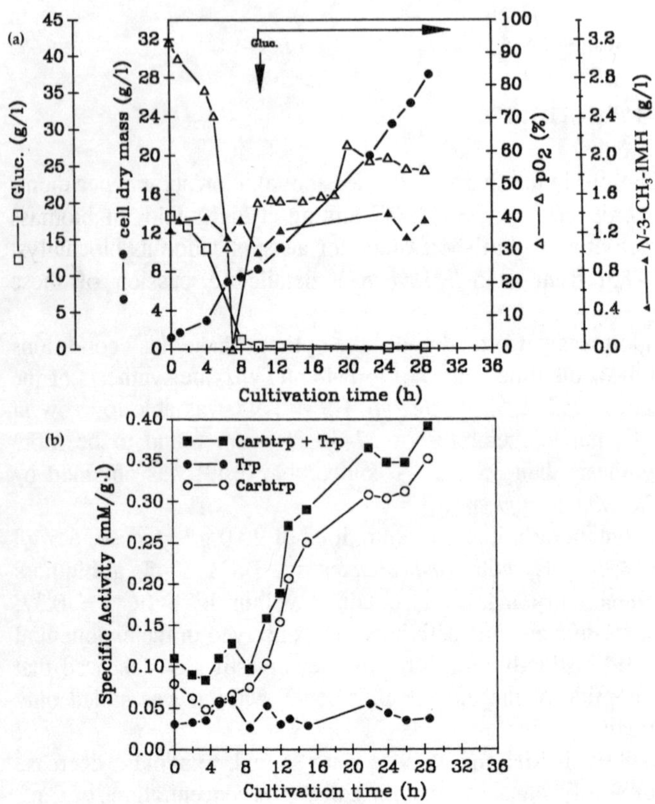

Figure 6.11. Growth of *Arthrobacter* sp. DSM 3747 in a 20-l Bioreactor with *N*-3-CH_3-IMH as Inducer and Feeding of Glucose [17]; (a) growth; (b) enzymatic activities.

In further investigations it was possible to reduce the concentration of DL-5-IMH by starting the cultivation without inducer and Mn^{2+} and beginning the feeding of both after a minimum of 8 to 10 h of growth. This procedure resulted in a much shorter cultivation time since inhibition of cell growth caused by Mn^{2+}, DL-5-IMH, and degradation products could be overcome under Mn^{2+}- and inducer-free conditions: using this novel procedure, the cultivation time to obtain 25 g/l cell dry mass could be reduced from 22 h to 13 h as compared to cultivations with Mn^{2+} and DL-5-IMH from the onset, continuously enhanced feeding of DL-5-IMH and pO_2-controlled feeding of glucose. The quantity of inducer could be reduced from 8.5 g/l DL-5-IMH to only 1.0 g/l, and the specific L-Trp-forming activity of the biomass was enhanced from 13 mmol·g^{-1}·h^{-1} (Trp per unit cell dry mass per unit time) to 20 mmol mmol·g^{-1}·h^{-1} (Trp per unit cell dry mass per unit time).

In parallel experiments a chemically modified inducer was tested: DL-5-IMH was methylated at the N-3 position of the hydantoin ring [44]. *Arthrobacter* sp. DSM 3747 was not able to hydrolyze this compound. Using 1 g/l *N*-3-CH_3-IMH as inducer in bioreactor experiments, a biomass concentration of 28 g/l cell dry mass was obtained within 28 h (see Figure 6.11). The specific enzyme activity of the biomass was significantly enhanced from 0.20 mmol·g^{-1}·h^{-1} to 0.34 mmol·g^{-1}·h^{-1} (Trp per unit cell dry mass per unit time). Additionally, no drop in the specific enzyme activities of the biomass could be observed during the cultivation [17].

6.5 L-SELECTIVE MICROBIAL BIOCONVERSION OF 5-MONOSUBSTITUTED HYDANTOIN DERIVATIVES WITH WHOLE MICROBIAL CELLS

6.5.1 Influence of the pH Value and Temperature

The goal of these investigations was to find optimal reaction conditions for the total conversion of DL-5-IMH to 100% L-Trp by resting cells of *Arthrobacter* sp. DSM 3747. In the first experiments, which were performed with DL-5-IMH–induced cells, the effects of the pH value and of different buffer systems on this reaction and the *N*-Carbtrp/Trp ratio were investigated at 27 °C (for details see [19, 20, 43]. The highest yield of L-Trp from DL-5-IMH was obtained at pH values between pH 8.5 and 9.0. The influence of the various buffer systems was quite different: especially Tris-HCl and diethanolamine-HCl buffer led to an unfavorable ratio of L-Trp to *N*-Carbtrp, while 0.1 M potassium phosphate buffer, glycine-NaOH buffer and carbonate buffer gave good results for the conversion yield in favor of the production of L-Trp.

Comparable results were obtained by Nishida *et al.* [10] and Yokozeki *et al.* [34] for resting cells of *Flavobacterium* sp. I-3 and AJ 3940: best results for L-Trp production were obtained at alkaline pH values of about pH 8.5. Interestingly, no *N*-Carbtrp was detected in the reaction supernatant of the latter microorganism, as reported before by Sano *et al.* and the same working group [32] for the bacterial strain T-523. In contrast, in *Arthrobacter* sp. DSM 3747 [19, 20, 43] as well as in *Flavobacterium* sp. I-3 [10], *N*-carbamoyl-L-amino acid amidohydrolase activity in the resting cells was lower than hydantoinase activity and depended on the reaction conditions (see below).

The conversion of DL-5-IMH by *Arthrobacter* sp. DSM 3747 was optimal at a temperature of 50 °C, the same result being found by Yokozeki *et al.* [34] for *Flavobacterium* sp. AJ 3940.

The latter two working groups also found that the hydantoin-cleaving activity of resting cells of *Arthrobacter* sp. DSM 3747 and *Flavobacterium* sp. AJ 3940 had a high pH stability and was resistant to temperatures up to 45 °C [19, 20, 34, 43] under certain conditions: when cells of *Arthrobacter* sp. DSM 3747 were preincubated at different pH values at 27 °C for 24 h, best stability of enzymatic activities was observed at pH 8.0–10.0 with a residual activity of 50–60% for hydantoinase and 40–50% for *N*-carbamoyl-L-amino acid amidohydrolase activity. Preincubating cells of *Arthrobacter* sp. DSM 3747 at various temperatures for 24 h, before measuring the remaining L-Trp-forming activity, gave the result that without addition of the substrate DL-5-IMH or the reaction products (see Section 6.5.2) both enzymatic activities were very labile. At 30 °C and 40 °C, a drop in activity of 50% and 70%, respectively, was observed. At 50 °C both enzymatic activities were negligibly low. In contrast, high activities were obtained after preincubation of the resting cells in the presence of DL-5-IMH (see Section 6.5.3).

6.5.2 Influence of Medium Components

Sometimes high substrate or product concentrations can cause an inhibition of the bioconversion; in other cases they can have a stabilizing effect on the enzymes involved in the bioconversion. So for *Arthrobacter* sp. DSM 3747 the influence of NH_4^+ and HCO_3^- ions as well as of L-*N*-Carbtrp and L-Trp was investigated in order to prevent a possible inhibition (for experimental details see [20] and [43]).

NH_4Cl, $NaHCO_3$ and NaCl were added to the reaction solution to detect possible effects of the counterions and the ionic strength. No inhibition by these compounds at 1, 10 and 50 mM could be detected, but a negative influence on the ratio of L-Trp/*N*-Carbtrp could be observed.

Adding either L-*N*-Carbtrp or L-Trp to the reaction solution, up to a concentration of 20 mM, resulted in an increase in the bioconversion rate after 24 h of incubation. For the 20 mM concentration of L-*N*-Carbtrp, the yield was double that of the control containing no L-*N*-Carbtrp. The hydantoinase seemed to be activated especially beyond an L-*N*-Carbtrp concentration of 10 mM. The increase in the L-Trp concentration after 24 h could be caused either by activation or by a shift in the reaction equilibrium, caused by the high substrate concentration for L-*N*-carbamoyl-L-amino acid amidohydrolase. The strong effect of L-*N*-Carbtrp concentrations greater than 10 mM may be due to improved diffusion through the cell membrane [21], which results in a sufficient intracellular concentration.

For the complete range of L-Trp concentrations the product yield was 30% higher after 24 h than that of the control to which no L-Trp had been added. L-Trp seemed to stimulate the entire bioconversion of DL-5-IMH to L-Trp by resting cells of *Arthrobacter* sp. DSM 3747. The increased yield was correlated with a higher ratio of L-Trp/*N*-Carbtrp.

Yokozeki *et al.* [34] observed an inhibitory effect on the L-Trp forming activity of *Flavobacterium* sp. I-3 AJ 3940, when working with substrate concentrations greater than 30 g/l L-5-IMH [34]. They were able to show by adding 18 g/l L-Trp to the reaction solution that this effect was not caused by L-5-IMH, but by high product concentrations. To relieve this end-product inhibition, they added to the reaction mixture an equimolar amount of inosine, which formed an insoluble adduct with L-Trp.

6.5.3 Stability of the Resting Cell Biocatalyst

In contrast to the results described in Section 6.5.1, both hydantoinase and *N*-carbamoyl-L-amino acid amidohydrolase activities of resting cells of *Arthrobacter* sp. DSM 3747 could be restored at 30 °C, 40 °C, and even at 50 °C during the preincubation period by adding DL-5-IMH to the reaction solution at a concentration that could not be converted completely [20, 43]. Even an activation of both enzymatic activities was observed at 30 °C, which may be explained by the permeabilization of the cells during the incubation period, which resulted in higher specific conversion rates. The hydantoinase activity could be stabilized at higher temperatures by the addition of Mn^{2+} ions and by Zn^{2+} ions up to 40 °C, too. For a detailed discussion of the stability of the isolated and purified enzymes hydantoinase and *N*-carbamoyl-L-amino acid amidohydrolase, see Sections 6.6.2 and 6.6.3.

6.5.4 Permeabilization of Cells

The bioconversion substrate DL-5-IMH has a solubility of only 0.5 g/l in water at 25 °C [43]. Thus, one problem to be investigated was the uptake of this compound by the microorganisms. Another problem was the transport limitation for *N*-carbamoylamino acids through the bacterial cell wall [21]. Nishida *et al.* [10] added 0.26 mM cetyltrimethylammonium bromide to the reaction solution of *Flavobacterium* sp. I-3 to study these problems.

In order to enable free exchange between the intracellular enzymes of *Arthrobacter* sp. DSM 3747 and the surrounding reaction solution, several organic solvents and detergents were added to the reaction mixture to permeabilize the cells [19, 20, 43]. The results of these experiments are summarized in Table 6.4.

Freezing and thawing of the biomass proved to be slightly successful: 136% activity was obtained in comparison with fresh cells, but the *N*-Carbtrp/Trp ratio was negatively influenced. The addition of 1.7% isobutylmethylketone to the reaction solution resulted in 172% conversion in comparison with the reaction without addition, but had a negative effect on the *N*-carbamoyl-L-amino acid amidohydrolase activity of the resting cells. Except for Triton X-100, nonionic detergents proved to be inhibitory. Triton X-100 caused a distinctly higher DL-5-IMH-cleaving activity of the resting cells, but a negative *N*-Carbtrp/Trp ratio. Sodium deoxycholate (NaDOC) led to a 2.6-fold increase in the DL-5-IMH-cleaving activity of resting cells. Therefore, in further experiments, the conditions of permeabilization with NaDOC relating to the concentration, incubation time and temperature were optimized. The optimal concentration of NaDOC proved to be 0.75–1.0 g/g (amount of detergent per unit cell wet mass). The permeabilization of biomass came to completion after 1 h of incubation with NaDOC. At this point 12% of the protein of the biomass employed could be detected in the supernatant. A rather low constant amount of *N*-Carbtrp and a slowly increasing amount of L-Trp were also detectable in the supernatant. This L-Trp was probably derived from the intracellularly accumulated inducer. But neither the hydantoinase nor the *N*-carbamoyl-L-amino acid amidohydrolase activity of the resting cells decreased during a preincubation period of 3 h with NaDOC. It was supposed that the enzymes therefore were not released from the cells during this time.

When using NaDOC-permeabilized cells for bioconversions, the optimal bioconversion temperature decreased from 50 °C to 30 °C, in contrast to the reaction without NaDOC (see Section 6.5.1). Therefore, the temperature optimum for intact cells of 50 °C was hypothesized to be not only a result of an increasing reaction rate, but of increasing membrane permeability.

6.5.5 Bioconversions under Optimal Reaction Conditions

The goal of these experiments was to get results on the bioconversion of 10 g/l (43.7 mM) and 20 g/l (87.4 mM) DL-5-IMH to L-Trp, in a 5-l bioreactor under optimized reaction conditions [20, 43]. At first, bioconversions of 43.7 mM DL-5-IMH were performed with DL-5-IMH–induced intact cells at 50 °C and DL-5-IMH–induced permeabilized cells at 30 °C, respectively (Figure 6.12a and 6.12b). For

TABLE 6.4 Influence of Several Detergents on Cell Permeabilization

Addition	Concentration (%)	*N*-Carbtrp (mM)	L-Trp (mM)	Activity (%)
Fresh cells	–	0.65	8.28	100
3h at –200 °C	–	1.17	10.99	136
Toluene	0.05	0.81	8.82	108
Isobutyl methyl ketone	1.7	4.66	10.73	172
Ethyl methyl ketone	5.0	0.85	10.54	127
$CHCl_3$	0.81	1.42	8.13	107
Dimethylformamide	5.0	0.77	7.99	98
Ethanol	5.0	0.73	10.44	125
n-Propanol	5.0	1.74	11.42	147
Dimethylsulfoxide	5.0	0.73	10.05	121
Diethylene glycol diethyl ether	5.0	1.01	8.62	108
Polyethylene glycol 200	10.0	1.30	13.38	164
Triton X-100	1.0	4.41	6.66	124
Triton X-405	1.0	0.53	7.20	87
Tween 20	1.0	0.20	2.50	30
Tween 40	1.0	0.28	3.53	43
Tween 60	1.0	0.61	4.70	60
Span 80	1.0	0.65	4.46	57
Na deoxycholate	1.0	6.20	17.05	260
Cetyltrimethyl-ammonium bromide	1.0	9.56	8.18	199

Conditions: 43.7 mM DL-5-IMH; 0.1 mM glycine-NaOH pH 8.5; 4.2 g/l cell dry mass; 27 °C; 24 h [43]

DL-5-IMH-induced cells a permeabilization with NaDOC and a temperature of 30 °C seemed to be more sufficient for the *N*-carbamoyl-L-amino acid amidohydrolase activity than 50 °C.

The bioconversion of 87.4 mM DL-5-IMH to L-Trp at 30 °C took more than twice as long as the conversion of 43.7 mM DL-5-IMH at the same temperature. While

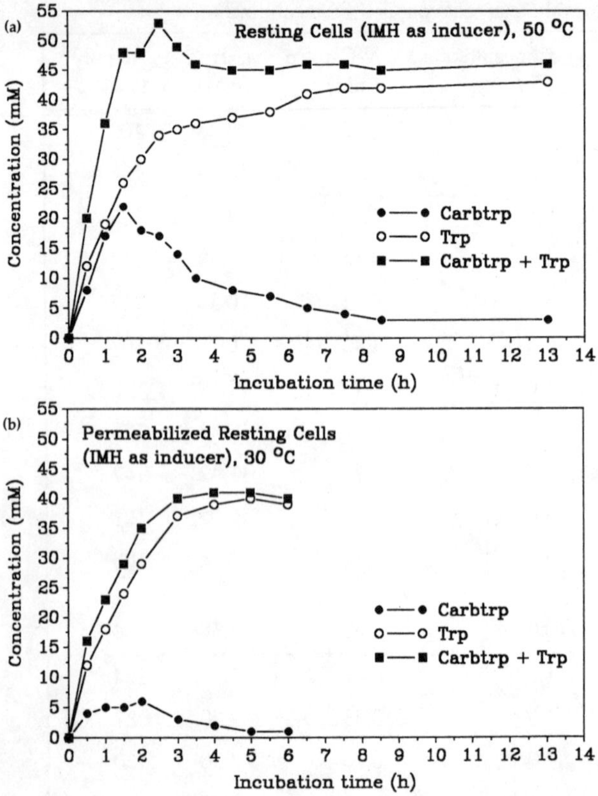

Figure 6.12. Bioconversion of 43.7 mM DL-5-IMH; (a) with resting cells of *Arthrobacter* sp. DSM 3747 at 50 °C; and (b) with permeabilized resting cells at 30 °C [20, 43].

with intact as well as with permeabilized cells the cleavage of DL-5-IMH was completed in comparable times, that of the residual *N*-Carbtrp took a longer time in the reaction mixture containing 87.4 mM DL-5-IMH (see Figure 6.13a and 6.13b, and compare to Figure 6.12). In the case of intact DL-5-IMH–induced cells at 50 °C, more than 25% of the product was *N*-Carbtrp, while only 5% remained with permeabilized cells. Product inhibition caused by L-Trp seemed unlikely, because with permeabilized cells the reaction velocity remained nearly constant until the end of the bioconversion (see Section 6.5.6, too).

In parallel experiments *N*-3-CH_3-IMH–induced cells of *Arthrobacter* sp. DSM 3747 were used for bioconversions. Figures 6.14a and b show typical results of bioconversions of 43.7 mM and 87.4 mM DL-5-IMH, respectively (cell dry mass concentration of the resting cells: 15 g/l). In contrast to the results obtained with DL-5-IMH–induced cells (Figures 6.12 and 6.13) the reaction velocities with *N*-3-CH_3-IMH–induced cells were much higher and only very small amounts of *N*-Carbtrp were observed during the reaction.

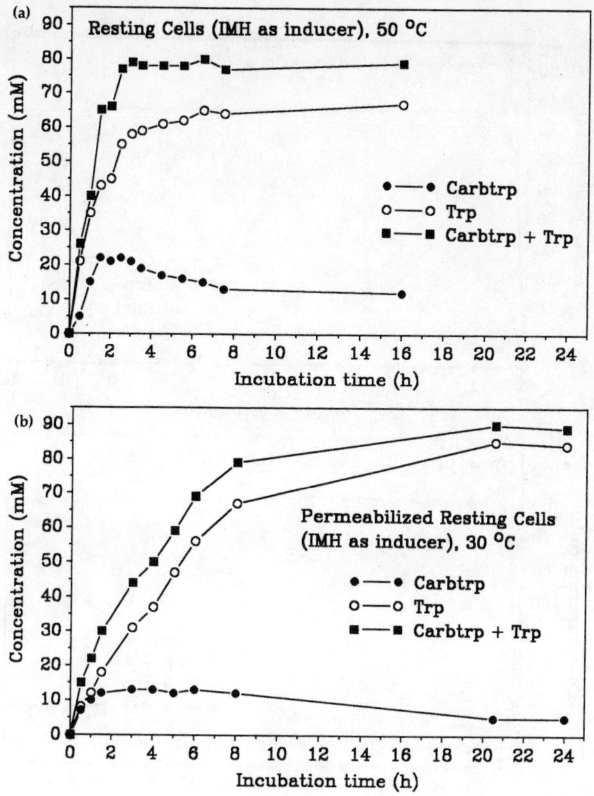

Figure 6.13. Bioconversion of 87.4 mM DL-5-IMH;
(a) with resting cells of *Arthrobacter* sp. DSM 3747 at 50 °C; and
(b) with permeabilized resting cells at 30 °C [20, 43].

6.5.6 Stereoselectivity of Resting Cells

The cleavage of optically pure D- and L-5-IMH by DL-5-IMH–induced resting cells of *Arthrobacter* sp. DSM 3747 was studied to determine the influence of the enantiomers on the reaction [20, 43]. The reaction courses of both enantiomers of 5-IMH leading to L-Trp were the same [24]. In both cases the conversion of 43.7 mM 5-IMH was complete after 1 h, while another 3 h were necessary to complete the transformation of the residual *N*-Carbtrp to L-Trp. Thus the reaction catalyzed by the *N*-carbamoyl-L-amino acid amidohydrolase represented the rate-limiting step. Determination of the configuration of the *N*-Carbtrp formed revealed that both L-*N*-Carbtrp and D-*N*-Carbtrp were produced from L-5-IMH as well as from D-5-IMH. In both cases the ratio of L-*N*-Carbtrp to D-*N*-Carbtrp was rather constant at about 70:30 until the conversion of 5-IMH was complete. The hydantoinase reaction of resting cells therefore proved not to be stereoselective. L-Trp was produced predominantly from L-*N*-Carbtrp:

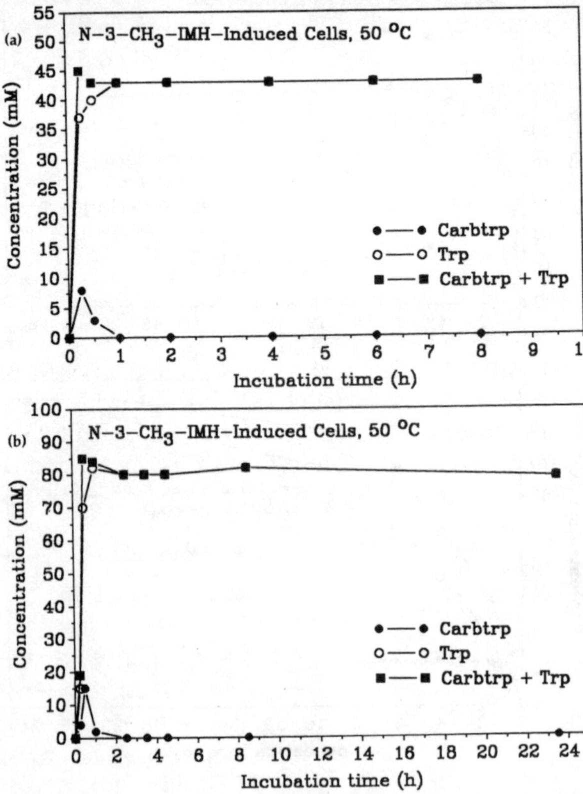

Figure 6.14. Bioconversion of (a) 43.7 mM and (b) 87.4 mM DL-5-IMH with Resting Cells of *Arthrobacter* sp. DSM 3747 at 50 °C after Induction with *N*-3-CH_3-IMH [16].

Arthrobacter sp. DSM 3747 was able not only to produce L-Trp from both enantiomers of 5-IMH, but also to produce L-Trp from *N*-Carb-L-Trp and from *N*-Carb-D-Trp [21], although with a much lower reaction velocity. So during the bioconversion of D- and L-5-IMH the optical purity of the D-enantiomer increased to nearly 100% after a reaction time of 5 h.

All these results were strong evidence that in resting cells of *Arthrobacter* sp. DSM 3747, either:

(i) a racemase, probably at the level of 5-IMH, or
(ii) a nonstereoselective hydantoinase, or
(iii) a D- and L-hydantoinase of different reaction velocities also able to catalyze the formation of 5-IMH from *N*-Carbtrp,

were involved in this process.

While chemical racemization of *N*-Carbtrp could be excluded [46], the ratio of chemical and enzymatic racemization of optically pure 5-IMH had to be determined to prove this hypothesis. The chemical racemization rate of D-5-IMH and L-5-IMH at pH 8.5 in 0.1 M potassium phosphate buffer at 30 °C indicated that the racemization was enzymatic [20, 43] (see Section 6.6.1). Chemical racemization proceeded more rapidly at 50 °C, but again in comparison with the enzymatic conversion rates of D-5-IMH at 50 °C it could be assumed that the racemization was mainly enzymatic (for a detailed discussion of D- and L-5-IMH racemization, see Section 6.6.1). All other hypotheses had to be examined with isolated and enriched or purified enzymes (see Section 6.6).

6.5.7 Reuse of Resting Cells

In order to determine the operational stability of the DL-5-IMH cleavage with DL-5-IMH–induced intact and permeabilized cells of *Arthrobacter* sp. DSM 3747, a semicontinuous procedure was studied under optimized conditions using a substrate concentration of 43.7 mM to prevent an accumulation of D-*N*-Carbtrp (see Section 6.5.6). The results of these experiments are summarized in Figure 6.15a and 6.15b [43]. When bioconversions were complete after 2.8 h, the cells were harvested by centrifugation and new buffer and substrate were added. This procedure was repeated four times.

No decrease of the enzymatic activities could be observed during the investigated time period of 11.3 h with NaDOC-treated cells at 30 °C (see Figure 6.15b). A reaction yield of 146 mM L-Trp was obtained after this period of incubation. In contrast, intact cells at 50 °C distinctly lost activity after 4 cycles (see Figure 6.15a).

6.6 ISOLATION AND CHARACTERIZATION OF THE ENZYMES

6.6.1 Isolation and Characterization of a Hydantoin Racemase

With regard to an economical production of optically pure amino acids from racemic mixtures, the racemization of the unreacted enantiomer of the amino acid or amino acid precursor is of great interest.

Racemizing activity is a widespread characteristic of microorganisms. A comprehensive survey is given by Adams [47], including some reflections about the physiological functions of amino acid racemases.

Several cofactors have been demonstrated to be present or active in racemases, for example, pyridoxal 5'-phosphate and ATP [47, 48]. The action of FAD as a possible hydrogen acceptor is not yet fully understood. On the other hand, there are some racemases which do not require any cofactors for their catalytic activity [49, 50].

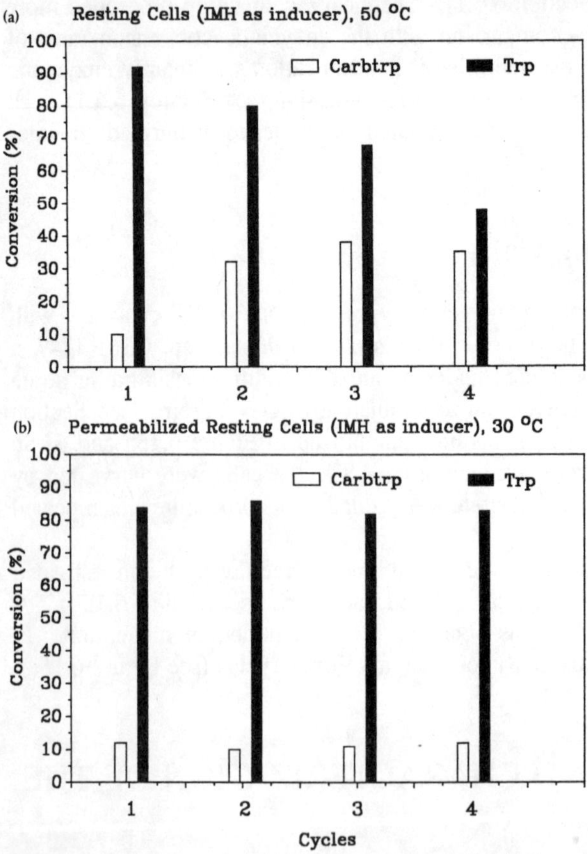

Figure 6.15. Reuse and Stability of *Arthrobacter* sp. DSM 3747 for Bioconversion of DL-5-IMH; (a) resting cells at 50 °C; and (b) after permeabilization at 30 °C [43].

Investigations on the nonenzymatic racemization of amino acid esters catalyzed by pyridoxal 5'-phosphate were carried out, and reported to be a possible step in the production of L-amino acids [51]. This nonenzymatic racemizing activity can cooperate with a stereoselective enzymatic esterase activity for the continuous optical resolution of DL-amino acids. The free amino acids formed are highly stable with respect to racemization under the reaction conditions. Thus, tedious and expensive changes of production conditions, which are necessary if unreacted enantiomer is racemized in anhydrous medium, can be avoided.

The method described above does not work if cyclic substrates, for example, amino acids, amino acid precursors or other cyclic structures have to be racemized, since pyridoxal 5'-phosphate cannot act as cofactor in an enzymatic process. Therefore neither the purified hydroxyproline 2-epimerase [52, 53] nor the partially purified proline racemase [54] require any cofactors.

Various attempts to explain the mechanism of racemization of hydroxyproline have been made, including the hypothesis of a decarboxylation-recarboxylation sequence, but the absence of cofactors makes it very difficult to propose a mechanism unless more information about protein structure is obtained.

Another enzyme acting on a cyclic substrate has been identified as an allantoin racemase (EC 5.1.99.3) which enables several bacteria to use both optical antipodes of allantoin as substrates [55, 56] (see Section 5.3.1). The only naturally occurring enantiomer is the *S*-(+)-allantoin produced by uricase (EC 1.7.3.3.) from uric acid. Thus racemic mixtures of allantoin formed by spontanous racemization in dead materials can be totally metabolized by various bacterial strains using a sequence of racemase and *S*-(+)-specific allantoinase.

Several racemases act on cyclic precursors of amino acids. For example, the total conversion of D-α-amino-ε-caprolactam into L-lysine carried out by several yeasts, such as *Cryptococcus laurentii*, and bacteria, such as *Achromobacter obae* nov. sp., indicates the presence of an aminocaprolactam racemase [57].

Various bacteria including the strains *Pseudomonas* sp., *Arthrobacter globiformis*, *Agrobacterium radiobacter*, and *Arthrobacter* sp. are found to racemize different 5-monosubstituted hydantoins (for general structure see Figure 5.1) [5, 58, 59, 60].

Before starting any investigations on enzymatic racemization, chemical racemization has to be examined. The velocity of the chemical racemization of hydantoins is influenced by the electronic nature of the substituent in the 5-position. In Figure 6.16 the chemical racemizations of (a) D-hydantoin-5-propionic acid [8], (b) L-5-(*p*-hydroxyphenyl)hydantoin [61], and (c, d, e) 5-indolylmethylenehydantoin [60, 62] are compared.

As can be determined from Figure 6.16 a–c, either electronegativity or stabilization of the enolate structure due to formation of mesomeric border structures, or both, is responsible for the velocity of racemization [63]. However, an increasing racemization rate can be also obtained by increasing the pH value (a, b, d) and the temperature (e) of the reaction system.

The velocity of chemical racemization of L-5-isopropylhydantoin was investigated by Battilotti *et al.* [58] and shown to be rather slow. So the fast and total conversion of DL-isopropylhydantoin to D-valine led them to the suggestion that a hydantoin racemase must be responsible for the racemization of the L-enantiomer.

Resting cells of *Arthrobacter* sp. DSM 3747 were able to hydrolyze D-5-indolylmethylenehydantoin to L-tryptophan very rapidly [24]. At 37 °C and pH 8.5, only poor chemical racemization was observed (see above). Neither addition of metal ions nor catalytic proteins had a significant effect on racemization [60].

First investigations on enzyme purification using liquid-liquid extraction after cell disruption [64] led to a first bottom phase which could only catalyze the conversion from L-5-indolylmethylenehydantoin to *N*-carbamoyl-L-tryptophan and could not act on the D-hydantoin enantiomer. In contrast, the first top phase was able to catalyze both the conversion of *N*-carbamoyl-L-tryptophan into L-tryptophan and the fast

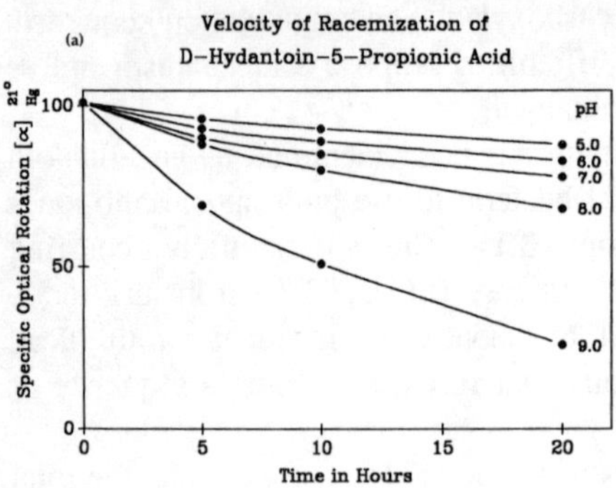

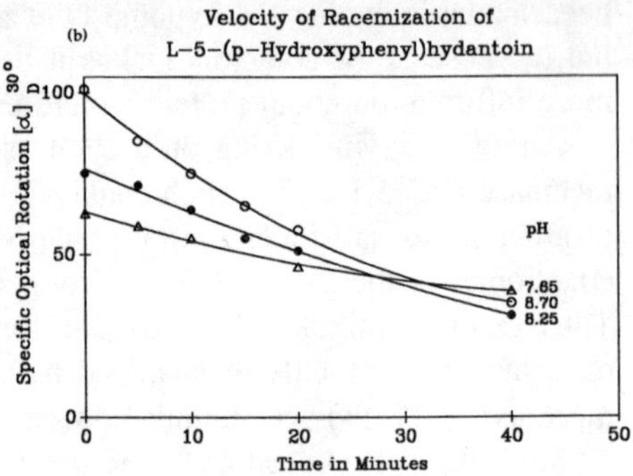

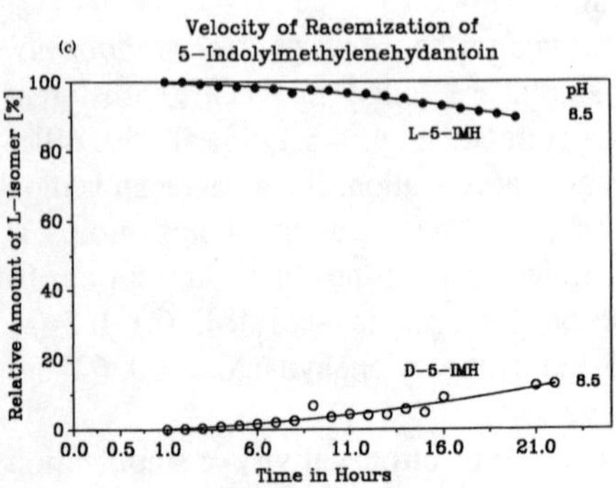

Figure 6.16. Chemical Racemization of 5-Monosubstituted Hydantoins—Influence of pH and Temperature.
a: D-hydantoin-5-propionic acid (from Tsugawa *et al.* [8]);
b: L-5-(*p*-hydroxyphenyl)hydantoin (from Yokozeki *et al.* [61]);
c: 5-indolylmethylenehydantoin;
d: D-5-IMH: influence of pH;
e: D-5-IMH: influence of temperature [62].

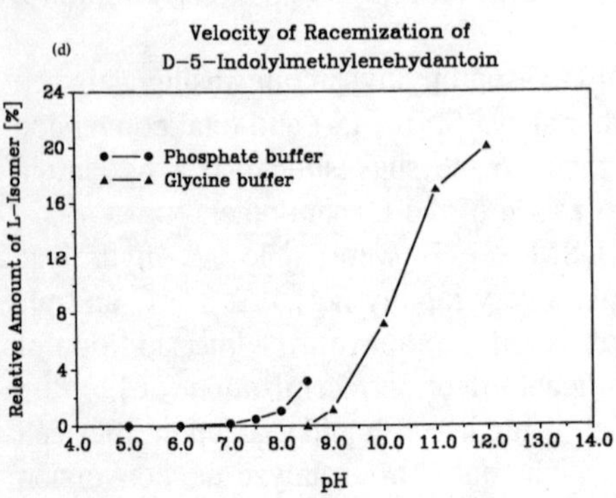

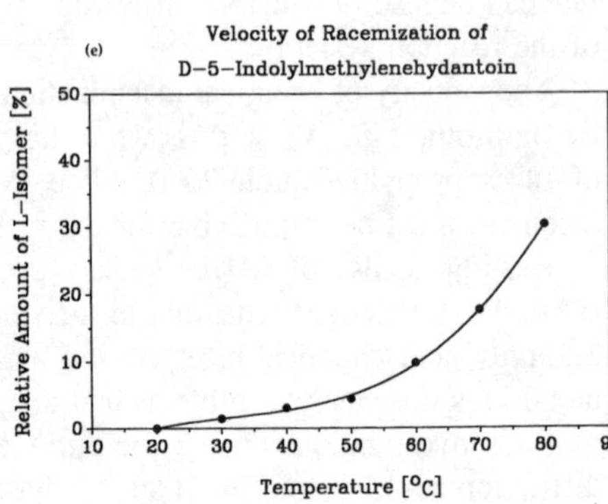

transformation of the optically pure enantiomers of 5-indolylmethylenehydantoin into the racemate. In Figure 6.17 the time dependence of the enzymatic reaction is shown in comparison with that of the chemical one.

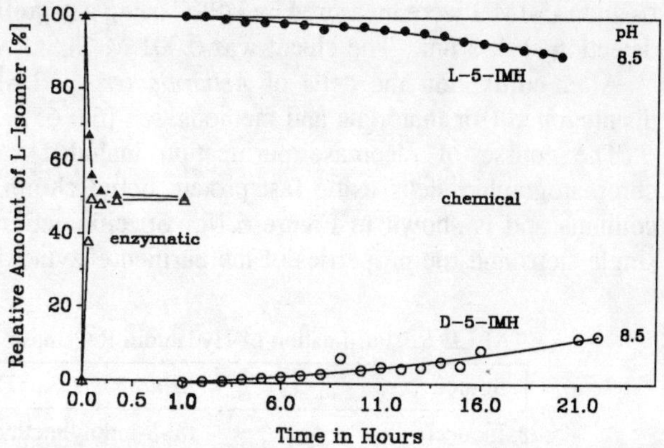

Figure 6.17. Chemical and Enzymatic Racemization of 5-Indolylmethylenehydantoin. (The enzyme fraction used in the assay was obtained after cell disintegration, liquid-liquid extraction, and dialysis.) [62].

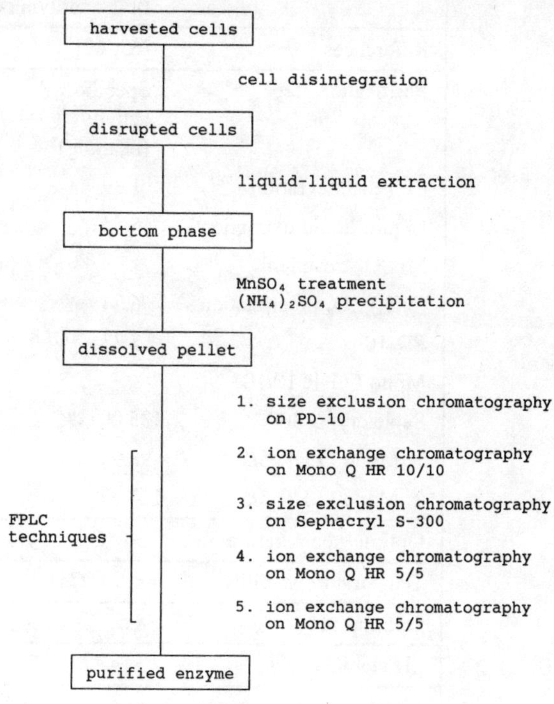

Figure 6.18. Purification Scheme of Hydantoin Racemase from *Arthrobacter* sp. DSM 3747 [62].

6.6.1.1 Purification of Hydantoin Racemase

The standard enzyme assays were carried out for 10 min at 37 °C using a 0.05 M glycine/NaOH buffer, pH 8.5, containing 0.2 g/l D-5-indolylmethylenehydantoin. The assays were stopped by adding methanol. After centrifugation the concentrations of D- and L-5-IMH were measured by HPLC using a D-proline–copper-II column and UV detection at 280 nm. The eluent was 0.001 M $CuSO_4$ solution.

After cultivation the cells of *Arthrobacter* sp. DSM 3747 were harvested and disintegrated (for materials and methods, see [62, 65]).

The course of racemase purification included various salt precipitation and chromatographic steps using fast protein liquid chromatography on high resolution columns and is shown in Figure 6.18. Specific activity, purification factors of the single steps and the properties of the purified enzyme are listed in Table 6.5.

TABLE 6.5 Purification of Hydantoin Racemase

Source	*Arthrobacter* sp. DSM 3747	
Inducers	DL-5-Indolylmethylene-*N*-3-methylhydantoin and DL-5-Indolylmethylenehydantoin	
References	[65, 62]	
Purification steps	Specific activity (nkat/mg)*	Purification factor
Cell disintegration	1.39	1.0
Liquid-liquid extraction	5.55	4.0
$MnSO_4$ treatment		
$(NH_4)_2SO_4$ precipitation	6.34	4.6
PD-10	8.33	6.0
Mono Q HR 10/10		
Sephacryl S 300	135.00	97.3
1. Mono Q HR 5/5	98.93	71.3
2. Mono Q HR 5/5	127.37	91.8
Optimal temperature	50 °C	
Temperature stability	< 50 °C	
Optimal pH	8.5	
pH stability	5.0–7.0	
Metal ion requirements	none	

* nkat = nanokatal = 10^{-9} mol/s

For investigations on the substrate specificity of the hydantoin racemase, the optically pure substrates were dissolved in a 0.05 M glycine/NaOH buffer, pH 8.5. The optical rotation was measured using a polarimeter at 365 nm and a temperature of 37 °C.

Various hydantoins, amino acids, amino acid esters, and amides and other compounds containing a keto function in a cyclic structure in the vicinity of a chiral center were tested for their availability to the racemase. As can be seen from Table 6.6, only aliphatic and aromatic hydantoin derivatives were substrates.

TABLE 6.6 Substrate Specificity of Hydantoin Racemase

Source	Purified from *Arthrobacter* sp. DSM 3747
Reference	[65]
Racemization of:	
D-5-Indolylmethylenehydantoin	+
D-5-Indolylmethylene-*N*-3-methylhydantoin	+
D-5-Benzylhydantoin	+
D-5-(2-Methylthioethylene)hydantoin	+
D-5-(*p*-Hydroxybenzyl)hydantoin	+
D-5-*i*-Butylhydantoin	+
D-Fructose	–
N-Acetyl-L-tryptophanamide	–
D-Tryptophan methyl ester	–
D-Tryptophan	–
L-Proline	–
L-Alanyl-L-alanine	–
Dihydro-L-orotic acid	–
(+)-6-Aminopenicillanic acid	–
Cephalexin	–
Cephalosporin C	–
Cephalothin	–

The influence of various chemicals including SH- and OH-blocking agents on the enzymatic activity was investigated by incubating the enzyme with a solution of the potential effector, obtaining a 0.01 M final concentration. The results are shown in Figure 6.19. The enzyme activity was totally destroyed by thiol-blocking agents such as *p*-chloromercurobenzoic acid (PCMB) and iodoacetamide. A reversible oxidation of the enzyme could be overcome by adding dithiothreitol (DTT) as reducing agent.

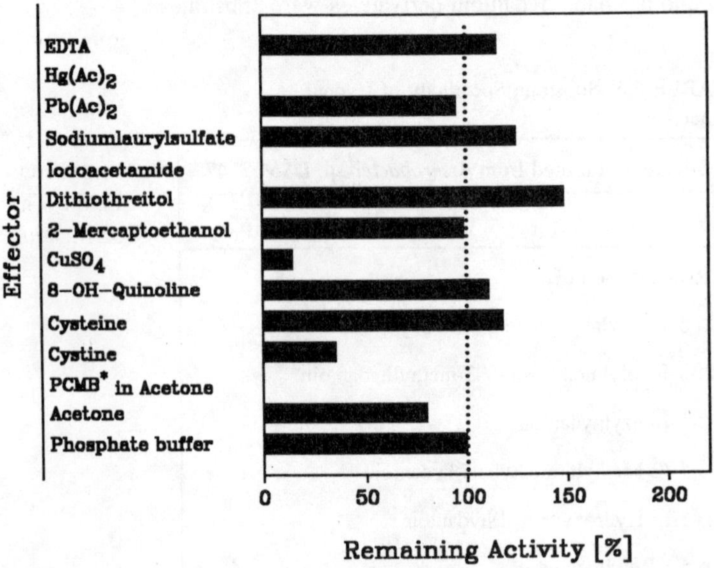

Figure 6.19. Racemase Activity Remaining after Incubation with Various Effectors [62]. PCMB* = *p*-Cl-mercuribenzoate.

6.6.2 Isolation and Characterization of L-Hydantoinases

6.6.2.1 Purification of L-Hydantoinase from *Arthrobacter* sp. DSM 3747

Like the other two enzymes which are involved in the microbial bioconversion of DL-5-monosubstituted hydantoins to L-amino acids, the L-hydantoinase is located in the cytosol of *Arthrobacter* sp. DSM 3747. Therefore cell disruption was necessary as the first step. Several methods had been tested, including treatment with lysozyme or disintegration by X-press [22, 67]. However, the best results were obtained using glass bead mills (Dyno Mill, W.A.B, Basel, Switzerland, and LME4, Netzsch, Selb, FRG) [67, 68].

After cell disruption, a clear L-hydantoinase solution free of cell debris was prepared by a two-step liquid-liquid extraction. Aqueous two-phase systems consisting of polyethylene glycol 6000 (PEG 6000) and ammonium sulfate in phosphate buffer were utilized [69, 70]. In the first extraction step the L-*N*-carbamoylase was separated, while cell debris, nucleic acids, and some other proteins were removed in the second step.

After dialysis of this enzyme solution, the retentate was further purified by a batch adsorption ion exchange chromatography using DEAE-Sepharose Fast Flow (Pharmacia, Uppsala, Sweden). The pooled active fractions were applied to a Sephacryl S-200 column (Pharmacia) for desalting and preparative size exclusion chromatography. After a second ion exchange chromatography step using Q-Sepharose Fast Flow, the solution containing the L-hydantoinase was purified by adsorption of other proteins on Blue Sepharose CL-6B. A homogeneous L-hydantoinase fraction was obtained using a Mono Q HR 10/10 FPLC column. The purification scheme is illustrated in Figure 6.20. Table 6.7 presents the purification parameters [70].

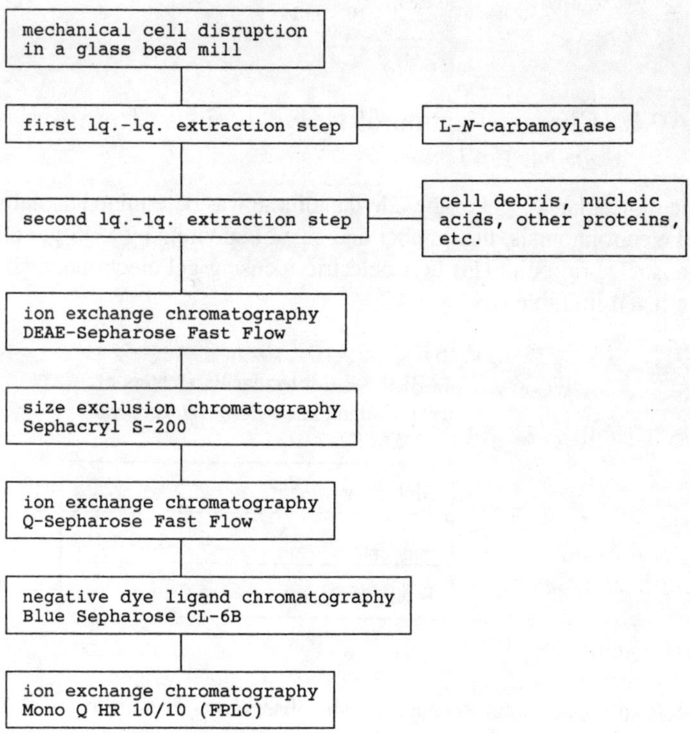

Figure 6.20. Purification of the L-Hydantoinase from *Arthrobacter* sp. DSM 3747 [70].

TABLE 6.7 Purification Table for the L-Hydantoinase from *Arthrobacter* sp. DSM 3747

Purification Step	Spec. activity (nkat/mg)*	Yield (%)	Purification factor
Cell disruption	1.92	100.0	1.0
Lq.-lq. extraction	11.87	79.7	6.2
Dialysis	4.88	75.8	2.5
DEAE Sepharose	5.70	33.8	3.0
Sephacryl S-200	11.22	33.0	5.8
Q-Sepharose	30.79	28.0	16.0
Blue Sepharose	54.35	13.5	28.3
Mono Q	422.00	9.3	219.6

* nkat = nanokatal = 10^{-9} mol/s

6.6.2.2 Characterization of L-Hydantoinase from *Arthrobacter* sp. DSM 3747

The molecular weight of the L-hydantoinase was determined by native polyacrylamide gel electrophoresis, the number and size of subunits by SDS gel electrophoresis, and the isoelectric point (Pi) by isoelectric focusing gel electrophoresis [70]. The results are listed in Table 6.8.

TABLE 6.8 Molecular Parameters of the L-Hydantoinase from *Arthrobacter* sp. DSM 3747

Molecular weight	232 kd
Subunits	4 × 55.4 kd
Isoelectric point	5.9

Total loss of catalytic activity was observed after addition of EDTA. The chelating agent was removed by gel filtration and various metal ions added. The enzyme could be reactivated by Mn^{2+} and Co^{2+} and partially by Ca^{2+} [67]. The optimal parameters for ring cleavage and reverse reaction are presented in Table 6.9.

TABLE 6.9 Optimal Conditions and Stability for the L-Hydantoinase from *Arthrobacter* sp. DSM 3747

	Ring Cleavage	Reverse Reaction
Optimal pH	8.8–9.25	6.6–7.3
Optimal Temperature	50 °C	30 °C
Temperature Stability	< 46 °C	n.d.

Various compounds were tested as substrates. As shown in Table 6.10, only those hydantoins were accepted which contained large residues and a methylene spacer. The L-hydantoinase accepted the L-enantiomers exclusively of both the hydantoin derivatives and the *N*-carbamoylamino acids.

6.6.2.3 L-Hydantoinases from Other Sources

L-Hydantoinases were isolated and partially purified also from *Flavobacterium* sp. AJ 3912 by Yokozeki *et al.* [40] and *Bacillus brevis* AJ 12299 by Yamashiro *et al.* [29]. The three hydrolases are different enzymes:

- The *Bacillus* L-hydantoinase requires ATP and Mg^{2+}, Mn^{2+} or K^{+} as cofactors and acts L-specifically. The optimal parameters for the ring cleavage are pH 8.0 and 50 °C. Only a few substrates have been investigated, as shown in Table 6.10.
- The hydantoinase from *Flavobacterium* is only L-selective. The optimal pH is remarkably high at 9.7. Its optimal temperature is 40 °C. Substrates containing aromatic residues are hydrolyzed with the highest rates.

6.6.3 Isolation and Characterization of *N*-Carbamoyl-L-Amino Acid Amidohydrolases

N-Carbamoyl-L-amino acid amidohydrolases (L-*N*-carbamoylases) were isolated and characterized from several microorganisms. In the following subsections, the enzyme isolated from *Arthrobacter* sp. 3747 by our working group [71] will be discussed first and after that a survey of the other microbial L-*N*-carbamoylases will be given.

TABLE 6.10 Substrate Specificity of L-Hydantoinases

Source	*Arthrobacter* sp. DSM 3747	*Flavobacterium* sp. AJ 3912	*Bacillus brevis* sp. AJ 12299
Reference	[70]	[40]	[29]
Inosine 5'-monophosphate	–	o	o
Allantoin	–	o	o
Barbituric acid	–	o	o
Creatinine	–	o	o
Dihydroorotic acid	–	o	o
Oxindole	–	o	o
Hydantoin	–	+	o
1-Methylhydantoin	–	o	o
5-Methylhydantoin	–	–	+
5-*i*-Propylhydantoin	–	+	+
5-*i*-Butylhydantoin	+	+	+
5-*sec*-Butylhydantoin	–	o	+
5-(3'-Aminopropylene)hydantoin	–	o	o
5-(4'-Aminobutylene)hydantoin	–	–	o
5-(2'-Methylthioethylene)hydantoin	o	+	+
5-Methoxymethylenehydantoin	o	+	o
5-Carboxymethylenehydantoin	–	–	o
5-Carboxyethylenehydantoin	–	–	o
5-Carbamoylethylenehydantoin	o	+	o
5-Carbamoylmethylenehydantoin	–	+	o
5,5-Dimethylhydantoin	–	o	o
5,5-Dimethyl-1-hydroxymethylhydantoin	–	o	o
Proline-derived hydantoin	–	o	o
5-Phenylhydantoin	–	o	o
5-*p*-Hydroxyphenylhydantoin	–	o	o
5-Benzylhydantoin	+	+	+
5-*p*-Hydroxybenzylhydantoin	+	+	+
5-(3,4-Dihydroxybenzyl)hydantoin	+	o	o
5-(3,4-Dimethoxybenzyl)hydantoin	o	+	o
5-Benzyloxymethylenehydantoin	+	+	o
5-(*S*-Benzylmercaptomethylene)hydantoin	+	o	o
5-(3',4'-Methylenedioxybenzylhydantoin	o	+	o
5-(*p*-Hydroxymethylenephenyl)-5-phenyl-hydantoin	–	o	o
5-Indolylmethylenehydantoin	+	+	o
5-(4-Imidazolmethylene)hydantoin	o	–	o
5-(2-Thienyl)hydantoin	–	o	o
N-Carbamoylglycine	–	o	o
N-Carbamoyltryptophan	+	o	o
Tryptophanamide	–	o	o
Tryptophan methyl ester	–	o	o

+ = conversion; – = no conversion; o = not determined

6.6.3.1 Purification of L-*N*-Carbamoylase from *Arthrobacter* sp. DSM 3747

After cell disruption by a glass bead mill (Netzsch LME4 or Dyno Mill) the L-*N*-carbamoylase was extracted into the top phase of the first partitioning using the aqueous two-phase system described above. The viscid enzyme solution was desalted by dialysis and then applied to a special ion exchange device named Zetaprep 60 disk QAE (Atlanta, Heidelberg) which could be run with acceptable flow rates. The active fractions were desalted by gel filtration (PD-10 columns, Pharmacia) and loaded onto a preparative Mono Q column (HR 10/10, Pharmacia) on the FPLC system. After concentrating with the analytical Mono Q HR 5/5, 0.2 ml of the L-*N*-carbamoylase was separated by size exclusion chromatography (Superose-12 HR 10/30, Pharmacia). The fraction with the highest catalytic activity was homogeneous (IEF 3-9 gel electrophoresis) [71]. Figure 6.21 shows the purification scheme; yields and purification factors are listed in Table 6.11.

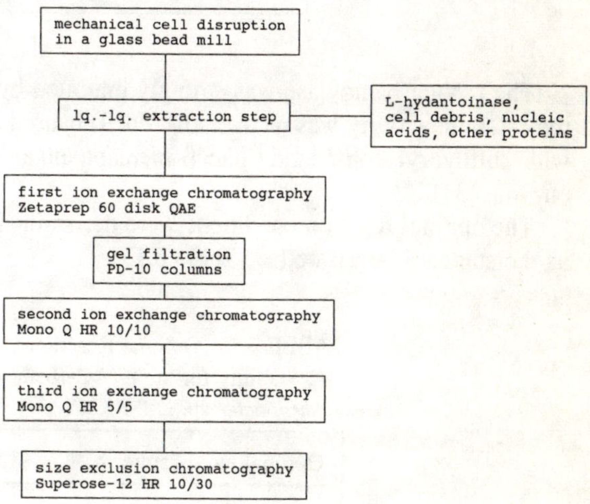

Figure 6.21. Purification of the *N*-Carbamoyl-L-amino Acid Amidohydrolase from *Arthrobacter* sp. DSM 3747 [69, 71].

TABLE 6.11 Purification Table for the L-*N*-Carbamoylase from *Arthrobacter* sp. DSM 3747

Purification Step	Spec. Activity (nkat/mg)*	Yield (%)	Purification Factor
Cell disruption	0.17	100	1.0
Lq.-lq. extraction	1.70	180	10.3
Zetaprep 60 QAE	0.85	52	5.1
Mono Q HR 10/10	9.21	13	55.8
Superose-12 HR 10/30	29.32	3	177.7

* nkat = nanokatal = 10^{-9} mol/s

6.6.3.2 Characterization of L-*N*-Carbamoylase from *Arthrobacter* sp. DSM 3747

The molecular weight of the holoenzyme as well as the number and molecular weight of the subunits (see Table 6.12) were determined by gel electrophoresis (native PAGE, IEF, SDS PAGE) [31, 69].

TABLE 6.12 Molecular Parameters of the L-*N*-Carbamoylase from *Arthrobacter* sp. DSM 3747

Molecular weight	70 kd
Subunits	2 × 32 kd
Isoelectric point	4.6

The L-*N*-carbamoylase was strongly inhibited by EDTA. After addition of Fe(II) the catalytic activity was recovered. Mn(II) caused a reactivation of 60%. Treatment with sulfhydryl compounds like β-mercaptoethanol and dithiothreitol activated the enzyme [31].

The optimal reaction conditions were determined using *N*-carbamoyl-L-tryptophan as the standard substrate (see Table 6.13).

TABLE 6.13 Optimal Reaction Conditions and Stability for the L-*N*-Carbamoylase from *Arthrobacter* sp. DSM 3747

Optimal temperature	45 °C
Optimal pH	~9.0
Temperature stability	< 35 °C
pH stability	6.0–6.5

The substrate specificity of this hydrolase is listed in Table 6.14 in comparison with the other L-*N*-carbamoylases discussed in the literature. No peptidase, amidase, aminoacylase or esterase activity was found. Only those amino acid derivatives which contain an aromatic residue coupled via a spacer group to the *S*-configured alpha carbon atom were accepted as substrates.

TABLE 6.14 Substrate Specificities of L-*N*-Carbamoylases

Source	*Arthrobacter* sp. DSM 3747	*Bacillus brevis* AJ 12299	*Flavobacterium* sp. AJ 3912	*Pseudomonas* sp. AJ 11220	*Zymobacterium oroticum*
Reference	[71]	[29]	[40]	[61]	[31]
N-Carbamoylglycine	o	o	+	–	o
N-Carbamoylalanine	–	+	+	+	o
N-Carbamoyl-β-alanine	o	o	+	o	o
N-Carbamoylsarcosine	o	o	o	o	o
N-Carbamoylvaline	–	+	+	+	o
N-Carbamoylleucine	–	+	+	+	o
N-Carbamoylisoleucine	o	+	+	o	o
N-Carbamoylproline	o	o	–	o	o
N-Carbamoylserine	–	o	+	o	o
N-Carbamoylmethionine	–	+	+	+	o
N-Carbamoyllysine	o	o	–	o	o
N-Carbamoylarginine	–	o	o	o	o
N-Carbamoylglutamic acid	–	o	–	+	–
N-Carbamoylaspartic acid	o	o	–	o	+
N-Carbamoylasparagine	o	o	+	+	o
N-Carbamoylglutamine	o	o	+	o	o
N-Carbamoylphenylglycine	–	o	o	o	o
N-Carbamoylphenylalanine	+	+	+	+	o
N-Carbamoyl-3,4-dihydroxyphenylalanine	o	o	o	–	o
N-Carbamoyl-3,4-dimethoxyphenylalanine	o	+	+	o	o
N-Carbamoyl-3,4-methylenedioxyphenylalanine	o	+	+	o	o
N-Carbamoyltryptophan	+	o	+	–	o
N-Carbamoyltyrosine	+	+	+	+	o
N-Carbamoyl-*O*-benzylserine	o	o	+	o	o
N-Carbamoyl-*O*-methylserine	o	o	+	o	o
N-Carbamoylhistidine	o	o	–	o	o
L-Citrulline	o	o	o	o	–
N-Acetyltryptophan	–	o	o	o	o
Glycyltryptophan	–	o	o	o	o
Tryptophylglycine	–	o	o	o	o
Tryptophanamide	–	o	o	o	o
Phenylalanineamide	–	o	o	o	o
Tyrosineamide	–	o	o	o	o
Tryptophan methyl ester	–	o	o	o	o
N-Formyltryptophan	-	o	o	o	o
N-Formylphenylalanine	-	o	o	o	o
N-Formylmethionine	–	o	o	o	o
tert-Butyloxycarbonyltryptophan	–	o	o	o	o

+ = conversion; – = no conversion; o = not determined

6.6.3.3 L-*N*-Carbamoylases from Other Microorganisms

There are four other microbial L-*N*-carbamoylases described in the literature. Table 6.14 gives a survey of the substrate specificities.

In 1955 Lieberman and Kornberg [31] reported on the isolation of the Mn(II)-/Fe(II)-dependent ureidosuccinase from *Clostridium oroticum* (also known as *Zymobacterium oroticum*). This hydrolase works best at pH 7.8 to 8.5 and its biological function is the conversion of *N*-carbamoylaspartic acid to L-aspartic acid (orotic acid degradation pathway).

Another L-specific enzyme, which also requires Mn(II) or Fe(II), was isolated from *Bacillus brevis* by Yamashiro *et al.* [29]. The spectrum of substrates includes aliphatic as well as aromatic *N*-carbamoylamino acids, and pH 7.5 and 50 °C are the optimal reaction conditions.

The *N*-carbamoylase from *Flavobacterium* sp. AJ 3912 described by Yokozeki *et al.* [40] is L-specific, too, and shows the highest relative rates of hydrolysis with aromatic *N*-carbamoyl derivatives. Optimal pH and temperature are 8.0 and 40 °C.

Finally, Yokozeki *et al.* [61] discovered L-*N*-carbamoylase activity in the crude extract of *Pseudomonas* sp. AJ 11220. The most rapid hydrolysis was observed with *N*-carbamoylvaline and -leucine as substrates.

All five L-*N*-carbamoylases discussed in this review have different properties, and their biological function is still unknown with the exception of the ureidosuccinase.

6.7 INDUSTRIAL APPLICATIONS

6.7.1 Survey

Since the 1970s there has been especially great interest in industry in producing essential L-amino acids from chemically synthesized precursors by bioconversion [1, 2, 3]. Until now the most successful method using a hydrolase is the L-aminoacylase method, which was investigated in detail by Chibata *et al.* (for a practical approach see [72]). Other methods of interest are discussed in this book.

To establish a new bioconversion method, this process must have distinct advantages over other reactions. For producing L-amino acid, the advantages of the hydantoin method, which were discussed in this chapter and in Chapter 5, are:

(i) the substrates can be easily synthesized by chemical methods from low-priced non-amino acid precursors (see Chapter 5);

(ii) the hydantoin method can be applied to a wide range of both proteinogenic and unnatural amino acids (see Section 6.6); and

(iii) the racemic substrates can be totally converted to 100% L-amino acid by either chemical or enzymatic racemization of the hydantoin derivatives (see Chapter 5 and Section 6.6.1).

In spite of these advantages, D-*p*-hydroxyphenylglycine is the only product which is industrially produced from a hydantoin substrate by bioconversion at this time (see Chapter 5). Although there are no reports in the literature on L-amino acids produced industrially by the L-hydantoin method at the moment, it must be noted that most of the publications discussed in this chapter are from research groups of industrial companies or of working groups which are working together with industrial companies: Guivarch *et al.* [5] are from the French company Rhône-Poulenc, Klages *et al.* [35] from the German company Schering, Miyoshi *et al.* [6] from the Japanese company Denki, Nishida *et al.* [10] from the Japanese company Tanabe, Sano *et al.* [32] as well as Yamashiro *et al.* [28, 29] and Yokozeki *et al.* [33, 34, 40] from the Japanese company Ajinomoto, and our investigations were done together with the German company Rütgerswerke AG, and are documented in common publications and patents [17, 27, 44, 73, 74].

When investigations on the L-selective hydantoin cleavage started in the late 1970s, most of the working groups were interested in producing the essential amino acid L-Trp from DL-5-IMH, but did not apply this method due to the reduction of the world market price for L-Trp in the 1980s. But with the increasing interest in nonproteinogenic L-amino acids of high value for peptide synthesis, as chiral synthons and for medical uses—for example L-*O*-benzylserine [15, 40]—the L-selective hydantoin method seems to be the most interesting one. So it is supposed that the L-selective hydantoin method can be successfully applied to most of the substrates used for the L-aminoacylase by Chenault *et al.* [75], too.

6.7.2 Process Design

The industrial application of resting free or immobilized microbial cells as well as of immobilized enzymes for the production of D-*p*-hydroxyphenylglycine from the corresponding DL-5-hydantoin derivative was discussed in detail in Chapter 5. All these methods cannot be directly applied to the production of L-Trp or other aromatic L-arylalkylamino acids by resting cells of *Arthrobacter* sp. DSM 3747 or isolated enzymes from this microorganism: in contrast to 5-*p*-hydroxyphenylhydantoin, these substrates do not racemize quickly under reaction conditions suitable for enzymes, as discussed in detail in Section 6.6.1. So it would not be economic to use only the L-hydantoinase reaction. Additionally, this reaction would not be stereoselective using resting cells of *Arthrobacter* sp. DSM 3747 (see Section 6.5). For a total conversion of DL-5-IMH to 100% L-Trp the hydantoin racemase is absolutely necessary. The reaction scheme is shown in Figure 6.22.

In addition, for good process economics the biocatalyst should be reused. So only two strategies are useful for the production of L-Trp from DL-5-IMH:

(i) the use of resting immobilized cells of *Arthrobacter* sp. DSM 3747 with high activities of all three enzymes, or
(ii) the use of immobilized enzymes.

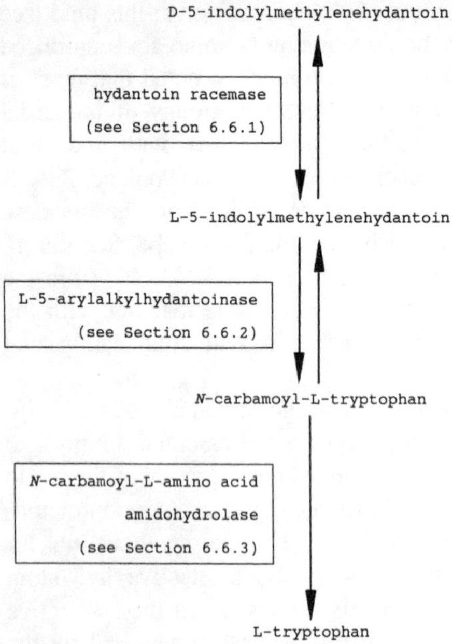

Figure 6.22. Schematic Summary of the Enzymatic Production of L-Tryptophan in Resting Cells of *Arthrobacter* sp. DSM 3747 [26].

In the second case it will be advantageous to work in a two-step process (L-hydantoinase and hydantoin racemase in the first step, L-*N*-carbamoylamino acid amidohydrolase in the second step) because of the different stabilities of and reaction parameters optimal for the three enzymes. Both methods are under investigation at the moment.

6.7.3 Other Applications and Future Outlook

In the future the L-selective hydantoin method will become of interest especially for the production of nonproteinogenic L-amino acids of high value for peptide synthesis, as chiral synthons and for medical uses (for a detailed discussion of the substrate specificities of all three enzymes involved in this process, see Section 6.6). The hydantoin racemase can become an interesting enzyme in combination with enzymes from the D-selective process, too, for substrates which only slowly racemize under reaction conditions, for example, for L-5-isopropypylhydantoin in the production of D-valine [58]. The isolated hydantoinase can be used not only for the L-selective cleavage of DL-5-monosubstituted hydantoin derivatives, but for the synthesis of optically pure L-hydantoin derivatives from chemically synthesized *N*-carbamoyl-DL-amino acids (see Section 6.6.2 and [40]). This application could be of interest for medical uses (for a discussion of the medical application of hydantoin derivatives see [63]). Experiments on genetic engineering, as described for the D-hydantoinase [76], can enable the production of high amounts of all three enzymes in the L-selective process.

6.8 CONCLUSIONS

Comparable to the D-selective hydantoin method, the L-selective reactions are of high stability and can be applied to a wide range of substrates. In *Arthrobacter* sp. DSM 3747, the most potent biocatalyst isolated in screening, a hydantoin racemase enabled the total conversion of DL-5-IMH to 100% L-Trp. High L-Trp-producing activity could be induced especially, when using the chemically modified substrate *N*-3-CH_3-IMH as inducer. This compound could not be hydrolyzed by the L-hydantoinase. Successful experiments in reusing resting cells of *Arthrobacter* sp. DSM 3747 or immobilized enzymes are a good starting point for further investigations, for example, on the production of nonproteinogenic L-amino acids of high value from chemically synthesized hydantoin derivatives in the future.

6.9 ACKNOWLEDGMENTS

Parts of this work were supported by grants of the “Bundesministerium für Forschung und Technik”, Bonn, FRG, and the Rütgerswerke AG, Castrop-Rauxel, FRG. We thank Dr. Albrecht Läufer and Dr. Hartmut Höke for good cooperation in this project, and Professor Karsten Krohn and his co-workers from the Institute of Organic Chemistry, Technical University of Braunschweig, FRG, for the synthesis of most of the hydantoin derivatives used in these experiments.

6.10 REFERENCES

1. Soda, K., Tanaka, H., Esaki, N. In *Biotechnology*; Rehm, H.J., Reed, G., Eds.; Verlag Chemie, Weinheim, 1983; Vol. 3, pp 480-530.
2. Schmidt-Kastner, G., Egerer, P. In *Biotransformations, Biotechnology*; Kieslich, K., Rehm, H. J., Reed, G., Eds.; Verlag Chemie, Weinheim, 1984; Vol. 6a, pp 387-419.
3. Yamada, H. In *Proc. 4th Europ. Congr. Biotechnol.*; Neijssel, O. M., van der Meer, R. R., Luyben, K. C. A. M., Eds.; Elsevier Science Publishers B. V., Amsterdam, 1987; Vol. 4, pp 689-692.
4. Yamada, H., Shimizu, S., *Angew. Chem. 100*, 640 (1988).
5. Guivarch, M., Gillonier, C., Brunie, J.-C., *Bull. Soc. Chim. Fr. 1-2*, 91 (1980).

6. Miyoshi, T., Hironoshin, K., Masaaki, K., Susuma, C., Eur. Pat. 0159 866 (1985).
7. Syldatk, C., Cotoras, D., Dombach, G., Groß, C., Kallwaß, H., Wagner, F., *Biotechnol. Lett. 9*, 25 (1987).
8. Tsugawa, R., Okumura, S., Ito, T., Katsuga, N., *Agric. Biol. Chem. 30*, 27 (1966).
9. Lieberman, I., Kornberg, A., *J. Biol. Chem. 207*, 911 (1954).
10. Nishida, Y., Nakamichi, K., Nabe, K., Tosa, T., *Enzyme Microb. Technol. 9*, 721 (1987).
11. Yokozeki, K., Nakamori, S., Yamanaka, S., Eguchi, C., Mitsugi, K., Yoshinaga, F., *Agric. Biol. Chem. 51*, 715 (1987).
12. Syldatk, C., Müller, R., Dombach, G. In *Abstracts of Papers*, 8th Int. Biotechnol. Symposium, Paris, 1988; p 154.
13. Müller, C., Syldatk, R., Wagner, F. In *BMFT - Statusreport "Stoffmwandlung mit Biokatalysatoren"*; Bundesministerium für Forschung und Technik, FRG, 1988; in press.
14. Takahashi, S., Ohashi, T., Kii, Y., Kumagai, H., Yamada, H., *J. Ferment. Technol. 57*, 328 (1979).
15. Cotoras, D., Wagner, F. In *Proc. 3rd Europ. Congress Biotechnol., München*; Verlag Chemie, Weinheim, 1984; Vol. 3, pp 351-354.
16. Groß, C., Syldatk, C., Wagner, F., *Biotechnol. Techniques 1*, 85 (1987).
17. Syldatk, C., Mackowiak, V., Höke, H., Groß, C, Dombach, G., Wagner, F., *J. Biotechnol. 14*, 345 (1990).
18. Mackowiak, V., Syldatk, C., Röhrmann, A., Wagner, F. In *Abstracts of Papers*, 8th Int. Biotechnol. Symposium, Paris, 1988; p 275.
19. Groß, C., Syldatk, C., Wagner, F. In *Proc. 4th Europ. Congr. Biotechnol.*; Neijssel, O. M., van der Meer, R. R., Luyben, K. C. A. M., Eds.; Elsevier Science Publishers B. V., Amsterdam, 1987; Vol. 2, pp 249-252.
20. Groß, C., Syldatk, C., Mackowiak, V., Wagner, F., *J. Biotechnol. 14*, 363-376 (1990).
21. Syldatk, C., Mackowiak, V., Dombach, G., Groß, C., Röhrmann, A., Wagner, F. In *Proc. 4th Europ. Congr. Biotechnol.*; Neijssel, O. M., van der Meer, R. R., Luyben, K. C. A. M., Eds.; Elsevier Science Publishers B. V., Amsterdam, 1987; Vol. 2, pp 245-248.
22. Dombach, G., Syldatk, C., Wagner, F. In *Proc. 4th Europ. Congr. Biotechnol.*; Neijssel, O. M., van der Meer, R. R., Luyben, K. C. A. M., Eds.; Elsevier Science Publishers B. V., Amsterdam, 1987; Vol. 2, pp 577-580.
23. Müller, R., Dombach, G., Syldatk, C., Wagner, F. In *Abstracts of Papers*, 8th Int. Symp. on HPLC of Proteins, Peptides, and Polynucleotides; Copenhagen, Denmark, 1988; p 34.
24. Syldatk, C, Dombach, G., Müller, C., Groß, R., Wagner, F. In *Enzyme Engineering 9*; Blanch, H.W., Klibanov, A.M., Eds.; Ann. N. Y. Acad.Sci., 1988; Vol. 542, pp 323-329.
25. Müller, R., Syldatk, C., Wagner, F. In *Dechema Biotechnology Conferences*; DECHEMA Frankfurt a.M., FRG, 1989; Vol. 3, pp 163-166.
26. Syldatk, C., Wagner, F., *J. Food Biotechnol. 4*, 87-95, (1990).
27. Syldatk, C., Läufer, A., Müller, R., Höke, H. In *Adv. Biochem Eng./Biotechnol.*; Fiechter, A., Ed.; Springer-Verlag, Berlin, FRG, 1990; Vol. 41, pp 29-76.
28. Yamashiro, A., Yokozeki, K., Kano, H., Kubota, K., *Agric. Biol. Chem. 52*, 2851 (1988).
29. Yamashiro, A., Kubota, K., Yokozeki, K., *Agric. Biol. Chem. 52*, 2857 (1988).
30. Lieberman, I., Kornberg, A., *Biochim. Biophys. Acta 12*, 223 (1953).
31. Liebermann, I., Kornberg, A., *J. Biol. Chem. 212*, 909 (1955).

32. Sano, K., Yokozeki, K., Eguchi, C., Kagawa, T., Noda, I., Mitsugi, K., *Agric. Biol. Chem. 41*, 819 (1977).
33. Yokozeki, K., Sano, K., Eguchi, C., Yamada, K., Mitsugi, K., *Agric. Biol. Chem. 51*, 819 (1987).
34. Yokozeki, K., Sano, K., Eguchi, C., Iwagami, K., Mitsugi, K., *Agric. Biol. Chem. 51*, 729 (1987).
35. Klages, U., Weber, A., Wilschowitz, L., Ger. Pat. Applic. DE 3 702 384 (1988).
36. Hassall, H., Greenberg, D. M., *J. Biol. Chem. 238*, 3325 (1963).
37. Akamatsu, N., *J. Biochem. 47*, 809 (1960).
38. Syldatk, C., Cotoras, D., Dombach, G., Groß, C., Kallwaß, H., Andree, H., Gebinoga, M., Wagner, F., *Hoppe-Seyler's Z. Physiol. Chem. 367*, 340 (1986).
39. Syldatk, C., Cotoras, D., Möller, A., Wagner, F., *BTF-Biotechforum 3*, 9 (1986).
40. Yokozeki, K., Hirose, Y., Kubota, K., *Agric. Biol. Chem. 51*, 737 (1987).
41. Cotoras, D. *Mikrobielle, enantioselektive Hydrolyse von DL-5-substituierten Hydantoinen und N*-Carbamoyl-DL-Aminosäuren; Ph. D. Thesis; Technical University of Braunschweig, FRG, 1985.
42. Yamada, H., Takahashi, S., Kii, Y., Kumagai, H., *J. Ferment. Technol. 56*, 484 (1978).
43. Groß, C. *Optimierung der Herstellung von L-Tryptophan aus DL-5-Indolylmethylhydantoin mit einer Mutante des Stammes Arthrobacter* spec. DSM 3747; Ph. D. Thesis; Technical University of Braunschweig, FRG, 1988.
44. Wagner, F., Syldatk, C., Mackowiak, V., Krohn, K., Höke, H., Läufer, A., Europ. Pat. Appl. 89118193.5 (1989).
45. Murata, M., Miyagawa-Koshima, K., Nakanishi, K., Naya, Y., *Science 234*, 585 (1986).
46. Weethall, H. H. In *Enzyme Engineering 3*, Dinelli, D., Marconi, W., Cecere, F., Galli, G., Morisi, F., Pye, E. K., Eds.; Plenum, New York, 1978; pp 477-481.
47. Adams, E. In *Enzymes*; Dixon, M., Webb, E. C., Eds.; 3rd ed., Academic Press, San Diego, 1979; Vol. 3.
48. Yamada, M., Kurahashi, K., *J. Biochem. 66*, 529 (1969).
49. Nakajima, N., Tanizawa, K., Tanaka, H., Soda, K., *Agric. Biol. Chem. 52*, 3099 (1988).
50. Soda, K., Osumi, T., *Agric. Biol. Chem. 31*, 1097 (1967).
51. Pugnière, M., Commeyras, A., Previero, A., *Biotechnol. Lett. 5*, 447 (1983).
52. Adams, E., Norton, I. L., *J. Biol. Chem. 239*, 1525 (1964).
53. Finlay, T. H., Adams, E., *J. Biol. Chem. 245*, 5248 (1970).
54. Stadtman, T. C., Elliott, P., *J. Biol. Chem. 228*, 983 (1957).
55. van der Drift, L., Vogels, G.D., van der Drift, C., *Biochim. Biophys. Act. 391*, 240 (1975).
56. Okumura, I., Kondo, K., Miyake, Y., Itaya, K., Yamamoto, T., *J. Biochem. 79*, 1013 (1976).
57. Fukumura, T., *Agric. Biol. Chem. 41*, 1327 (1977).
58. Battilotti, M., Barberini, U., *J. Mol. Cat. 43*, 343 (1988).
59. Knabe, J., Wumm, W., *Arch. Pharm. (Weinheim) 313*, 538 (1980).
60. Pietzsch, M. *Stereochemische Untersuchung zur enzymatischen Hydrolyse von DL-5-monosubstituierten Hydantoinen*; Diploma Thesis; Technical University of Braunschweig, FRG, 1989.
61. Yokozeki, K., Kubota, K., *Agric. Biol. Chem. 51*, 721 (1987).

62. Pietzsch, M. In *Dechema Biotechnology Conferences*; DECHEMA Frankfurt a.M., FRG, 1990; Vol. 4, part A, pp 259-262.
63. Ware, E., *Chem. Rev. 46*, 403 (1950).
64. Müller, R. *Anreicherung und Charakterisierung einer* L-*N*-Carbamoylase; Diploma Thesis; Technical University of Braunschweig, FRG, 1987.
65. Wagner, F., Pietzsch, M., Syldatk, C. In *Enzyme Engineering 11; Ann. N. Y. Acad. Sci.*, in press.
66. Schwedes, J., Bunge, F. In *Biotechnology Focus 3*; Finn, R. K., Präve, P., Eds.; Hanser-Verlag, Munich, New York, 1992; pp 185-233.
67. Dombach, G. *Isolierung und Charakterisierung einer Hydantoinase aus Arthrobacter* spec. DSM 3747; Ph. D. Thesis; Technical University of Braunschweig, FRG, 1989.
68. Bunge, F., Pietzsch, M., Müller, R., Syldatk, C., *Chem. Eng. Science 47*, 225 (1992).
69. Müller, R. *Untersuchungen zur Charakterisierung und Immobilisierung der N*-Carbamoyl-L-Aminosäureamidhydrolase aus *Arthrobacter* spec. DSM 3747; Ph. D. Thesis; TU Braunschweig (1990).
70. Kölle, R., Nöthe, C., Siemann, M., Syldatk, C., in preparation.
71. Wohlfahrt, G., Müller, R., Wagner, F., Syldatk, C., in preparation.
72. Chibata, I., Tosa, T., Sato, T., Mori, T., *Methods Enzymol. 44*, 747 (1976).
73. Höke, H., Wagner, F., Höltmann, W., Cotoras, D., Syldatk, C., Groß, C., Dombach, G., Wagner, T., German Pat. DE 3712539 C2 (1988).
74. Syldatk, C., Läufer, A., H., Höke, Wagner, F., *Forum Mikrobiologie 11*, 224 (1988).
75. Chenault, H. K., Dahmer, J., Whitesides, G. M., *J. Am. Chem. Soc. 111*, 6354 (1989).
76. Lungershausen, R., Martin, C., Marcinowski, S., Siegel, H., Kuesters, W., German Pat. DE 3031151 A1 (1982).

PRODUCTION OF NATURAL AND SYNTHETIC L- AND D-AMINO ACIDS BY AMINOPEPTIDASES AND AMINO AMIDASES

J. Kamphuis, E. M. Meijer, W. H. J. Boesten, Q. B. Broxterman, B. Kaptein, H. F. M. Hermes, and H. E. Schoemaker

CONTENTS

7.1 GENERAL INTRODUCTION

Amino acids hold an important position in biochemistry and chemistry. They are the building blocks of proteins and play an essential role in the regulation of the metabolism of living organisms. In general it can be stated that microbial processes (fermentations) are the industrial production methods of choice for the large-scale production of naturally occurring proteinogenic L-α-*H*-amino acids, whilst for the small-scale production of synthetic D- or L-amino acids, chemoenzymatic methods are highly competitive with classical chemical resolution procedures.

Recently there has been a growing interest in rare and nonproteinogenic amino acids, since these compounds are valuable synthons for peptide-derived chemotherapeutics. D-Phenylglycine and D-*p*-hydroxyphenylglycine are well-known chiral building blocks for the broad-spectrum antibiotics ampicillin and amoxycillin, respectively. Other examples are D-valine, an intermediate in the production of the pyrethroid insecticide fluvalinate, and L-homophenylalanine, which is a general constituent of a number of recently developed Angiotensin-Converting Enzyme (ACE) inhibitors.

At present, a number of enzymatic kinetic resolution processes for the production of amino acids is known, namely, the application of aminoacylases, hydantoinases, and carbamoylases. Whilst the biochemistry of amidases and aminopeptidases has been studied for many years—mainly to understand their role in the degradation of various compounds—it was only recently that their application in the kinetic resolution of amino acid amides was commercialized.

In this chapter the nature and occurrence of different types of aminopeptidases and amino amidases are discussed. The commercial application of an L-specific aminopeptidase from *Pseudomonas putida* for the stereoselective hydrolysis of a broad structural range of amino acid amides is described in detail. Emphasis will be placed on the specific characteristics of this process, that is, the chemical preparation of the substrates, the pH profile, the substrate specificity, and the reaction kinetics of the enzymatic resolution step.

Amino acids are versatile chiral building blocks for a whole range of fine chemicals. Some applications of amino acids are given.

In recent years medicinal chemists have shown increasing interest in bioactive peptide analogs containing α-alkyl amino acids (such as neuropeptides, chemotactic peptides, enkephalins, etc.), mainly because they tend to freeze specific conformations, dramatically slow down enzymatic degration processes, or both. A novel chemoenzymatic production method developed at DSM based on the stereoselective hydrolysis of α-alkyl amino acid amides with an amidase from *Mycobacterium neoaurum* is discussed in detail. Some applications of α-alkyl amino acids are given as well.

An important disadvantage of the use of hydrolases like amino amidases and aminopeptidases in kinetic resolution processes is the product yield, which is intrinsically limited to 50%. One of the exceptions is the production of L-lysine from DL-α-amino-ε-caprolactam. This process is based on the combination of two enzymatic reactions: the enantiospecific hydrolysis of L-α-amino-ε-caprolactam to L-lysine and the simultaneous racemization of D-α-amino-ε-caprolactam into the DL-enantiomer. In this way a yield of nearly 100% L-lysine is obtained. Another exception is the use of hydantoinases with aromatic hydantoins as substrates, because under the reaction conditions the unhydrolyzed hydantoin will racemize, and therefore a 100% conversion is observed.

For the future, researchers involved in the development of new biocatalytic procedures are challenged to develop new systems resulting in a 100% yield of an optically pure product in either enantiomeric form. However, these new developments will not be restricted to the field of biocatalysis alone. Many interesting results have already been obtained in the field of homogeneous asymmetric catalysis, for example, the asymmetric hydrogenation of *N*-acetamidocinnamic acid derivatives to form L-phenylalanine derivatives (developed by Knowles) and the Sharpless epoxidation of allylic alcohols, as well as the use of BINAP complexes (Noyori) in the synthesis of chiral alcohols, etc. These research fields are complementary and should be seen as a set of powerful tools in the (bio)chemical synthesis of optically pure compounds.

7.2 AMINOPEPTIDASES (EC 3.4.11.–) AND AMINO AMIDASES

7.2.1 Introduction

In the cell, peptides can be generated from the processing of leader sequences of exported proteins, the turnover of denaturated or abnormal proteins, and the postulated translation of attenuator sequences in amino acid biosynthetic operons. It has been observed that these peptides do not normally accumulate in Gram-negative bacteria, and this can be attributed to the presence of numerous peptidase activities [1]. A firm grasp of the exact role of all these peptidases is essential to understanding the regulation and metabolism of living organisms. On the other hand, the capacity of proteolytic enzymes to catalyze the synthesis of peptide bonds has been studied for a long time, with the aim of elucidating the potential use of these enzymes in the production of biologically active peptides [2]. One of the classes of these peptidases is the *aminopeptidases*, a group of (zinc-containing) metalloenzymes (exopeptidases) with specificity for cleavage at the amino terminus of a polypeptide chain. They are of biochemical as well as medicinal importance because of their putative involvement in the degradation of biologically active peptides, such as the opioid peptide enkephalins, and in certain pathological conditions such as human eye cataracts [3, 4].

Arylamidases readily hydrolyze β-naphthylamides of amino acids. In the literature [5] there is a suggestion that those enzymes should be treated differently from aminopeptidases. However, enzymes that readily hydrolyze β-naphthylamides of amino acids are sometimes also active towards amino acid amides [6].

It is also confusing that several authors identify and characterize arylamidases without regard to their activities towards amino acid amides or dipeptides [7, 8]. In general, amidases are capable of hydrolyzing amides to acids. This class of enzymes is outside the scope of this chapter. However, a mutant strain of the amidase of *Brevibacterium* sp R312 is capable of hydrolyzing not only formamide, urea, and nicotinamide, but also α-amino acid amides, and was referred to as an α-amino amidase [9].

In this section only those enzyme systems that are reported to be able to hydrolyze α-amino acid amides are discussed; others, if appropriate, are briefly mentioned.

7.2.2 Occurrence

The role of aminopeptidases in the degradation of various compounds of peptidic character, although not yet well understood, has stimulated continuous interest in these enzymes.

The aminopeptidases comprise an apparently large and diverse group of enzymes, widely distributed in animal and plant tissue and in microorganisms [10, 11].

By far the best known is leucine aminopeptidase (EC 3.4.11.1), which has been highly purified from swine kidney and bovine eye lens. The main microbial aminopeptidases described are from *E. coli* B [12], *Talaromyces duponti* [13], *Brevibacterium linens* [14], *Aspergillus oryzae* [15], *Micrococcus lysodeikticus* [16], yeast hydrolysate [17], *Bacillus subtilis* [6, 18, 24], *Streptomyces soyaensis* [19], *Bacillus stearothermophilus* [20], *Keratomyces ajelloi* [21], *Aspergillus japonica* [22], *Streptomyces peptidofaciens* [23], *Brevibacterium* species A4 [9], *Lactobacillus casei* [25], and *Aeromonas* species such as *Aeromonas proteolytica* [26].

7.2.3 Characteristics

Aminopeptidases are a group of zinc-containing exopeptidases with specificity of cleavage at the amino terminus of a polypeptide chain. They show a wide range of substrate specificities. Although the mechanism of action is still unclear, the primary amino acid sequence of leucine aminopeptidase from bovine eye has been reported [27] and some x-ray crystallographic studies have been published [28]. Studies on the mechanism of action of aminopeptidases have been reported [29, 30]. Most of these aminopeptidases have an alkaline optimum pH, usually above 8.0. They are usually inhibited by heavy metals, but are activated by Mg^{2+}, Mn^{2+}, Zn^{2+}, or Co^{2+}.

Prescott *et al.* [31] in 1966 gave a detailed description of the purification and properties of *Aeromonas* (*proteolytica*) aminopeptidase (Tables 7.1 and 7.2). The enzyme hydrolyzes di- and tripeptides, amino acid amides, amino acid *p*-nitroanilides, and β-naphthylamides. Only those peptides having an unsubstituted amino group were attacked. Some remarkable results with amino acid amides as substrates were obtained (zero-order kinetics).

TABLE 7.1 First-Order Proteolytic Coefficients (C_1) of *Aeromonas* Aminopeptidase with Some Amino Acid Amides *

Amide of	C_1
L-leucine	29
L-norvaline	7
L-norleucine	5
L-isoleucine	1
L-phenylalanine	1
D-leucine	0

* Assayed at substrate concentrations of 0.05 M at 40 °C

TABLE 7.2 Comparative Activities of Leucine Aminopeptidase and *Aeromonas* Aminopeptidase

Enzyme	Substrates	
	L-Leucine amide	LNA°, specific activity toward *p*-nitroanilides
Leucine aminopeptidase	25	0.4
Aeromonas aminopeptidase	29	23

The activity towards D-amino acid amides was zero. To the best of our knowledge this is one of the first reports of an L-specific aminopeptidase. The activities of the *Aeromonas* aminopeptidase were also compared with the activities of leucine aminopeptidase (Table 7.2). The enzyme was activated by the addition of Zn^{2+}, Mn^{2+}, or Co^{2+}.

Some characteristics of other reported aminopeptidases are summarized in Table 7.3.

7.2.4 Applications

Relatively few reports have appeared in the literature on aminopeptidases which have been tested with respect to their substrate specificity on amino acid amides; those that have been described are leucine aminopeptidase, and aminopeptidases from *Aeromonas proteolytica*, *Lactobacillus acidophilus* R-26, *Brevibacterium* A-4, *Rhodococcus erythropolis*, *Pseudomonas putida*, and *Ochrobactrum anthropi*. Many other aminopeptidases have been described, but have been tested only on amino acid *p*-nitroanilides or amino acid β-naphthylamides, with no mention of their possible activity towards amino acid amides. The possible use of *Brevibacterium* in the stereospecific synthesis of amino acids has been reviewed before [9]. The L-aminopeptidase activity was used in combination with the nitrilehydratase activity to convert aminonitriles into amino acids. To the best of our knowledge this process has not been commercialized. The main reason for this could be that most aminonitriles are very unstable in water. They are always in equilibrium with the aldehyde or ketone and ammonia/HCN. Polymerization of hydrogen cyanide gives an equilibrium shift resulting in the loss of aminonitrile. Therefore a low overall yield in amino acids is to be expected, which makes this method less attractive for the industrial synthesis of amino acids.

Other L-aminopeptidases are reported in the patent literature by DSM, Mitsubishi, and Nitto Chemicals. The use of an aminopeptidase from *Pseudomonas putida* was described by Boesten *et al.* in 1976 [50, 51]. The commercial use of this aminopeptidase is discussed in Section 7.3. D-Specific aminopeptidases were recently discovered by DSM and Sagami Research Institute [47–49, 52–54].

At the Sagami Research Institute [47], an enrichment culture technique was used to select D-stereospecific aminopeptidase from an *Achromobacter* species SCRC C1-38 (later renamed *Ochrobactrum anthropi*) from a soil sample.

The molecular weight of the native enzyme was estimated to be approximately 122,000, with two identical subunits having molecular weights of 56,000.

The substrate characteristics are illustrated in Table 7.4.

The enzyme has a higher affinity towards peptide substrates than towards amino acid amides. The enzyme exhibites a mode of action typical of aminopeptidases (EC 3.4.11).

TABLE 7.3 Characteristics of Aminopeptidases from Different Sources

α-Aminoacyl peptide hydrolases	Type (EC 3.4.11.–)	M.W.	pH	T	Inhibitors	Refs.
Aeromonas hydrophila	extracellular zinc protease		8.0	70 °C	EDTA, 1,10-phenanthroline	[26]
Aeromonas proteolytica	extracellular	29,000	8.0	70 °C	activated by Cu^{2+} & Ni^{2+}	[31, 32]
Alteromonas B207	periplasmic	450,000 tetrameric			Zn^{2+}	[43]
Bacillus subtilis	L-aminopeptid. D-aminopeptid.	220,000 ± 20,000				[33, 34]
Bovine lens aminopeptidase	exopeptidase	326,000 hexameric	8.0			[36]
Bovine leucocyte aminopeptidase						[40]
Brevibacterium A-4						[9]
Clostridium histolyticum ($AlaNH_2$)	extracellular	hexameric	8.6	40 °C	activated by Mn^{2+} & Co^{2+}	[35]
Peptidase N from *E. coli* K-12						[41]
Lactobacillus acidophilus R-26	exopeptidase	156,000 4 mol Zn^{2+} tetrameric			Zn^{2+}, Co^{2+}, Ni^{2+}, Mn^{2+}	[45]
Lactobacillus lactis	surface-bound	78,000–81,000	6.2–7.2	47.5 °C		[37]
Lyophyllum cinerascens		150,000 tetrameric	5.7–7.7	40 °C		[39]
Ochrobactrum anthropi D-aminopeptidase	exopeptidase	122,000 dimeric			thiolpeptides	[47–49]
Porcine liver peptidase						[40]
Rat kidney Ala aminopeptidases		210,000 dimeric				[44]
Streptococcus cremoris	membrane-bound	130,000 trimeric				[46]
Thermus aquaticus YT-1		108,000 dimeric	8.5–9.0	80 °C		[38]
Tissue leucine aminopeptidase		40,000 tetrameric				[36]

TABLE 7.4 Substrate Specificity of Aminopeptidase from *Achromobacter* sp. SCRC C1-38

Amide	Relative velocity (%)	K_m (mM)
D-Alanine amide	100	0.65
Glycine amide	44	22.3
D-α-Aminobutyric acid amide	30	18.3
D-Serine amide	29	27.0
D-Ala-3-aminopentane amide	32	2.27
D-Ala-*p*-nitroanilide	96	0.51
D-Alanine methylester	75	–
D-Alanine dimer	21	10.2
D-Alanine trimer	92	0.57
D-Alanine tetramer	89	0.32
D-Alanylglycine	95	0.98
D-Alanylglycylglycine	45	0.37
D-Alanyl-L-alanyl-L-alanine	100	0.65

The immobilized enzyme was also active in water-saturated organic solvents (benzene, butyl acetate, 1,1,1-trichloroethane). D-Alanine 3-aminopentane amide was quantitatively synthesized from D-alanine methylester and 3-aminopentane in one hour. The catalytic activity, K_{cat}, was as high as 7700 min^{-1}, which is several hundred to ten thousand times greater than those of α-chymotrypsin- or subtilisin-catalyzed peptide synthesis. The reactivity of L-alanine amide was less than 0.01% that of D-alanine amide.

Another D-stereospecific amino acid amidase has been partially purified from *Ochrobactrum anthropi* SCRC SV3 [49], which was isolated and selected from soil. The substrate specificity is shown in Table 7.5.

TABLE 7.5 Substrate Specificity of D-Amino Acid Amidase from *Ochrobactrum anthropi* SCRC SV3

Substrate*	Relative activity (%)	Substrate*	Relative activity (%)
D-Phenylalanine amide	100	D-Phenylglycine amide	15
D-Tyrosine amide	97	D-Proline amide	9.7
D-Tryptophan amide	96	D-Lysine amide	2.5
D-Leucine amide	46	D-Histidine amide	1.5
D-Norleucine amide	40	D-Asparagine	1.4
D-Alanine amide	33	D-Glutamine amide	1.1
D-Methionine amide	28	D-Phenylalanine methylester	192
D-Norvaline amide	15		

* D-Threonine amide, glycine amide, D-glutamine, D-asparagine amide, D-α-aminobutyric acid amide, D-isoglutamine, D-arginine amide, D-valine amide, and D-isoleucine amide were hydrolyzed at 0.1% to 0.6% the rate for D-phenylalanine amide.

The following compounds were not substrates for the enzyme: 2-phenylacetamide, acetamide, propionamide, Boc-D-alanine amide, D-isoasparagine, D-alanine-*p*-nitroanilide, D-alanylglycine, D-alanyl-D-alanine, D-alanyl-D-alanyl-D-alanine, D-alanyl-L-alanine, D-alanyl-L-alanyl-L-alanine, L-alanyl-D-alanine, L-phenylalanine amide, L-tyrosine amide, L-tryptophan amide, L-leucine amide, L-alanine amide, L-methionine amide, L-phenylglycine amide, L-proline amide, L-lysine amide, L-histidine amide, L-alanine-*p*-nitroanilide, L-alanyl-L-alanine, and L-phenylalanine methylester.

7.3 USE OF AMINOPEPTIDASES IN THE PRODUCTION OF α-*H*-AMINO ACIDS

7.3.1 Introduction

Amino acids are an important class of chiral compounds belonging to the so-called *chiral pool*. They also play an important role in biology, biochemistry, and chemistry. Amino acids are the building blocks of proteins and they play an essential role in the regulation of the metabolism of living organisms.

Commercial production of amino acids dates back to 1908, with the production of monosodium glutamate, used as a seasoning agent. Since 1957, microbial processes have been successfully used for the production of natural proteinogenic amino acids like lysine, proline, threonine, phenylalanine, etc. In general, when chemoenzymatic methods are compared with microbial processes from a commercial standpoint, the latter are highly competitive for the large-scale production (thousands of tons per year) of naturally occurring proteinogenic L-amino acids, whilst for the production of natural amino acids or synthetic D- or L-amino acids on a small scale, chemoenzymatic methods (in addition to all-chemical procedures) are the method of choice.

Established chemoenzymatic kinetic resolution processes are based on the use of aminoacylases and hydantoinases. A novel and recently commercialized kinetic resolution process based on the use of an L-specific aminopeptidase for the stereospecific hydrolysis of DL-amino acid amides is described in this section.

7.3.2 Use of an Aminopeptidase from *Pseudomonas putida*

A universally applicable process for the production of optically pure L- as well as D-amino acids was commercialized by DSM in 1988. Pivotal in this process is the enantioselective hydrolysis of DL-amino acid amides. The stable substrates, DL-amino acid amides, are efficiently prepared under mild reaction conditions from simple, readily available raw materials (aldehydes) (Figure 7.1).

Aldehydes of a broad range of types are readily converted to the aminonitriles under Strecker reaction conditions. Hydrolysis of the aminonitriles under alkaline conditions in the presence of a catalytic amount of ketone affords in high yield the DL-amino acid amides [51]. The kinetic resolution step [52, 53] is completed with permeabilized whole cells of *Pseudomonas putida* ATCC 12633, with a nearly 100% enantioselectivity for the hydrolysis of only the L-amino acid amide. The biocatalyst accepts a broad structural variety of amino acid amides, as is illustrated in Figure 7.2 [54–58].

Not only aliphatic amino acids like valine and leucine, but also (substituted) aromatic amino acids, heterosubstituted amino acids (methionine, homomethionine [59], and thienylglycine) and other functionalized α-*H*-amino acids can be obtained in both optically pure L- and D- forms. No enzymic side effects are observed and substrate concentrations up to 20% by weight can be used without adverse effects on the enzyme activity. The biocatalyst is active in a broad pH range (8.0–10.5) and can be used in soluble form in a batchwise process, so poorly soluble amino acids can be resolved and obtained without technical difficulty. The presence of an α-hydrogen atom in the substrate is an essential structural feature for the enzymatic activity of the aminopeptidase of *P. putida*. Its enantioselectivity for α-*H*-amino acid amides is absolute, yet it accepts substrates having a wide range of structures and functionalities. This structural variety is emphasized and nicely illustrated in Figure 7.3.

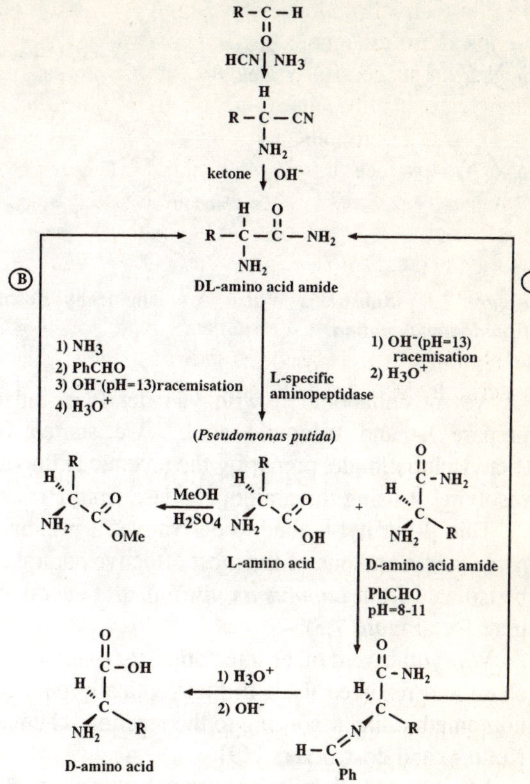

Figure 7.1. DSM Process for the Chemoenzymatic Synthesis of L- and D-α-*H*-Amino Acids.

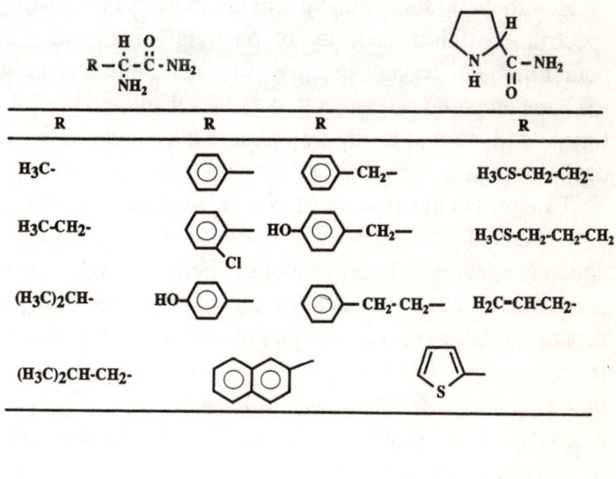

Figure 7.2. Substrate Specificity of the Aminopeptidase from *Pseudomonas putida* ATCC 12633 (selected examples).

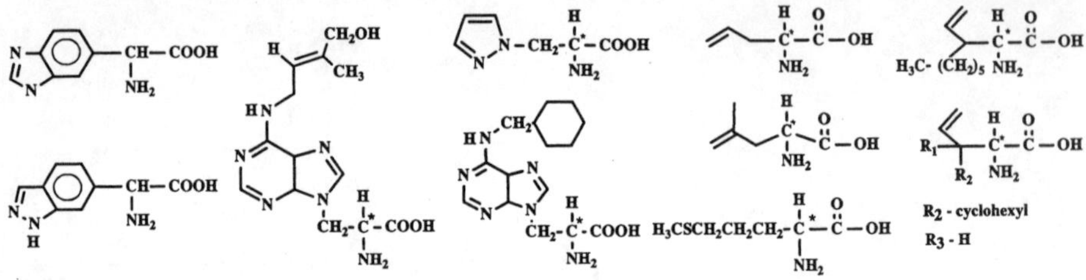

Figure 7.3. Substrates Which Are Kinetically Resolved Using the Aminopeptidase from *Pseudomonas putida*.

We, in collaboration with van der Plas and co-workers [60], were the first to prepare L- and D-lupinic acid. We started from *N*-(4-acetoxy-3-methyl-*E*-but-2-enyl)phthalimide, preparing the racemic amino acid amide in four steps, and readily resolving it using the aminopeptidase from *Pseudomonas putida* (Figure 7.4).

This β-purinyl-L-alanine derivative, a principal metabolite of the phytohormone *trans*-zeatin and one of the most effective natural stimulants of plant cell division, can be isolated from *Lupinus angustifolins*. Several analogs were also prepared optically pure (see Figure 7.3).

We synthesized other interesting α-*H*-amino acid amides functionalized in the side chain and resolved them in their optically pure form. For example, we synthesized homomethionine according to the reaction scheme in Figure 7.5, in collaboration with Kellogg and co-workers [59].

We obtained allylglycines through alkylation, starting from *N*-acetylmalonic diesters (Figure 7.6).

Heterocyclic substituted phenylglycines are interesting because of the known broad-spectrum antibiotic activity of phenylglycine- and *p*-hydroxyphenylglycine-derived penicillins and cephalosporins. In collaboration with Vanderhaeghe and co-workers [61], we prepared several heterocyclic substituted phenylglycines in their optically pure form, as illustrated in Figure 7.7, and used them in the synthesis of the corresponding cephalosporins.

To separate the mixture of D-α-*H*-amino acid amide and the L-α-*H*-amino acid, one equivalent of benzaldehyde (with respect to the D-amino acid amide) is added to the aqueous enzymatic hydrolysate; the benzaldehyde reacts selectively with the D-amino acid amide to form a Schiff base, which is insoluble in water and, therefore, easily separated [62]. Acid hydrolysis results in the formation of the D-amino acid (without racemization). Alternatively the D-amino acid amide can be hydrolyzed by whole-cell preparations of *Rhodococcus erythropolis* [63]. This option is very useful for amino acids, which are highly soluble in the neutralized reaction mixture obtained after acid hydrolysis of the amide.

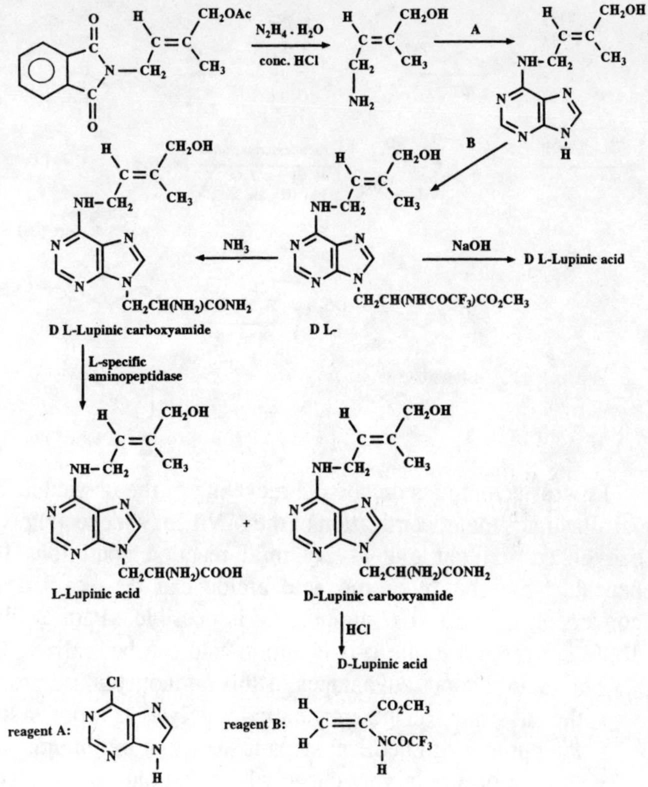

Figure 7.4. Synthesis of L(–) and D(+) Lupinic Acid.

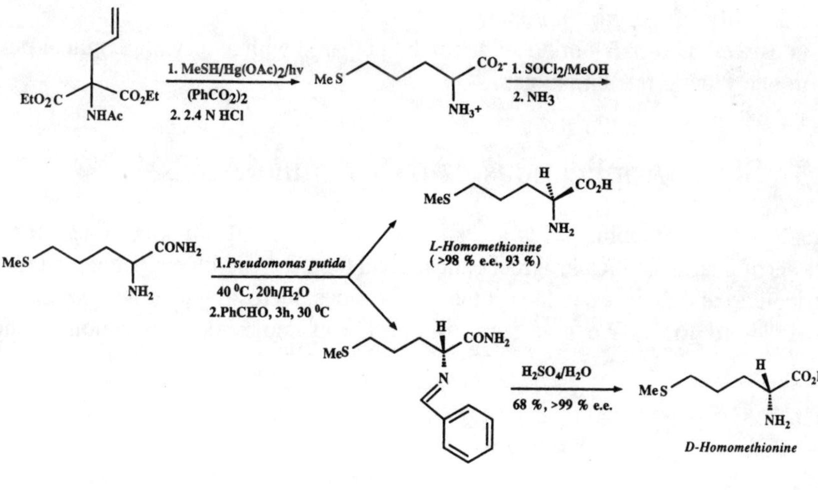

Figure 7.5. Synthesis of L- and D-Homomethionine.

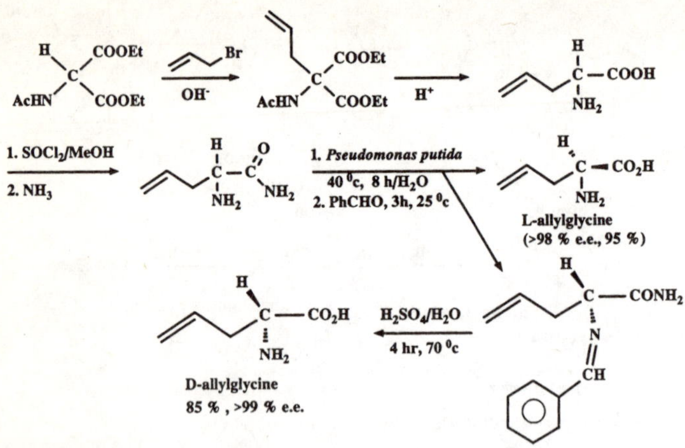

Figure 7.6. Synthesis of L- and D-Allylglycine.

Process economics dictate the recycling of the unwanted isomer. Path A in Figure 7.1 illustrates that racemization of the D-*N*-benzylidene amino acide amide is facile and can be carried out under very mild reaction conditions [64]. After removal of benzaldehyde the DL-amino acid amide can be recycled, and in principle 100% conversion to the L-α-*H*-amino acid is possible. Path B illustrates that in this way 100% conversion to the D-α-*H*-amino acid can be realized.

The main distinct advantages of this aminopeptidase process are [52, 53, 55–59]:

- the substrate for the enzymatic hydrolysis is a precursor of the amino acid;
- the number of chemical steps is kept to a minimum;
- the use of a relatively cheap whole-cell biocatalyst contributes to the economic feasibility of the procedure; and
- L- as well as D-α-*H*-amino acids can be prepared with a very high optical purity in one kinetic resolution step.

7.3.3 Some Applications of α-*H*-Amino Acids

Optically pure α-*H*-amino acids are known versatile chiral building blocks for the synthesis of pharmaceuticals, agrochemicals, and food/feed additives. Chemical routes to derivatives and their use in the synthesis of various interesting pharmaceuticals have been reviewed [65]. We give some illustrative examples in this section, without

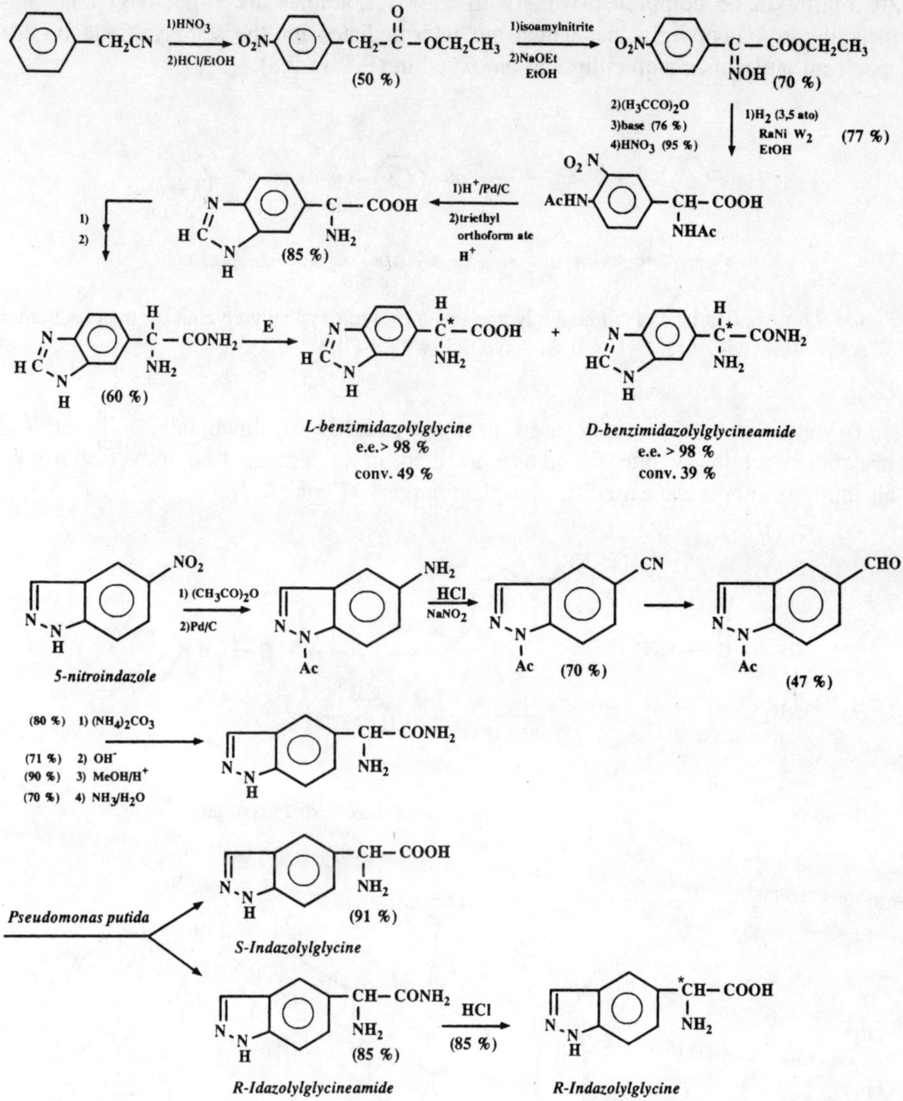

Figure 7.7. Synthesis of (A) Benzimidazolylglycine- and (B) Indazolylglycine-Derived Cephalosporins.

attempting to be comprehensive. Well-known examples are D-phenylglycine and D-parahydroxyphenylglycine, which are intermediates for the semisynthetic broad-spectrum antibiotics ampicillin and amoxycillin (Figure 7.8).

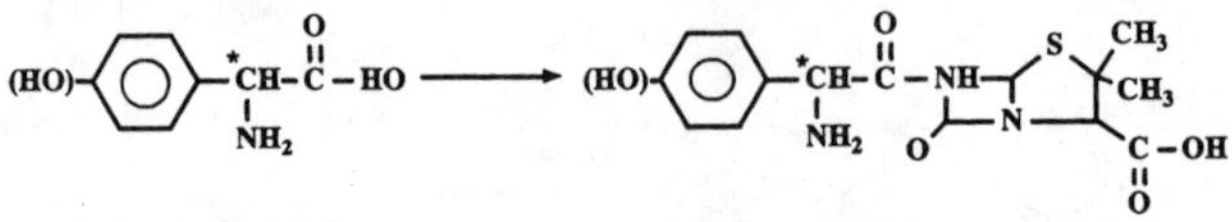

D-(p-hydroxy)phenylglycine — Semisynthetic penicillin Ampicillin (Amoxycillin)

Figure 7.8. Applications of D-Phenylglycine and D-*p*-Hydroxyphenylglycine for the Production of Ampicillin (without *p*-OH) and Amoxycillin (with *p*-OH).

D-Valine is commercially used in the synthesis of fluvalinate, a pyrethroid insecticide, while L-valine is used as a precursor in the fermentation of cyclosporin A, an immune suppressant used in transplant surgery (Figure 7.9).

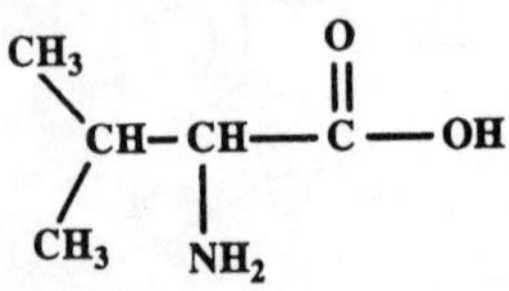

D-valine

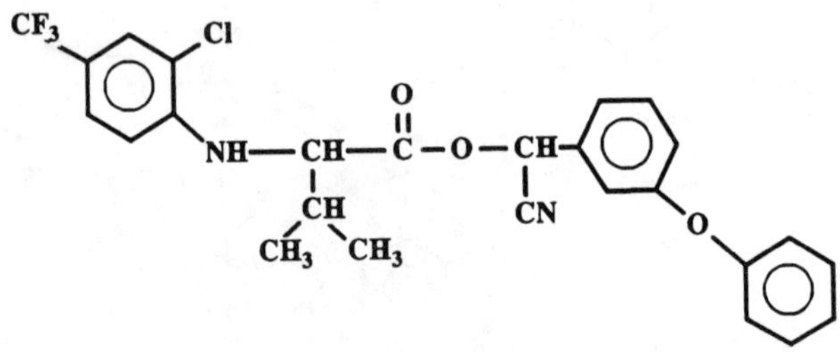

Pyrethroid insecticide Fluvalinate

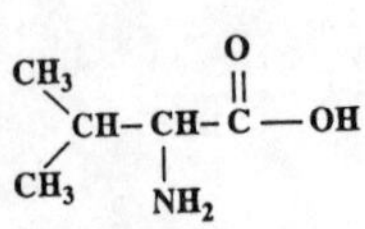

L-valine

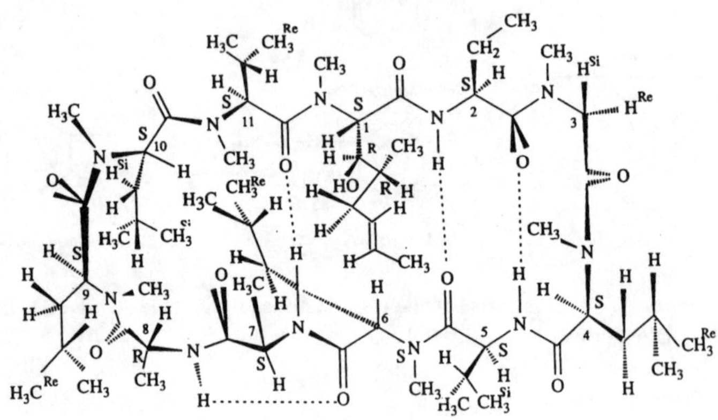

Figure 7.9. Applications of D- and L-Valine.

L-Phenylalanine is used in the commercial synthesis of Aspartame™, an artificial low-calorie sweetener 200 times sweeter than sucrose (Figure 7.10).

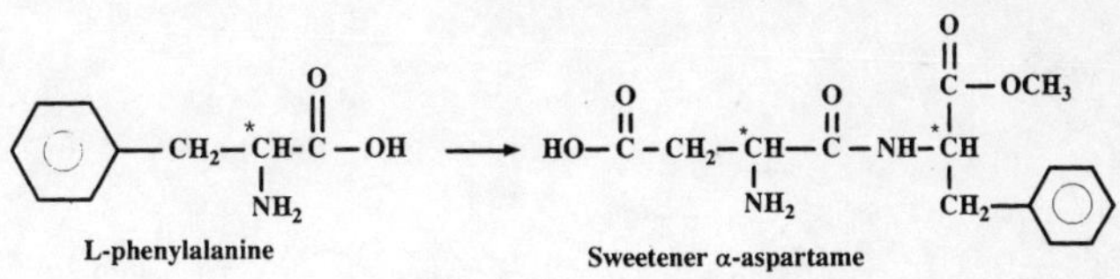

Figure 7.10. Application of L-Phenylalanine.

L-Homophenylalanine and also D-homophenylalanine are versatile intermediates for the newly developed class of ACE inhibitors like Enalapril, Benzapril, etc. (Figures 7.11 and 7.12).

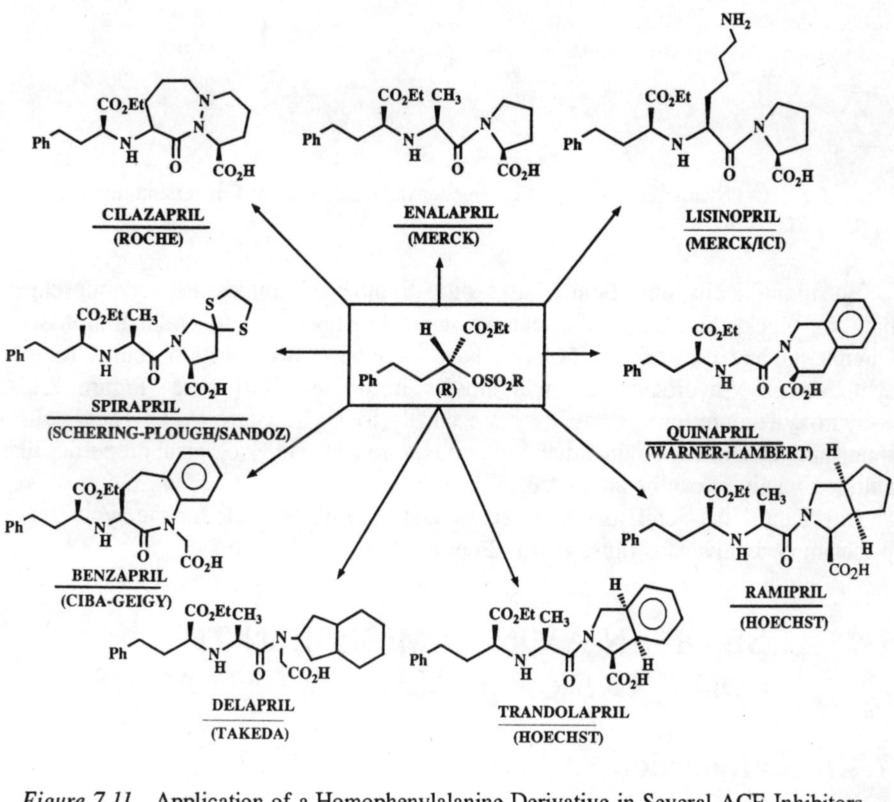

Figure 7.11. Application of a Homophenylalanine Derivative in Several ACE Inhibitors.

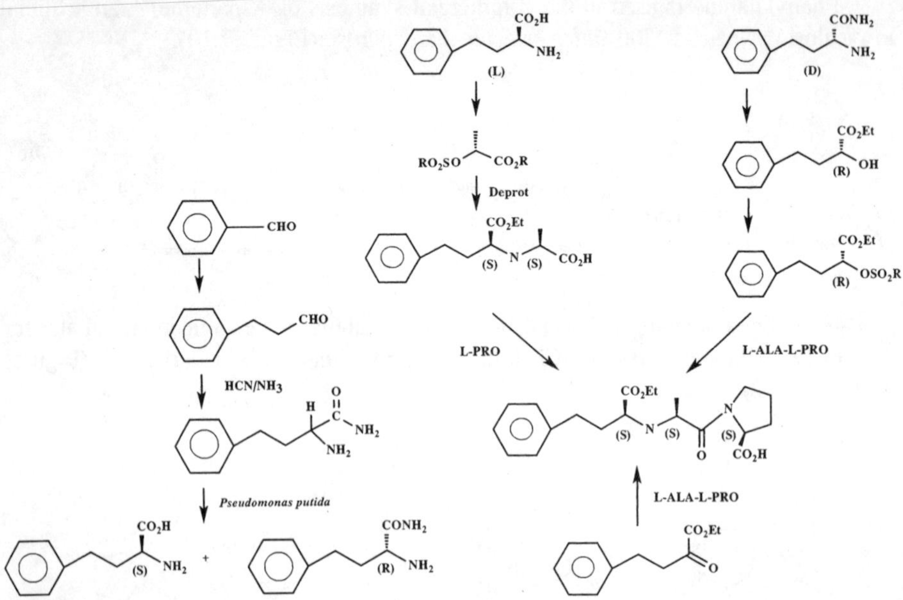

Figure 7.12. (A) Synthesis of L- and D-Homophenylalanine, and (B) Conversion into the ACE Inhibitor Enalapril.

Enantiomerically pure Schiff bases of D-amino acid amides are versatile chiral building blocks because of the reactive imine functionality. In collaboration with Ottenheijm and co-workers [66], we have developed two reaction routes for the synthesis of *N*-hydroxyamino acid amides from the Schiff base (Figure 7.13). *N*-hydroxy-α-amino acid derivatives are widely found in nature [67]. They can be found amongst others as constituents of peptides, to which physiological properties like antibiotic activity can be attributed [68].

In principle the Schiff bases are also versatile synthetic tools for the synthesis of β-lactam derivatives, as illustrated in Figure 7.14.

7.4 USE OF AN AMINO AMIDASE IN THE PRODUCTION OF α-ALKYLAMINO ACIDS

7.4.1 Introduction

Several routes to enantiomerically pure α-alkyl-α-amino acids have been elaborated during the past few years [69]. Schöllkopf and collaborators [70] have devised a versatile and useful method for preparing a large variety of amino acids, based on

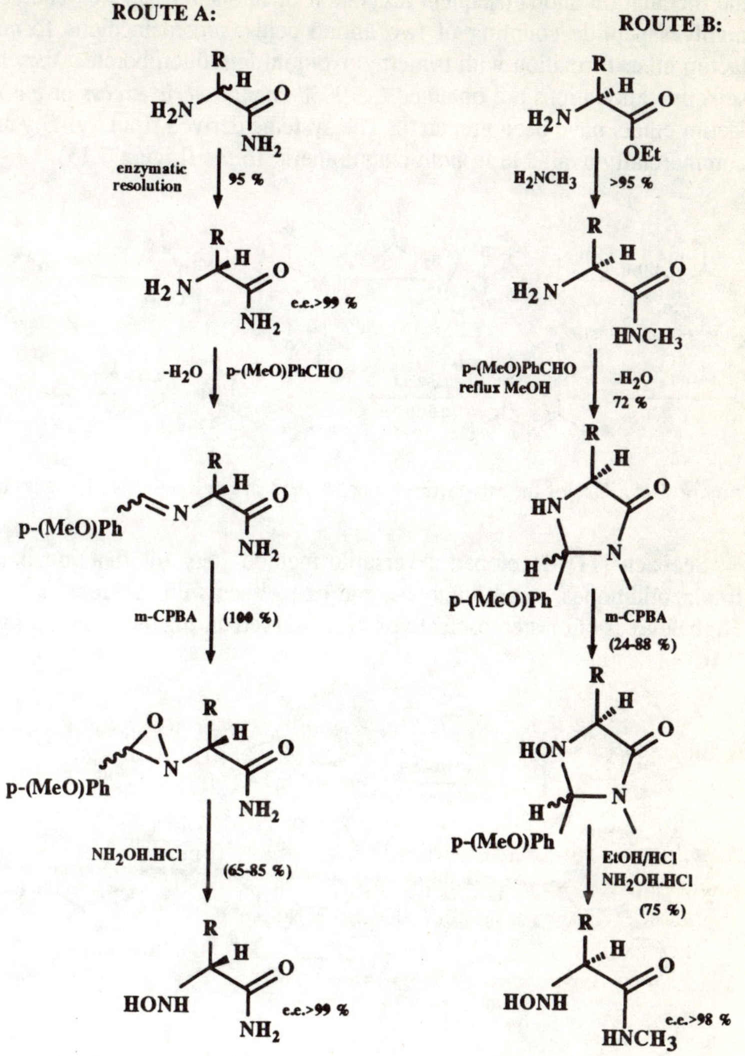

Figure 7.13. Oxidative Synthetic Routes to Optically Active *N*-Hydroxy-α-amino Acid Amides.

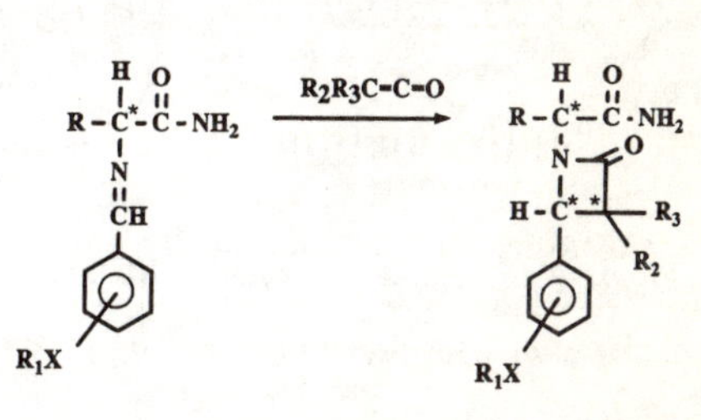

Figure 7.14. Synthesis of Lactams using Optically Pure Schiff Bases of Amino Acid Amides.

the metallation and subsequent alkylation of bis-lactim ethers. The general protocol involves peptide coupling of two amino acids, piperazinedione formation and bis-lactim ether formation with trimethyloxonium tetrafluoroborate. Very high diastereoselective alkylations are obtained (> 90% *enantiomeric excess* or *e.e.*). Several bis-lactim ethers have been prepared. The systems derived from Val-Gly and Val-Ala are commercially available in both enantiomeric forms (Figure 7.15).

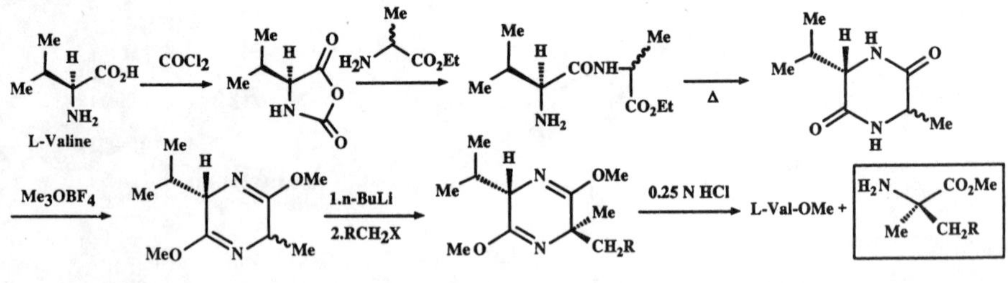

Figure 7.15. Use of the Bis-lactim Ether Method in the Synthesis of α-Alkyl-α-amino Acids.

Seebach [71] developed a versatile methodology for functionalizing enolates of imidazolidinones, oxazolidinones, and oxazolines with self-reproduction of chirality. High degrees of diastereoselectivity are observed during the enolate alkylations (Figure 7.16).

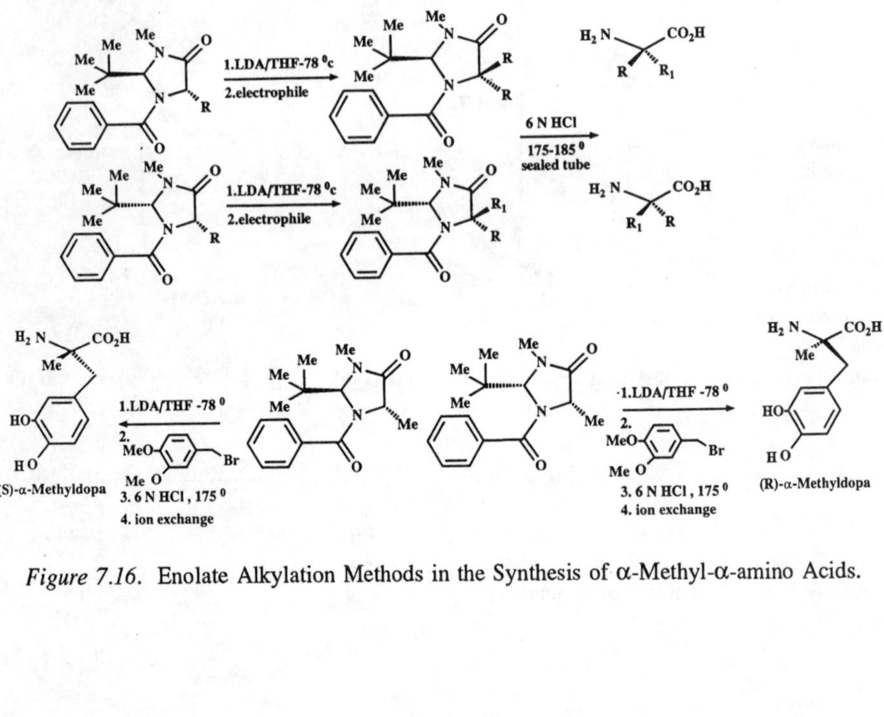

Figure 7.16. Enolate Alkylation Methods in the Synthesis of α-Methyl-α-amino Acids.

More chemical methods have been reported recently. For example, Ojima *et al.* [72] developed a clever and potentially useful way to make optically active amino acids and peptide derivatives by stereoselective alkylation of β-lactams (Figure 7.17).

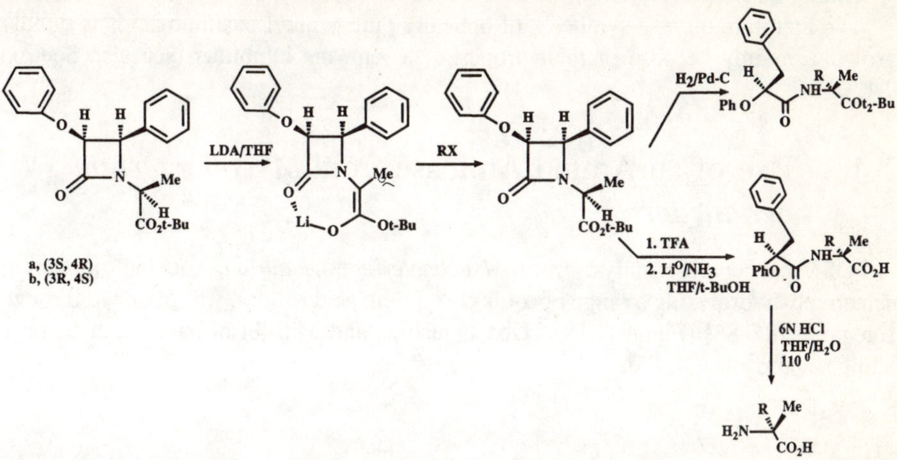

Figure 7.17. Stereoselective Alkylation of β-Lactams and Further Conversion into α-Methyl-α-amino Acids.

For an excellent overview of these methods, see Chapter 1 of reference [73].

Although most of these syntheses are scientifically very elegant they are far too expensive and difficult to scale up to produce α-alkyl-α-amino acids in bulk. Major problems arise in performing alkylations at very low temperatures (generally –70 °C) on a large scale.

Some chemoenzymatic approaches have been described also. Watanabe *et al.* [74] at Kanegafuchi described the stereoselective cyclization of *N*-carbamoyl-α-alkylated α-amino acids to the corresponding D-hydantoins using cultivated microorganisms belonging to the genera *Aerobacter*, *Bacillus*, *Corynebacterium*, etc. The D- and L-α-alkyl-α-amino acids can be prepared from the D-hydantoins or the remaining L-*N*-carbamoyl-α-alkylated amino acids, respectively.

Björkling *et al.* [75] reported using esterases on prochiral diesters to obtain α-methyl-α-amino acids. Only in the very special case of 3,4-dimethoxyphenylalanine, was a relatively high enantiomeric excess obtained (93% e.e.).

Anantharamaiah *et al.* [76] resolved some α-methyl-α-amino acid esters using α-chymotrypsin at pH 5.0, but the amino acids obtained had low optical purity.

α-Nitropropanoate esters were resolved by kinetic resolution procedures via chymotrypsin catalysis, as reported by Wong and co-workers [77]. Products with a relatively high optical purity were obtained (> 95% e.e.).

In principle, acylases (benzyl penicillinase, carboxypeptidase, hog kidney acylase) stereospecifically hydrolyze *N*-phenylacetyl-, *N*-trifluoracetyl-, and *N*-acetyl-α-methyl-α-amino acids. Most of these combinations, however, suffer from very low substrate specificity and relatively long reaction times [78–80].

The literature on new syntheses of optically pure α-alkyl-α-amino acids is rapidly growing, mainly because of their properties as enzyme inhibitors (see also Section 7.1.3).

7.4.2 Use of an Amino Amidase from *Mycobacterium neoaurum*

At DSM a new biocatalyst from *Mycobacterium neoaurum*, capable of stereoselectively hydrolyzing a range of α-alkyl-α-amino acid amides, was obtained through screening [81–83] (Figure 7.18). Ube Industries and Mitsubishi have reported on a similar approach [84, 85].

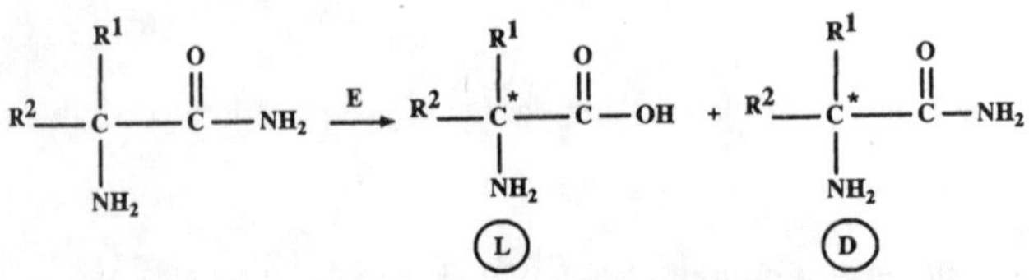

Figure 7.18. Kinetic Resolution Process for α-Alkyl-α-amino Acids.

The basis of this process is essentially the same as that discussed for α-*H*-amino acids (Section 7.2.2). The aminonitrile obtained after reaction of the ketone with hydrogen cyanide and ammonia is converted to the amino acid amide by hydrogen peroxide in ammonia [82]. Enzymatic hydrolysis affords L-α-alkyl-α-amino acid and D-α-alkyl-α-amino acid amide. The unwanted isomer can be recycled by converting the amino acid into the starting material (Figure 7.19).

The use of a relatively cheap whole-cell biocatalyst contributes to the economic feasibility of the procedure. L- as well as D-amino acids can be prepared in one enzymatic resolution with a very high optical purity (approximately 100% e.e.). The biocatalyst is versatile [83] because it is capable of stereoselectively hydrolyzing a broad range of structurally different α-alkyl-α-amino acid amides (Figure 7.20).

Some of the characteristics of the process are summarized below:

- Both L- and D-α-alkyl-α-amino acids can be produced.
- Permeabilized whole cells of *M. neoaurum* ATCC 25795 as crude enzyme preparations can be used.
- Very high stereoselectivity (> 98% e.e.) and broad substrate specificity can be obtained.
- The enzyme is active in a broad pH range (Figure 7.21), between 5.0 and 11.0.

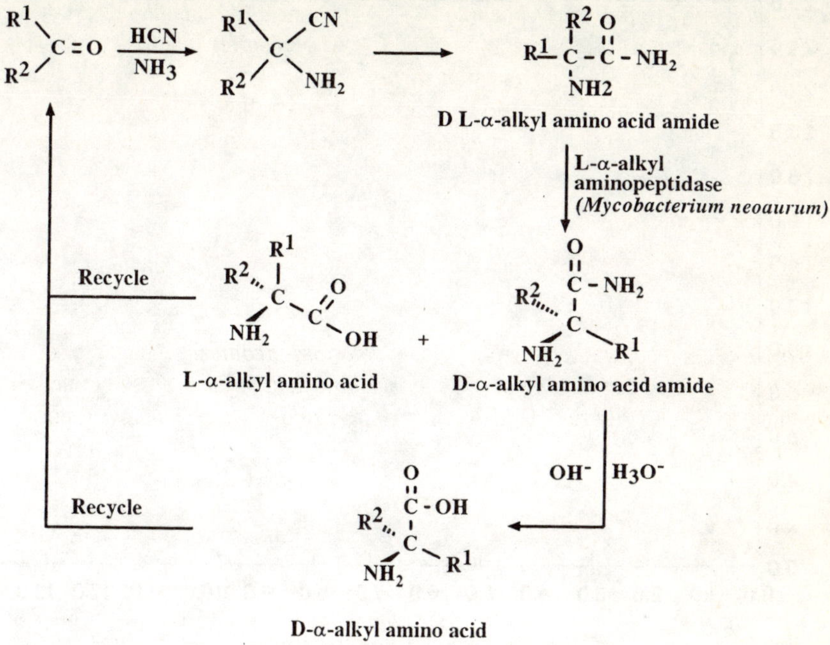

Figure 7.19. DSM Production Method for Optically Pure α-Alkyl-α-amino Acids using an Amino Amidase from *Mycobacterium neoaurum.*

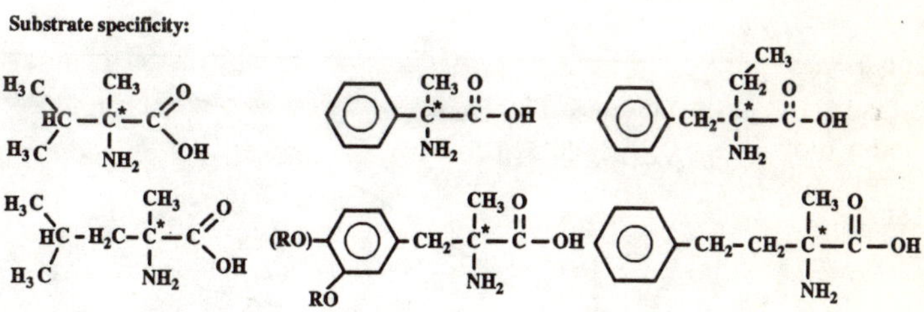

Figure 7.20. Substrate Specificity of the Amino Amidase from *Mycobacterium neoaurum.*

The kinetics of the enzymatic hydrolysis were studied with DL-α-methylleucine amide as the model substrate. In this case no substrate inhibition up to 10% (w/w) was observed (Figure 7.22).

From the Eadie-Hofstee plot, a V_{max} value of 30.2 mol·g^{-1}·min^{-1} (cell dry weight) and a K_m of 17.8 mM were determined (Figure 7.23). Other characteristics are now under investigation. The enzyme purification and physiology of this system (*M. neoaurum*) are currently being undertaken and will be published elsewhere [86].

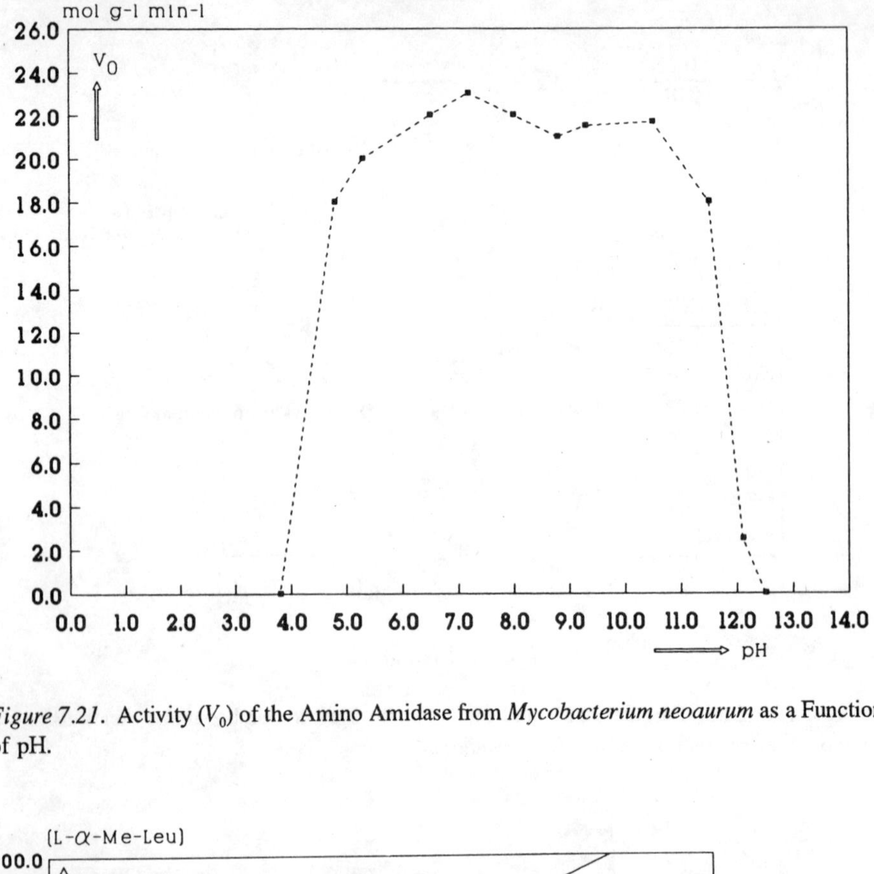

Figure 7.21. Activity (V_0) of the Amino Amidase from *Mycobacterium neoaurum* as a Function of pH.

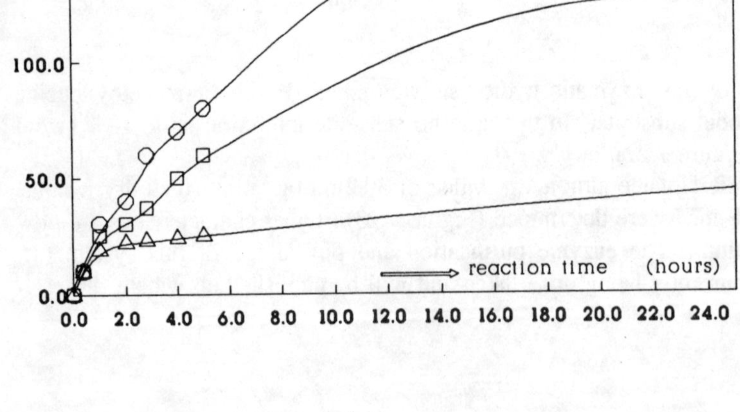

Figure 7.22. Activity of the Amino Amidase from *Mycobacterium neoaurum* as a Function of Different Substrate Concentrations. The enzymatic hydrolysis of DL-methylleucine amide is shown.

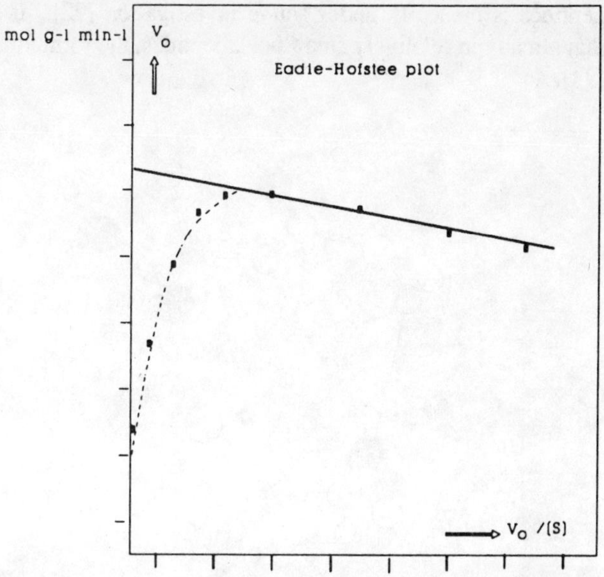

Figure 7.23. Eadie-Hofstee Plot of the Kinetics of the Enzymatic Hydrolysis of DL-α-Methylleucine Amide. The enzymatic hydrolysis can be described according to the Michaelis-Menten model. The Eadie-Hofstee plot gives a value for V_{max} of 30.2 mol·g^{-1}·min^{-1} (cell dry weight basis) and for K_m of 17.8 mM.

7.4.3 Some Applications of α-Alkyl-α-amino Acids

The best known application of the use of an α-alkyl-α-amino acid in the pharmaceutical industry is L-α-methyl-3,4-dihydroxyphenylalanine (L-MeDOPA, Aldomet™, [Figure 7.24]), which is produced by Merck, Sharp & Dohme [87] on a large scale and is used as such because of its antihypertensive activity. The use of nonproteinogenic amino acids having predetermined properties for the synthesis of new enzymes inhibitors, hormones, synthetic immunostimulants, drugs, and countless other valuable materials has been recognized and represents emerging exciting and creative fields [88]. The benefits of these endeavors will undoubtedly reach far into the future.

Figure 7.24. L-α-Methyl-DOPA (L-α-methyltyrosine), an Antihypertensive Drug.

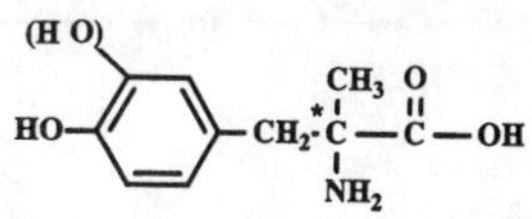

More recently, medicinal chemists became interested in bioactive peptides containing α-alkyl-α-amino acids, since the predetermined conformations of the molecules slow enzymatic processes down dramatically [89]. Nowadays, many α-alkyl-α-amino acids are being found in nature. For example L-isovaline is found in peptaibol antibiotics [90, 91]. Their influence on the conformational behavior of peptides is presently under active investigation [92]. It is remarkable (Figure 7.25) that already in relatively small pentapeptides, the formation of 3.10 helices is observed [93].

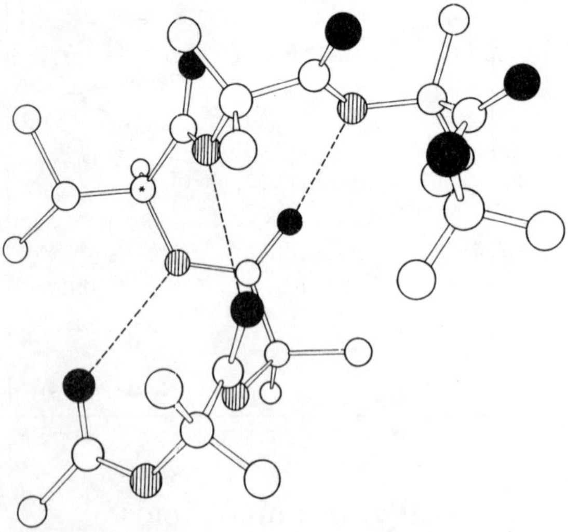

Figure 7.25. The Pentapeptide Ac-$(Aib)_2$-(α-Me)Val$(Aib)_2$O*t*Bu (CH_3CN solvate), Showing a 3.10 Helix [93].

In agrochemicals the potential benefit in using optically pure compounds is recognizable too [94]. For example, the broad spectrum herbicide Arsenal™ (Figure 7.26), on the market as the racemate, was introduced by American Cyanamid. From the patent literature [95] it can be inferred that the corresponding enantiomerically pure (D)-(2-imidazolin-2-yl) pyridine compound, containing D-α-methylvaline amide, is much more active than the racemic form in some applications.

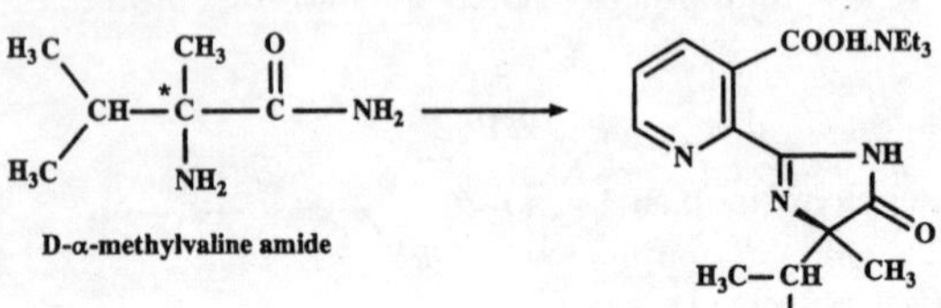

Figure 7.26. Potential Application of D-α-Methylvaline Amide in a Herbicidal Compound.

7.5 CONCLUSION AND FUTURE PROSPECTS

Optically pure proteinogenic and nonproteinogenic amino acids are increasingly interesting because they are versatile synthons for the synthesis of pharmaceuticals, agricultural products, and products for the food industry and materials sciences. The application of biocatalytic methods in the industrial synthesis of these compounds is still steadily expanding. Most of the methods are based on using "simple" non-cofactor–requiring enzymes, like the hydrolases (for example, aminopeptidase, hydantoinases, acylases, etc.).

In general, from a commercial viewpoint it can be stated that a microbial process (fermentation) is the method of choice for the production of proteinogenic amino acids on a large scale. For the production of synthetic D- or L-amino acids, chemoenzymatic methods are the method of choice, in addition to all-chemical procedures like diastereomeric resolution. In the future an ever-increasing number of enantiomerically pure (synthetic) amino acids will be produced both by the application of established procedures and by newly developed chemoenzymatic methods. Highly promising developments are the discovery of new enzyme activities, like D-specific aminopeptidases, which, if used in combination with the appropriate racemases, will dramatically improve the kinetic resolution processes in use today.

7.6 ACKNOWLEDGMENTS

We gratefully acknowledge the exceptionally valuable assistance of Mrs. C. J. M. Hendrix and Mr. W. Kortenoeven, in typing the manuscript and preparing the drawings, respectively.

7.7 REFERENCES

1. McCaman, M. T., Villargo, M. R., *Arch. Biochem. Biophys. 213*, 384 (1982).
2. Breddam, K., *Carlsberg. Res. Comm. 49*, 535 (1984).
3. Wagner, G. W., Tavianini, M. A., Herrmann, K. M., Dixon, J. E., *Biochemistry 20*, 3884 (1981).
4. Wolff, J. B., Resnik, R. A., *Biochim. Biophys. Acta 73*, 588 (1963).
5. Patterson, E. K., Hsiao, S.-H., Keppel, A., *J. Biol. Chem. 238*, 3611 (1963).
6. Ajabnoor, M. A., Ray, L. E., Wagner, F. W., *Arch. Biochem. Biophys. 202*, 540 (1980).
7. Marks, N., Datta, R. K., Lajtha, A., *J. Biol. Chem. 243*, 2882 (1968).
8. Mahadevan, S., Tappel, A. L., *J. Biol. Chem. 242*, 2369 (1967).
9. Maestracci, M., Bui, K., Thiéry, A., Arnaud, A., Galzy, P. In *Advances in Biochemical Engineering/Biotechnology*; Fiechter, A., Ed.; Springer Verlag, Heidelberg, 1988; Vol. 36, pp 69–111.
10. Delance, R. J., Smith, E. L. In *The Enzymes*; Boyer, P. D., Ed.; 3 ed., Academic Press, New York, London, 1971; Vol. 3, pp 81–118.
11. Hill, R. L., Smith, E. L., *J. Biol. Chem. 235*, 2332 (1960).
12. Brown, J. L., *J. Biol. Chem. 248*, 409 (1973).
13. Chapuis, R., Zuber, H., *Methods Enzymol. 19*, 552 (1970).

14. Foissy, H., *Milchwissenschaft 33*, 221 (1978); Foissy, H., *Z. Lebensm.—Unters. Forsch. 166*, 164 (1978).
15. Ivanov, N. M., Vaganova, I. I., Strognin, A. Y., Stepanov, V. M., *Biochemistry (USSR) 42*, 652 (1977).
16. Koelsch, R., Hanson, H., *Acta Biol. Med. Ger. 26*, 1117 (1971).
17. Matsuda, T. M., Hayashi, R., Hata, T., *Agric. Biol. Chem. 39*, 499 (1975); Nowak, J., Tsai, H., *Can. J. Microbiol. 34*, 118 (1988).
18. Mimamiura, N., Matsumura, Y., Yamamoto, T., Fukumoto, J., *Agric. Biol. Chem. 33*, 653 (1969).
19. Morihara, K., Tsuzuki, H., *Ann. Rep. Shionogi Res. Lab. 13*, 91 (1963).
20. Rocari, G., Stoll, E., Zuber, H., *Methods Enzymol. 45B*, 522 (1976); Gamal, R. F., Attia, R. M., *Zentralbl. Bakteriol., Parasitenkd., Infektionskrankh. Hyg., Abt. 2, Naturwiss.: Mikrobiol. Landswirtsch., Technol. Umweltschutzes 134*, 733 (1979).
21. Ruffin, P., Vanbrussel, E., Biguet, J., Biserle, G., *Biochemie 61*, 495 (1979).
22. Sugiura, M., Ishikawa, M., Sasaki, M., *Chem. Pharm. Bull. 26*, 3103 (1978).
23. Uwajima, T., Yoshikawa, M., Osamu, T., *Agric. Biol. Chem. 37*, 2727 (1973).
24. Wagner, F. N., Chung, A., Ray, L. E., *Can. J. Microbiol. 18*, 1883 (1972).
25. El-Soda, M., Desmazeaud, M., Bergere, J. L., *J. Dairy Res. 45*, 445 (1978).
26. Pasare, A. C., Venugopal, L. N. F., *Agric. Biol. Chem. 50*, 1743 (1986).
27. de Jong, W. W., Bloemendal, H., *J. Biol. Chem. 257*, 7077 (1982).
28. Jurnak, F., Rich, A., van Loon-Klaassen, L. A. H., Bloemendal, H., Taylor, A., Carpenter, F. H., *J. Mol. Biol. 112*, 149 (1977).
29. Allen, M. P., Yamada, A. H., Carpenter, F. H., *Biochemistry 22*, 3778 (1983).
30. Lin, S. H., van Wart, H. E., *Biochemistry 21*, 5528 (1982).
31. Prescott, J. M., Wilkes, S. H., *Arch. Biochem. Biophys. 117*, 328 (1966).
32. Prescott, J. M., Wilkes, S. H., Wagner, F. W., Wilson, K. J., *J. Biol. Chem. 256*, 1756 (1971).
33. Gogshtein-Gandman, L. V., Keynan, A., Milner, Y., *J. Bacteriol. 170*, 5895 (1988).
34. Desmond, E. P., Starnes, W. L., Behal, F. J., *J. Bacteriol. 124*, 353 (1975).
35. Kessler, E., Yaron, A., *Eur. J. Biol. 63*, 271 (1976).
36. van Loon-Klaassen, L. A. H., Cuypers, T., van Westreenen, H., de Jong, W. W., Bloemendal, H., *Biochem. Biophys. Res. Comm. 95*, 334 (1980).
37. Eggimann, B., Bachman, M., *Appl. Environ. Microbiol. 40*, 876 (1980).
38. Minagawa, E., Kaminogawa, S., Matsuzawa, H., Ohta, T., Yamanuchi, K., *Agric. Biol. Chem. 52*, 1755 (1988).
39. Abdus-Satter, A. K. M., Yashimoto, T., Tsuru, D., *Arch. Biochem. Biophys. 271*, 241 (1989).
40. Tsunematsu, H., Aratani, H., Mizusaki, K., Hatanaka, Y., Kawata, S., Yamamoto, M., Makisumi, S., *Chem. Pharm. Bull. 36*, 1205 (1988).
41. McCaman, M. T., Villargo, M. R., *Arch. Biochem. Biophys. 213*, 384 (1982).
42. Cuypers, H. T., van Loon-Klaassen, L. A. H., Egberts, V. W. T. M., de Jong, W. W., Bloemendal, H., *J. Biol. Chem. 257*, 7077 (1982).
43. Lee, C. C. T., Merkel, J. R., *Biochem. Biophys. Acta 661*, 39 (1981).
44. Sansot, J. L., Philippon, C. V., Colle, A., Prevot, D., Manuel, Y., *Enzyme 35*, 18 (1986).
45. Machuga, E. J., Ives, D. H., *Biochem. Biophys. Acta 789*, 26 (1984).

46. Exter, F. A., de Veer, G. J. C. M., *Appl. Environ. Microbiol. 53*, 577 (1987).
47. Asano, Y., Nakazawa, A., Kato, Y., Kondo, K., *Angew. Chem. 28*, 450 (1989).
48. Kato, Y., Asano, Y., Nakazawa, A., Kondo, K., *Tetrahedron 45*, 5743 (1989).
49. Asano, Y., Mori, T., Hanamoto, Y., Kato, Y., Nakazawa, A., *Biochem. Biophys. Res. Comm. 162*, 470 (1989).
50. Boesten, W. H. J., Brit. Pat. 1,548,032 (1976).
51. Boesten, W. H. J., U.S. Pat. 3,971,000 (1976).
52. Kamphuis, J., Hermes, H. F. M., van Balken, J. A. M., Schoemaker, H. E., Boesten, W. H. J., Meijer, E. M. In *Amino Acids, Synthesis, Biology and Medicine*; Lubeck, G., Rosenthal, A. E., Eds.; ESCOM Science Publishers, Leiden, 1990; p 119.
53. Sheldon, R. A., Schoemaker, H. E., Kamphuis, J., Boesten, W. H. J., Meijer, E. M. In *Stereoselectivity of Pesticides*; Ariëns, E. J., van Rensen, J. J. S., Welling, W., Eds.; Elsevier, Amsterdam, 1988; pp 409–451.
54. DSM/Stamicarbon-NOVO/Nordisk, Eur. Pat. Applic. 0 307 023 (1989).
55. Kamphuis, J., Kloosterman, M., Schoemaker, H. E., Boesten, W. H. J., Meijer, E. M. In *Proc. 4th Eur. Congr. Biotechnol.*; Neijssel, O. M., Van der Meer, R. R., Luyben, K. C. A. M., Eds.; Elsevier, Amsterdam, 1987; Vol. 4, pp 331–348.
56. Meijer, E. M., Boesten, W. H. J., Schoemaker, H. E., van Balken, J. A. M. In *Biocatalysts in Organic Syntheses*; Tramper, J., van der Plas, H. C., Linko, P., Eds.; Elsevier, Amsterdam, 1985; pp 135–156.
57. Boesten, W. H. J., Dassen, B. H. N., Kerkhoffs, P. L., Roberts, M. J. A., Cals, M. J. H., Peters, P. J. H., van Balken, J. A. M., Meijer, E. M., Schoemaker, H. E. In *Enzymes as Catalysts in Organic Syntheses*; Schneider, M. P., Ed.; NATO ASI Series C, Reidel Publishing Co., Dordrecht, Boston, Lancaster, Tokyo, 1986; Vol. 178, pp 355–360.
58. Meijer, E. M., Kamphuis, J., van Balken, J. A. M., Hermes, H. F. M., van den Tweel, W. J. J., Kloosterman, M., Boesten, W. H. J., Schoemaker, H. E. In *Trends in Drug Research*; Claassen, V., Ed.; Elsevier, Amsterdam, Oxford, New York, Tokyo, 1990; pp 363–382.
59. Vriesema, B. K., ten Hoeve, W., Wijnberg, H., Kellogg, R. M., Boesten, W. H. J., Meijer, E. M., Schoemaker, H. E., *Tetrahedron Lett. 26*, 2045 (1986).
60. Shadid, B., van der Plas, H. C., Boesten, W. H. J., Kamphuis, J., Meijer, E. M., Schoemaker, H. E., *Tetrahedron 46*, 913 (1990).
61. Vanrobays, M., Vanderhaeghe, H., Boesten, W. H. J., Schoemaker, H. E., Meijer, E. M., Kamphuis, J., *J. Org. Chem.*, submitted.
62. Boesten, W. H. J., U.S. Pat. 4,172,846 (1979).
63. Cals, M. J. H., Boesten, W. H. J., Dutch Pat. 8.403.093 (1984).
64. Boesten, W. H. J., Dassen, D. H. N., Schoemaker, H. E., Dutch Pat. Applic. 8.501.093 (1985).
65. Coppola, G. M., Schuster, H. F. *Asymmetric Synthesis: Construction of Chiral Molecules Using Amino Acids*; Wiley, New York, 1987.
66. Feenstra, R. W., Stokkinggreef, E. H. M., Reichwein, A. M., Lousberg, W. B. H., Ottenheijm, H. C. J., Kamphuis, J., Boesten, W. H. J., Schoemaker, H. E., Meijer, E. M., *Tetrahedron 46*, 1745 (1990).
67. Maehr, H., *Pure Appl. Chem. 28*, 603 (1971).
68. Keller-Schierlein, W., Prelog, V., Zähner, H., *Forschr. Chem. Org. Naturstoffe 57*, 328 (1964).

69. Beck, A. K., Seebach, D., *Khimiya 42*, 142 (1988) and references cited therein.
70. Schöllkopf, U., Tolle, R., Egert, E., Nieger, M., *Liebigs Ann. Chem.* 399 (1987); Schöllkopf, U., *Top. Curr. Chem. 109*, 65 (1983).
71. Naef, R., Seebach, D., *Helv. Chim. Acta 68*, 135 (1985).
72. Ojima, I., Qui, X., *J. Am. Chem. Soc. 109*, 6537 (1987).
73. Williams, R. M. *Synthesis of Optically Active α-Amino Acids*; Pergamon Press, Oxford, 1989.
74. Kanegafuchi, Eur. Pat. Appl. 0.175.312 (1985).
75. Björkling, F., Boutelje, J., Gatenbeck, S., Hult, K., Norin, T., *Tetrahedron Lett. 40*, 4957 (1985).
76. Anantharamaiah, E. M., Roeske, L. W., *Tetrahedron Lett. 23*, 3335 (1982).
77. Lalonde, J. J., Bergbreiter, D. E., Wong, C. H., *J. Org. Chem. 53*, 2323 (1989).
78. Rossi, D., Lucente, G., Romeo, A., *Experientia 33*, 1517 (1977).
79. Turk, J., Panse, G. T., Marshall, G. R., *J. Org. Chem. 40*, 953 (1975).
80. Keller, J. W., Hamilton, B. J., *Tetrahedron Lett. 27*, 1249 (1986).
81. DSM/Stamicarbon, Eur. Pat. 0.150.854 (1984).
82. DSM/Stamicarbon, Eur. Pats. 0.231.548 & 0.236.591 & 0.232.562 (1986).
83. Kruizinga, W. H., Bolster, J., Kellogg, R. M., Kamphuis, J., Boesten, W. H. J., Meijer, E. M., Schoemaker, H. E., *J. Org. Chem. 53*, 1826 (1988).
84. Ube Industries, Ger. Pat. DE 3 217 908 (1989).
85. Mitsubishi, Jpn. Pats. 88-123398, 88-157990, 88-160599, 88-008551, 88-089304, 88-094539, 88-207437 (1988).
86. Hermes, H. F. M., Tandler, R. F., Sonke, T., van Balken, J. A. M., Meijer, E. M., in preparation.
87. Pandey, R. C., Meng., H., Cook, J. C., Rinehart, K. L., *J. Am. Chem. Soc. 99*, 5203 (1977).
88. Bollinger, F. W., *J. Med. Chem. 14*, 373 (1971); Amato, J. S., Weinstock, L. M., Karady, S., Eur. Pat. 0 169 300 (1985).
89. Fox, R. O., Richard, F. M., *Nature 300*, 225 (1982).
90. Goodman, M., *Biopolymers 24*, 137 (1985).
91. Goodman, D. R. In *Chemical Regulation of Biological Mechanisms*; Lubitt, A. G., Creighton, A. M., Eds.; The Chemical Society, London, 1982.
92. Toniolo, C., Benedetti, E., *ISI Atlas of Science: Biochemistry 1*, 225 (1988).
93. Toniolo, C., Crisma, M., Bonora, G. M., Klajc, B., Lelj, F., Grimaldi, P., Rosa, A., Polinelli, S., Boesten, W. H. J., Meijer, E. M., Schoemaker, H. E., Kamphuis, J., *Int. J. Peptide Res. 38*, 242 (1991).
94. *Farm Chem. Handbook*; Hall, F. R., Knake, E. L., McCarty, R. H., Mortvedt, J., Terry, D. L., Eds.; 75th ed., Meister Publ. Co., Willoughby, OH, 1989.
95. Am. Cyanamid Corp., Eur. Pat. 0.123.830 (1984); 0.169.300 (1985).

PART II

BIOCATALYTIC PRODUCTION OF DERIVATIVES OF AMINO ACIDS

SYNTHESIS OF PEPTIDES BY PROTEOLYTIC ENZYMES

Kiyotaka Oyama

CONTENTS

Abbreviations Used in This Chapter:

Abbreviations used are those generally used in peptide chemistry: Bz = benzoyl; Z = benzyloxycarbonyl; OBzl = benzyl ester; OBu^t = ^{t}-butyl ester; OMe = methyl ester. Unless otherwise stated, all amino acids have the L- configuration.

8.1 INTRODUCTION

Proteins are one of the most important constituents of living organisms. Since peptides can be chemically synthesized in almost infinite numbers of structural combinations of more than 20 natural amino acids, each of which has specific properties as constituents of living organsisms, we may expect to synthesize peptides having unexpected properties. For such reasons, it might be said that the chemistry of peptides is a treasure-house in the field of life science. As conventional tools for the synthesis of peptides, chemical methods have been in use for a long time. The main problems with chemical methods have been racemization occurring during synthetic reactions, and a lack of suitable protecting groups capable of being easily introduced and removed. The efforts to overcome such problems in the past 40 years could consolidate the foundation of peptide synthesis today. Each chemical synthesis method has its own advantages and disadvantages; thus, when synthesizing peptides, the synthetic strategy must be planned with special care.

Apart from the chemical methods, peptides can be synthesized by the reverse reaction of hydrolysis of a peptide bond by proteolytic enzymes. Mohr and Strohschein reported in 1909 that the following reaction (eq 1) can take place in the presence of papain in high yield [1]:

$$\textbf{Bz-Leu-OH + H-Leu-NH}_2 \rightleftharpoons \textbf{Bz-Leu-Leu-NH}_2 \qquad (1)$$

Elucidation of the mechanism of protein synthesis in living systems is very important, so Bergmann and his co-workers actively studied enzymatic peptide synthesis in the late 1930s to the early 1940s chiefly from the mechanistic point of view, with the hypothesis that *in vivo* synthesis of proteins might be achieved by the action of proteolytic enzymes [2]. However, the interest in enzymatic peptide synthesis waned rapidly when the mechanism was clarified by the striking progress of molecular biology in the 1950s.

In recent years, the advantages of the enzymatic peptide synthesis have been drawing considerable attention as interest in biotechnology in general increases. In addition, peptide synthesis by genetic engineering also became practical; thus, now three methods, namely, the chemical methods, the enzymatic method, and the genetic engineering method, are available for peptide synthesis, and all of them are now used as complementary synthetic tools. In this chapter, such subjects as the basic principles of enzymatic peptide synthesis, the enzymes used for peptide synthesis, and some applications of the method will be described.

8.2 EQUILIBRIUM BETWEEN SYNTHESIS AND HYDROLYSIS OF A PEPTIDE BOND

The synthesis and hydrolysis of a peptide bond is a typical organic equilibrium reaction. The carboxylic and/or amine substrates and a product peptide exist as a complex mixture of non-ionized and ionized species under usual enzymatic reaction conditions. Therefore, equilibrium constants differ greatly depending on which equilibrium one considers [3]. The equilibrium constant K_I, shown by eq 2, is based on the equilibrium of non-ionic species. Although the probability of all substrates being in a non-ionic state is extremely small, this equilibrium is of importance when one considers the equilibrium from the theoretical point of view, since K_I is not affected by the pH of the medium [4].

$$K_{I} = \frac{[R^{1}COOH][H_{2}NR^{2}]}{[R^{1}CONHR^{2}]} \tag{2}$$

The equilibrium constant K_{II} is based on the equilibrium of ionic species, shown by eq 3. In this case, the probability of all substrates being in an ionic state is still small, though larger than in the non-ionic case. Here, K_{II} is also independent of the pH of the medium.

$$K_{II} = \frac{[R^{1}COO^{-}][H_{3}N^{+}R^{2}]}{[R^{1}CONHR^{2}]} \tag{3}$$

The apparent equilibrium constant K_{pH}, shown by eq 4, is based on the equilibrium of the total concentration of all ionic and non-ionic species of the substrates, and so it can be measured experimentally. Since K_{pH} is dependent on the pH of the medium, the pH of the medium at which the constant is determined must be indicated.

$$K_{pH} = \frac{[R^{1}COOH]_{tot}\,[H_{2}NR^{2}]_{tot}}{[R^{1}CONHR^{2}]_{tot}} \tag{4}$$

One can see in Table 8.1 how these equilibrium constants differ from each other in the α-chymotrypsin–catalyzed formation and hydrolysis of hydroxamic acids [5].

Carpenter [4] calculated $\Delta G_I°$ and $\Delta G_{pH}°$ of eqs 2 and 4, respectively, by using thermodynamic data of peptide and amide bonds (Table 8.2). It can be seen that

TABLE 8.1 Equilibrium Constants for Hydroxamic Acid Formation Catalyzed by α-Chymotrypsin [5]

Equilibrium constants	$CH_3CONHOH$	Ac-Trp-NHOH
K_I	402	363
K_{II}	16	0.49
K_{pH}	1.46	0.042

TABLE 8.2 Free Energy Changes in the Formation of Peptide or Amide Bonds[a, b] [4]

	$\Delta G_I°$ (kcal/mol)	$\Delta G_{pH}°$ (kcal/mol)[c]
H-Gly↓Gly-OH	–3.9	+4.3
DL-H-Ala↓Gly-OH	–3.9	+4.2
DL-H-Leu↓Gly-OH	–4.9	+3.4
Glutamine	–3.3	+3.4
Asparagine	–4.3	+3.0
Bz-Gly↓Gly-OH	–4.2	+3.1
Bz↓Gly-OH	–4.9	+1.9
Bz↓Gly-Gly-OH	–5.2	+0.8
Bz-Tyr↓Gly-NH_2	–5.4	+0.5

[a] At 25 °C.

[b] Arrows indicate the bonds which are involved in the reactions.

[c] At pH 7.0.

$\Delta G_I°$ is almost same for most of the compounds, whereas $\Delta G_{pH}°$ varies considerably depending on the compounds. This is probably because in the former the effects of substituents are canceled on both the reactant and product sides due to the non-ionic nature of $\Delta G_I°$, whereas in the latter the dissociation constants of substrates are affected by substituents. It should be noted that $\Delta G_{pH}°$, which is based on the actual system, shows positive values, whereas $\Delta G_{\mathrm{I}}°$ shows negative values, indicating that the equilibrium based on the non-ionic species, which is almost negligible in the actual system, lies rather on the synthesis side. From these it is suggested that one could shift the equilibrium toward the synthesis side if one could bring the system into a more non-ionic state. As will be discussed later, attempts to shift the equilibrium toward the synthesis side from this viewpoint were successful.

8.2.1 Effects of Substrate Concentration

As expected from the law of mass action in equilibrium reactions, conversion to the product is higher with higher initial concentration of substrates. The conversion to the product at the equilibrium point is expressed by eq 5, where α is the conversion, and a and b are the initial concentrations of the substrates ($a \leq b$).

$$K_{\mathrm{pH}} = \frac{a\alpha}{a(1-\alpha)(b-a\alpha)} \qquad (5)$$

The percent synthesis with different substrate concentration was calculated for the following equilibrium (eq 6), which has one of the largest equilibrium constants in Table 8.2 (K = 0.43 M^{-1} at pH = 7.0).

$$\text{Bz-Tyr-OH} + \text{H-Gly-NH}_2 \rightleftharpoons \text{Bz-Tyr-Gly-NH}_2 \qquad (6)$$

The results in Table 8.3 show that one can get only 24.5% of the product yield even when 1 M concentrations of each of the substrates are reacted. This clearly indicates that synthesis by a mere equilibrium in a homogeneous solution is not practical as a synthesis method, thus requiring some means to overcome the unfavorable situation.

8.2.2 Effects of Solubility of Product

One can shift the equilibrium toward either side by removing a component involved in the equilibrium out of the system. Therefore, if one can establish the system in which reactants are soluble and a product is insoluble, one can get the product in higher yield. Fortunately, in most peptide equilibria, the product is generally more insoluble in aqueous media than are the reactants, because the former is much more hydrophobic than the latter. Therefore, most enzymatic peptide syntheses effectively utilize this solubility difference. In order to shift the equilibrium further toward the synthesis side than the thermodynamically determined point, the concentration of the product needs to exceed its solubility. The synthesis reaction proceeds until concentration of the product at the equilibrium point becomes equal to its solubility.

TABLE 8.3 Extent of Peptide Formation with Different Substrate Concentrations and Product Solubilities[a]

Substrate concentration	Solubility[b]			
	Soluble	1×10^{-3} M	1×10^{-4} M	1×10^{-5} M
0.01 M	0.5%	(0.5%)[c]	(0.5%)[c]	51.7%
0.1 M	3.5%	50.8%	84.6%	95.2%
1.0 M	24.5%	95.1%	98.5%	99.5%

[a] Equilibrium of Bz-Tyr-OH + H-Gly-NH_2 ⇌ Bz-Tyr-Gly-NH_2 (K_{pH} = 0.43 M^{-1} [4]).

[b] Hypothetical solubility of Bz-Tyr-Gly-NH_2.

[c] The conversion to the product at the equilibrium point is less than the hypothetical solubility. Therefore the yield is the same as that of the homogeneous reaction.

Therefore, the conversion to the product, α, is correlated with solubility of the product, S, initial concentration of substrates (a and b, $a \leq b$), and K_{pH}, as shown by eq 7 [6]. As can be seen in Table 8.3, when the hypothetical solubility of the product is less than 1×10^{-5} M, the yield is almost quantitative even under relatively low concentrations of the reactants (0.1 M).

$$K_{pH} = \frac{S}{[a(1 - \alpha) - S][b - a\alpha - S]} \tag{7}$$

When $a = b$, α is calculated easily by eq 8, which clearly shows that one can get higher product yield with higher initial substrate concentration and lower product solubility.

$$\alpha = 1 - \frac{S + \sqrt{\dfrac{S}{K_{pH}}}}{a} \tag{8}$$

8.2.3 Effects of pH

The apparent equilibrium constant K_{pH} can be correlated with K_{II}, the constant based on ionic equilibrium, as shown by eq 9 (K_1 and K_2 are acid dissociation constants of carboxylic and amino substituents, respectively; thus, $K_1 >> K_2$) [7].

$$K_{pH} = \frac{K_{II}}{1 + \frac{[H^+]}{K_1} + \frac{K_2}{[H^+]}} \tag{9}$$

Therefore, from eq 9, the pH profile of K_{pH} can be obtained. The equilibrium of eq 10 in H_2O has a bell-shaped K_{pH}–pH profile with maximum K_{pH} at pH 4.5–6.5, in which calculated values and experimentally obtained values agree very well, as shown in Figure 8.1 [8].

$$\text{Z-Trp-OH} + \text{H-Gly-NH}_2 \rightleftharpoons \text{Z-Trp-Gly-NH}_2 \tag{10}$$

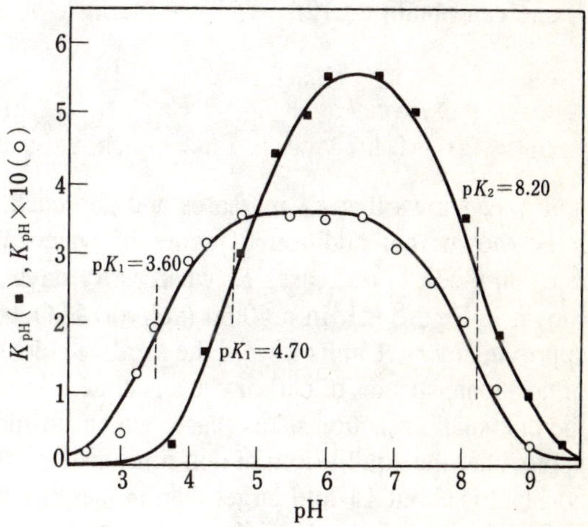

Figure 8.1. The pH–K_{pH} Profile for the Equilibrium of Z-Trp-OH + H-Gly-NH_2 ⇌ Z-Trp-Gly-NH_2. (○ in H_2O; ■ in 40:60 H_2O–triethylene glycol.) The solid lines were calculated using eq 9; the dashed lines indicate the *p*Ks of the amino acids [7].

Furthermore, the pH at the maximum K_{pH} can be calculated from eq 9, as $d[K_{pH}]/d[H^+] = 0$; thus:

$$[H^+] = \sqrt{K_1 K_2} \tag{11}$$

The above discussion is for the equilibrium in which only functional groups involved in the reaction undertake dissociation. Even if the components involve several ionization states, the equation which correlates between K_{pH} and pH can also be derived [9]. However, the equation is much more complicated than the above model, so it will not be discussed further.

8.2.4 Effects of Organic Solvents

As discussed above, the yield in peptide synthesis is very much dependent on the equilibrium constant, initial substrate concentration, and solubility of the product. Since organic solvents can affect all of these factors, they are added intentionally into the reaction medium. As can be seen in Table 8.2, K_I is almost constant, not much influenced by substituents or reaction conditions, and the equilibrium lies rather on the synthesis side. On the other hand, K_{II} is sensitive to these factors. From eqs 2 and 3, one can obtain eq 12:

$$K_{II} = K_I \frac{K_2}{K_1} \tag{12}$$

Eq 12 shows that K_{II} increases and approaches K_I with smaller K_1 and larger K_2. It is known that addition of some of water-miscible organic cosolvents into the reaction medium decreases K_1 whereas K_2 stays almost unchanged. For instance, as shown in Figure 8.1, in a 40:60 (vol/vol) H_2O–triethylene glycol solution, pK_1 shifts approximately 1.1 units toward the alkaline side. This indicates that the concentration of non-ionic species of carboxylic substrate is increased in the neutral pH region where the enzymatic reaction takes place, which, in turn indicates that the reaction system approaches the equilibrium of non-ionic species. In this case, the apparent equilibrium constant is about 15-fold larger than in the pure water system. Furthermore, in 15:85 H_2O–1,4-butanediol, pK_1 shifts 1.60 units toward the alkaline side, and the apparent equilibrium constant is approximately 80-fold larger than in the pure water system [8].

The above are examples of the effect of the addition of water-miscible organic cosolvents. It is also possible to increase the yield of the product by adding a water-immiscible organic cosolvent which can extract the product, thus shifting the equilibrium toward the synthesis side.

8.3 PEPTIDE SYNTHESIS BY AN IMMOBILIZED ENZYME

Industrial applications of enzymes to produce valuable materials have been facilitated by the recent progress of technology in immobilized enzymes. However, application to the enzymatic peptide synthesis in an aqueous solution, where the product coming out of solution is essential in order to shift the equilibrium toward the less favorable synthesis side, is quite difficult, since separation of the immobilized enzyme from the precipitated product is difficult. This difficulty may be overcome by the use of an organic cosolvent that can dissolve or extract the product. In selecting the organic cosolvent, one must consider the following aspects:

1. effects on activity and stability of the enzyme,
2. effect on equilibrium constant,
3. solubility of the product, and
4. partition of the reactants and the product between aqueous and organic phases when a water-immiscible cosolvent is used.

Generally, in the presence of a water-miscible cosolvent, the activity as well as the stability of enzymes are markedly impaired, although some cosolvents can significantly shift the equilibrium toward the synthesis side. On the other hand, with a water-immiscible cosolvent, the impairing effects are much smaller than with a water-miscible cosolvent, since the contact of an organic solvent with an enzyme is much smaller in the former case than in the latter case. In enzymatic peptide synthesis, aliphatic or aromatic hydrocarbons are in general unsatisfactory because of the poor solubility of the product in these solvents. Solvents such as chloroform, 1,2-dichloroethane, ethyl and isopropyl acetates, and methyl isobutyl ketone can satisfy the requirements to some extent, but ethyl acetate appears to be the most appropriate for the following reasons: (1), the impairing effects on the enzyme are smaller than with other solvents, especially chlorinated hydrocarbons; and (2), the solubility of products in the organic layer is quite high and partition of reactants into the organic layer is generally low [10]. These facts suggest that most of the reactants can stay in an aqueous layer where the enzymatic reaction takes place, and the product can smoothly transfer to the organic layer, thus shifting the equilibrium toward the synthesis side.

On immobilizing an enzyme, one must consider the following aspects:

1. high immobilization yield,
2. small deactivation during immobilization,
3. inexpensive immobilization cost,
4. high volumetric activity, and
5. small leakage of the immobilized enzyme from the support material.

There is no need to comment on the first three, but with regard to the fourth, as high a volumetric activity as possible is necessary because of the much slower reaction rate in the presence of an organic cosolvent than in a pure aqueous system. As to the fifth, enzymes immobilized by the physical adsorption method have a tendency to easily desorb from the support material in the presence of an organic cosolvent. Enzyme leakage also occurs in the ionic-binding method because of the ionic nature of amino acid substrates. Therefore, immobilization by covalent binding or entrapping appears to be preferable for the present case [10].

The immobilized enzyme can be applied in the organic-aqueous biphasic system as well as in the apparent single-phase system of water-immiscible organic solvents. In the apparent single-phase system, the pores of support material where enzymes are immobilized are filled with a minimum amount of water, and the outer phase is surrounded with an organic solvent, so it can be regarded as an extreme case of the biphasic system. From the practical point of view, the apparent single-phase system is more advantageous than the biphasic system, because the latter system, when applied to a continuous column reactor, causes channeling of the two layers in a packed bed of an immobilized enzyme, and also when applied to a continuous stirred tank reactor, it is difficult to discharge the biphasic reaction mixture homogeneously from the reactor. Matsuno *et al.* applied the apparent single-phase system to the synthesis of the aspartame precursor, Z-Asp-Phe-OMe, and found that continuous operation with the stirred tank reactor was better than the column reactor, since environmental conditions of the reaction could be more easily manipulated in the tank reactor [11].

In the apparent single-phase system, "immobilization" of an enzyme could also be done just by holding the aqueous enzyme solution inside the pores of supporting material, as shown in Figure 8.2. Therefore, even though the enzyme is not attached to the support by conventional immobilization techniques, one can "immobilize" the enzyme on such inert materials as glass beads [12]. This type of "immobilization" is possible only in the apparent single-phase system. A further attractive feature of this method is that when the activity of the "immobilized" enzyme is decreased, the regeneration can be done by simply washing out the enzyme with water and immersing it in a fresh enzyme solution.

8.4 KINETICS-CONTROLLED PEPTIDE SYNTHESIS

It is well known from mechanistic studies that thiol proteinases and serine proteinases catalyze a hydrolytic reaction via covalent bond formation between a substrate (an amide or ester, R^1COX) and an enzyme to lead to an acyl-enzyme intermediate. In

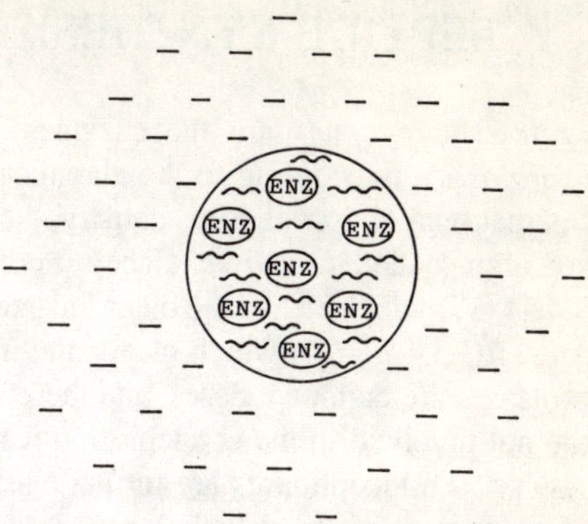

Figure 8.2. "Immobilized" Enzyme in the Apparent Single-phase System of a Water-immiscible Organic Solvent. (Outside of the pores in the supporting material is an organic phase; the inside is filled with an aqueous enzyme solution) [12].

a typical hydrolytic reaction, this intermediate is attacked by a water molecule to produce the carboxylic acid and regenerate the enzyme. If a nucleophile other than H_2O is present, this may competitively attack the intermediate to form a peptide linkage, as shown below [13]:

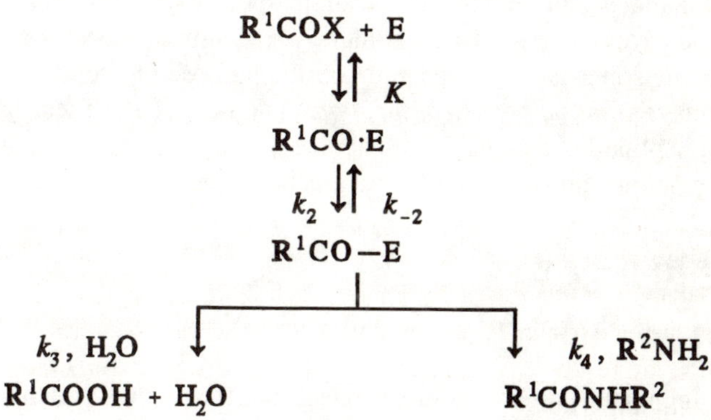

Although the peptide produced also undergoes hydrolysis via the acyl-enzyme intermediate, formation of the intermediate is much faster from esters than from peptides, so esters can be used as carboxylic substrates [14]. In this case, the yield of product is independent of the yield predicted by the thermodynamic equilibrium. Therefore, even if the product is soluble, it could give a high synthesis yield [15]. If the product is insoluble, its hydrolysis is almost nil, so the yield can be further increased.

8.5 THE ENZYMES AND THE STRATEGY FOR PEPTIDE SYNTHESIS

Proteolytic enzymes are the enzymes which can cleave a peptide bond (*peptide hydrolases*). According to the classification by the Nomenclature Committee of the International Union of Biochemistry, these enzymes belong to the group EC 3.4, and are often called *proteases*. They are divided into two groups; *exopeptidases* (EC 3.4.11–17) which cleave the outer linkages of proteins, and *proteinases* (or *endopeptidases*, EC 3.4.21–24) which cleave the inner linkages. At present, most widely used proteases are endopeptidases, and therefore, terminal carboxylic or amine groups that are not involved in the condensation reaction must be blocked by suitable protecting groups. Endopeptidases are further classified into four categories on the basis of the reaction mechanism at their active sites, that is, *serine proteinases* (EC 3.4.21), *thiol proteinases* (EC 3.4.22), *carboxyl proteinases* (or *acidic proteinases*, EC 3.4.23), and *metalloproteinases* (or *neutral proteinases*, EC 3.4.24). Usually, proteinases within the same group share many common properties, such as optimum pH, substrate specificities, etc. Much information on the enzyme's substrate specificity can be gleaned from hydrolysis studies [16], which one can consult when choosing a suitable enzyme for synthesis, because substrate specificities between hydrolysis and synthesis generally do not differ much. In addition to substrate specificities, one must also consider other enzyme properties, such as optimum pH, inhibitors, heat stability, resistance toward organic solvents, etc., depending on the purpose of use.

Many proteinases have *transamidase* (*transpeptidase*) activity [16]. Therefore, if a substrate contains a peptide linkage that is susceptible to the enzyme, for example, the B-C bond in the reaction shown in eq 13, side reactions could occur [17].

X-A-B-C-OH + **H-D-E-F-Y** →

X-A-B-C-D-E-F-Y + **X-A-B-OH**, **X-A-B-D-E-F-Y**, etc. (13)

In such a case, setting up the proper conditions to facilitate the product precipitation as soon as the product is formed can be workable. Furthermore, to accelerate the rate of the desired reaction, use of a large excess of another component or an ester substrate, when a thiol or a serine proteinase is used, can also be effective [18]. Since serine and thiol proteinases have esterase activity, one must also pay attention to undesirable side reactions due to this activity as shown in eq 14 (Y is a carboxyl-blocking lower alkyl ester group).

X-A-B-OH + **H-C-D-Y** →

X-A-B-C-D-Y + **X-A-B-C-D-OH**, **H-C-D-OH**, etc. (14)

When an enzyme with esterase activity is to be used, esters of bulky alkyl groups, for example, *t*-butyl or benzhydryl groups, are often used, since they are not enzymatically hydrolyzed and yet can be easily removed by acid [19].

Crude enzymes are often contaminated with other enzymes. Therefore, when they are used, it is also necessary to consider possible side reactions due to these contaminants. When the contaminating enzyme is identified, time-consuming and expensive purification may be avoided by carrying out the synthetic reaction in the presence of a specific inhibitor of the contaminating enzyme or by treating the crude enzyme with such an inhibitor prior to use. For instance, when one wishes to synthesize a peptide containing an ester linkage of a lower alkyl group by using crude metalloproteinases contaminated with a serine proteinase, one can completely suppress esterase activity of the latter enzyme by carrying out the reaction in the presence of an inhibitor from potato [20]. When trypsin is contaminated with undesired chymotrypsin, the latter can be inactivated by treating it with *N*-tosyl-L-phenylalanine chloromethyl ketone [21]. Thiol proteinases are easily oxidized by air to an inactive form, so it is preferable to use them in the presence of mild reducing agents such as mercaptoethanol [19].

Exopeptidases have not been so widely used as endopeptidases, but recently they are getting more attention and their use in peptide synthesis is increasing [22, 23]. Because they act on only the outer linkage of proteins, side reactions due to cleavage of the inner bonds are not significant. Furthermore, a terminal substituent does not need to be protected, thus stepwise elongation from the C-terminal (aminopeptidase) end or from the N-terminal (carboxypeptidase) end can be done without protection.

From the mechanistic point of view of enzymatic reactions, it is also important to consider the general environmental factors that can affect the reaction rate, for example, the pH of the medium, reaction temperature, ionic strength, or pressure. These are widely described in books on the enzymatic reaction mechanism [24].

8.6 SYNTHESIS OF BIOLOGICALLY ACTIVE PEPTIDES

As information on the enzymatic peptide synthesis increases and its usefulness as a synthetic method becomes prominent, studies have been extended from the methodological studies using model compounds to the synthesis of biologically active peptides. Many compounds have been partially or totally synthesized, including angiotensin [17], substance P [25], eledoisin [25, 26, 27], caerulein [28], Leu- and Met-enkephalins [23, 29, 30], salmon calcitonin [31], soybean trypsin inhibitor [14], human insulin [21, 32, 33, 34, 35, 36], bovine trypsin inhibitor [37], ribonuclease [38], aspartame [12, 39, 40], dynorphin [41], and mouse epidermal growth factor [22]. The enzymatic method can be applicable to the bond formation of certain structural settings. However, for the synthesis of enkephalins [29, 30] and 1-8-dynorphin [41], all of the peptide linkages are synthesized enzymatically. This may indicate that, if one could choose a proper enzyme and appropriate reaction conditions, the enzymatic synthesis could be as widely applicable as the chemical methods. Among the above examples, human insulin and aspartame have been produced enzymatically on an industrial scale, so they will be described in the following.

Insulin is an important drug for diabetes but supplying it from human organs or by chemical synthesis is extremely difficult. Therefore, insulins extracted from animals have been chiefly used, but they have a problem of side effects due to the antigen-antibody reaction. Porcine and human insulins differ from each other only at the 30th amino acid residue of the B chain, that is, the former has alanine while the latter has threonine. Firstly, Inoue *et al.* reported that using trypsin pretreated with *N*-tosyl-L-phenylalanine chloromethyl ketone, the octapeptide of B23-B30 is removed, and then the chemically synthesized human-type octapeptide is recombined [21]. Next, they found that using *Achromobacter* protease, which selectively cleaves at the carboxylic side of the lysine residue in proteins, the B30 alanine is removed and then H-Thr-OBu^t is introduced [32, 34]. Finally, porcine insulin is enzymatically transformed to the human insulin ester in a single-step reaction using a large excess of H-Thr-OBu^t [35, 36] (Figure 8.3). Peptide synthesis by the method involving genetic engineering has shown remarkable progress in the past decade, and human insulin is now commercially produced by the enzymatic method as well as by genetic engineering [42].

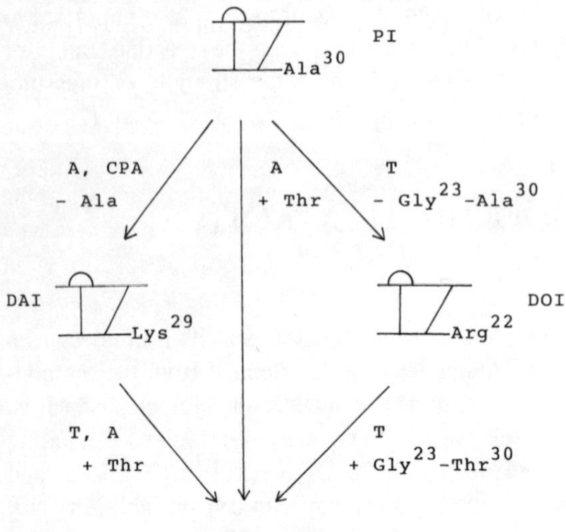

Figure 8.3. Enzymatic Conversion of Porcine Insulin (PI) to Human Insulin (HI).
(CPA = carboxypeptidase A;
A = *Achromobacter* protease;
DAI = desalanine-(B30)-insulin;
DOI = desoctapeptide-(B23–B30)-insulin; T = trypsin;) [21, 32–36].

Aspartame (H-Asp-Phe-OMe) is a dipeptide sweetener that is about 200 times sweeter than sugar. We found that in the presence of some metalloproteinases, X-Asp(OBzl)-OH and H-Phe-OMe react to give the fully protected aspartame (X is an N-protecting group) [43]. We further found that even though the side-chain carboxylic acid is not protected, X-Asp(OH)-OH can react under certain conditions to give the condensation product with an exclusive reaction at the α-carboxylate, which is then deposited as the insoluble addition compound with H-Phe-OMe [39]. It was also found that when racemic substrates are used, the reaction takes place only on L-isomers; the D-isomers remain unreacted. On the other hand, all of the chemical methods with industrial potential permit only L-isomers as feedstocks and produce 20–40% of the β-isomer, which has a bitter taste [42]. Therefore, this is an outstanding example of the enzymatic method not only in the peptide synthesis but also in enzymatic synthesis in general, since it contains many attractive features that cannot be achieved in chemical reactions, especially concerning such points as simultaneous peptide synthesis and optical resolution, and the regioselective peptide bond formation at the α-carboxylate. Based on these findings, we developed the industrial process by which aspartame is now commercially produced.

8.7 CONCLUSION

The enzymatic method for peptide synthesis has many advantages over the chemical ones. For instance, one does not need to worry about racemization during condensation or problems concerning introduction and removal of side-chain protecting groups. Furthermore, one can carry out the reactions in an aqueous solution under mild conditions. It is also possible to use racemic amino acids as feedstocks, indicating there is no need to be bothered with impurities of optical isomers. When using an exopeptidase, a terminal substituent may not be protected. Naturally, there are some drawbacks in the enzymatic method also. The enzymes cannot act on just any substrate, so one must find the appropriate enzyme for the desired reaction. In addition, one must also pay attention to possible side reactions due to transpeptidation and esterase activities. Modification of an enzyme sometimes may enhance its activity enormously and may also improve its properties. For instance, Blumberg and Vallee reported that a chemical modification of thermolysin caused increase of the enzyme activity by 70-fold [44]. Recent progress in protein engineering will without doubt enable the preparation of a tailor-made enzyme having improved properties. In the industrial production of aspartame and human insulin, the enzymatic method is now competitive with the chemical methods and the genetic engineering method, respectively, showing that it has taken its place as one of the basic methods in peptide synthesis. When working out the synthetic strategy, one should understand the characteristics of each method to select the proper one. At any rate, the enzymatic method will contribute greatly to the progress of peptide chemistry as one of the basic synthesis tools.

8.8 REFERENCES

1. Mohr, E., Strohschein, F., *Berichte 42*, 2521 (1909).
2. For example, Bergmann, M., Fraenkel-Conrat, H., *J. Biol. Chem. 124*, 321 (1937); Bergmann, M., Fruton, J. S., *Adv. Enzymol. 1*, 63 (1941).
3. Oyama, K., Kihara, K., *Jpn. Chem. Rev. 35*, 195 (1982).
4. Carpenter, F. H., *J. Am. Chem. Soc. 82*, 1111 (1960).
5. Jencks, W. J., Caplow, M., Gilchrist, M., Kallen, R. G., *Biochemistry 2*, 1313 (1963).
6. Oyama, K., Irino, S., Harada, T., Hagi, N. *Enzyme Engineering 7*; Laskin, A. I. *et al.*, Eds.; New York Acad. Sci., New York, 1984; p 95.
7. Dobry, A., Fruton, J. S., Sturtevant, J. M., *J. Biol. Chem. 195*, 149 (1952).
8. Homandberg, G. A., Mattis, J. A., Laskowski, M., Jr., *Biochemistry 17*, 5220 (1978)
9. Borsook, H., *Adv. Protein Chem. 8*, 127 (1953).
10. Oyama, K., Irino, S., Hagi, N., *Methods Enzymol. 136*, 503 (1987).
11. Nakanishi, K., Kamikubo, T., Matsuno, R., *Bio/Technol. 3*, 459 (1985).
12. Oyama, K., Nishimura, S., Nonaka, Y., Kihara, K., Hashimoto, T., *J. Org. Chem. 46*, 5241 (1981).
13. Oka, T., Morihara, K., *J. Biochem. 82*, 1055 (1977).
14. Laskowski, M., Jr. In *Semisynthetic Peptides and Proteins*; Offor, R. E., DiBello, D., Eds.; Academic Press, London, 1978; p 255.
15. Morihara, K., Oka, T., Biochem. J. 163, 531 (1977).
16. *Methods Enzymol. 19*, (1970).
17. Isowa, Y., Ohmori, M., Sato, M., Mori, K., *Bull. Chem. Soc. Jpn. 50*, 2766 (1977).
18. Oka, T., Morihara, K., *J. Biochem. 84*, 1277 (1978).
19. Isowa, Y., Ichikawa, T., Ohmori, M., *Bull. Chem. Soc. Jpn. 51*, 271 (1978).
20. Isowa, Y., Ichikawa, T., Ohmori, M., *Bull. Chem. Soc. Jpn. 51*, 271 (1978).
21. Inouye, K., Watanabe, K., Morihara, K., Tochino, Y., Emura, T., Sakakibara, S., *J. Am. Chem. Soc. 101*, 751 (1979).
22. Widmer, F., Bayne, S., Houen, G., Moss, B. A., Rigby, R. D., Whittaker, R. G., Johansen, J. T. *Peptide 1984*; Almquist & Wiksell, Stockholm, 1984; p 193.
23. Widmer, F., Breddam, K., Johansen, J. T. *Peptide 1980*; Scriptor, Copenhagen, 1981; p 46.
24. There are many books and reviews. See, for example, Dixon, M., Webb, E. C. *Enzymes*; Academic Press, New York, 1979.
25. Isowa, Y., *J. Synth. Org. Chem. Jpn. 36*, 195 (1978).
26. Kuhl, P., Doring, G., Neubert, K., Jakubke, H. D., *Monatsh. Chem. 115*, 423 (1984).
27. Kuhl, P., Konnecke, A., Doring, G., Daumer, H., Jakubke, H. D., *Tetrahedron Lett.*, 893 (1980).
28. Takai, H., Sakato, K., Nakamiso, N., Isowa, Y. *Peptide Chemistry 1980*; Protein Res. Foundation, Osaka, 1981; p 213.

29. Kullmann, W., *Biochem. Biophys. Res. Commun. 91*, 693 (1979).
30. Kullman, W., *J. Biol. Chem. 255*, 8234 (1980).
31. Gondo, M., Yamashita, H., Sakakibara, K., Isowa, Y. *19th Peptide Discussion*; Nagoya, 1981; p 34.
32. Morihara, K., Oka, T., Tsuzuki, H., Tochino, Y., Kanaya, T., *Biochem. Biophys. Res. Commun. 92*, 396 (1980).
33. Morihara, K., Oka, T., Tsuzuki, H., *Nature 280*, 412 (1979).
34. Masaki, T., Nakamura, K., Isono, M., Soejima, M., *Agr. Biol. Chem. 42*, 1433 (1978).
35. Morihara, K., Oka, T. *Peptide Chemistry 1982*; Protein Res. Foundation, Osaka, 1983; p 231.
36. Breddam, K., Widmer, F., Johansen, J. T., *Carlsberg Res. Commun. 46*, 361 (1981).
37. Chu, S. C., Wang, C. C., Brandenburg, D., *Z. Physiol. Chem. 362*, 647 (1981).
38. Homandberg, G. A., Komoriya, A., Chaiken, I. M., *Biochemistry 21*, 3385 (1982).
39. Isowa, Y., Ohmori, M., Ichikawa, T., Mori, K., Nonaka, Y., Kihara, K., Oyama, K., Satoh, H., Nishimura, S., *Tetrahedron Lett.*, 2611 (1979).
40. Oyama, K., Kihara, K., Nonaka, Y., *J. Chem. Soc. Perkin Trans. II*, 356 (1981).
41. Kullmann, W., *J. Org. Chem. 47*, 5300 (1982).
42. Oyama, K., Kihara, K., *CHEMTECH 14*, 100 (1984).
43. Isowa, Y., Oyama, K., Ichikawa, T. *35th Ann. Meeting Jpn. Chem. Soc.*; Sapporo, 1976; 4B23.
44. Blumberg, S., Vallee, B. L., *Biochemistry 14*, 2410 (1975).

ENZYMATIC PEPTIDE SYNTHESIS IN NONAQUEOUS MEDIA

Alexey L. Margolin

CONTENTS

Abbreviations Used in This Chapter:

AA = Amino acid; F = formyl; NHNph = β-naphthylamide; Ind = indolinyl; *tert*-Bu = *tert*-butyl ester; DMF = dimethylformamide; Cam = Carboxamidomethyl; Ph = phenyl; Boc = *tert*-butoxycarbonyl; Z = benzyloxycarbonyl.

9.1 INTRODUCTION

Enzyme-catalyzed formation of the peptide bond is becoming a valuable addition and, in some cases, a strong alternative to the classical techniques of peptide synthesis.

The greatest advantages of enzymatic peptide synthesis—absence of racemization and minimal protection and activation requirements—are well recognized [1–5]. Despite these promising features and several successful examples, enzymatic, namely protease-catalyzed, peptide synthesis has yet to reach the versatility of well-known chemical methods. Three major problems should be solved before enzymatic techniques become general routine procedure. These problems are: (1) unfavorable thermodynamics of peptide synthesis in water, (2) narrow substrate specificity and enantioselectivity of enzymes, and (3) undesirable proteolysis of the growing polypeptide chain.

Several methods that allow one to shift the unfavorable equilibrium of peptide bond formation have been developed [5]. These methods include the use of product precipitation, molecular traps, biphasic aqueous organic mixtures, mixtures of water with water-miscible organic solvents or nonaqueous media. The two other problems seem to be more difficult to handle. Indeed, it is impractical to use proteases with narrow substrate specificity, since, in the extreme this would require one enzyme per amino acid. On the other hand the use of proteases with broad substrate specificity usually leads to secondary hydrolysis of the existing polypeptide chain.

Recent research in the area of biocatalysis has unequivocally demonstrated that many enzymes are catalytically active in organic solvents [6–8]. The use of organic solvents as reaction media not only allows for the reversal of the equilibrium of many hydrolytic reactions, but also reveals some important and unexpected phenomena of enzymatic catalysis [9,10]. The most important of these is that substrate specificity and enantioselectivity of an enzyme in nonaqueous media can be changed and sometimes predictably controlled. In this chapter, we address the problems of secondary hydrolysis and strict substrate specificity and enantioselectivity of enzymes in peptide synthesis.

9.2 SUBSTRATE SPECIFICITY OF PROTEASES IN ORGANIC SOLVENTS

The substrate specificity of enzymes stems from their ability to utilize the free energy of bonding with substrate. It seems obvious that if one can somehow modify any part of the free energy of bonding (hydrophobic interactions, hydrogen bonds or electrostatic bonds) one will be able to change the substrate specificity of enzymes [11].

It has been discovered that the substrate specificity of several proteases in organic solvents is different from that in water. Zaks and Klibanov [12] studied the transesterification reaction between esters of *N*-Ac-Phe, *N*-Ac-His and *N*-Ac-Ser with primary alcohols in octane and the hydrolysis of the same esters in water. Upon transition from hydrophobic Phe to hydrophilic Ser in this series, k_{cat}/K_m for enzymatic hydrolysis drops by a factor of nearly 5×10^4. On the other hand, for transesterification in octane, the ester of *N*-Ac-Ser is 3 times more reactive than that of *N*-Ac-Phe. Furthermore, *N*-Ac-His-OMe, whose reactivity in water was 0.5% that of *N*-Ac-Phe-OEt, in octane becomes a 20-fold more reactive substrate. Qualitatively similar data were obtained for subtilisin. These data indicate that the main driving force of the enzyme-substrate binding in water—the hydrophobic interactions—do not play a major role in organic solvents.

This idea has been further supported by Gaertner and Puigserver for peptide synthesis in organic solvents [14]. The study of the synthesis of Bz-AA-PheNH$_2$ from the corresponding Bz-AA ester and PheNH$_2$ in benzene, catalyzed by poly(ethylene glycol)–modified chymotrypsin, has shown altered substrate specificity of the catalyst. The transition from hydrophobic Bz-Tyr-OEt to hydrophilic Bz-Lys-OMe leads to a dramatic decrease of k_{cat}/K_m by more than 8×10^5-fold. In peptide synthesis in benzene, both substrates exhibit equal reactivity. The effect of organic solvent on the peptide synthesis catalyzed by subtilisin was less dramatic but still profound. The ratio of k_{cat}/K_m for Tyr over k_{cat}/K_m for Lys in the hydrolytic reaction is 34; the same ratio for peptide synthesis in benzene is only 0.3.

The significant change in substrate specificity of enzymes in organic solvents seems to be a general phenomenon. For example, it is known that the hydrolysis of a peptide bond in water, catalyzed by thermolysin, is significantly enhanced by the presence of any hydrophobic amino acid as the carboxyl group donor of the scissile bond. In sharp contrast to what is observed for peptide synthesis in water, the hydrophilic amino acids Asp and Ser exhibit higher reactivity (as judged by k_{cat}/K_m) than hydrophobic Phe and Leu in the synthesis of dipeptides Z-AA-PheNH$_2$, catalyzed by poly(ethylene glycol)–modified thermolysin in benzene [14].

The broadening of substrate specificity of proteases in organic solvents opens new opportunities for peptide synthesis. It allows the use of a relatively small number of enzymes for a wide range of peptide syntheses that are impossible to carry out in water.

9.3 ENANTIOSELECTIVITY OF PROTEASES IN ORGANIC SOLVENTS AND SYNTHESIS OF PEPTIDES CONTAINING D-AMINO ACIDS

A number of biologically active peptides, including important antibiotics, synthetic vaccines, and enkephalins and other hormones, contain D-amino acid residues [15–18]. Proteases, on the other hand, exhibit a profound preference for L-amino acids [19]. This means that enzymatic methods (as well as recombinant DNA technology) are not generally applicable to peptides containing D-amino acids. The only exception has been the use of D-amino acid derivatives as nonspecific nucleophiles competing with water for the acyl enzyme formed in the reaction between a protease and a protected L-amino acid ester [20–23]. This approach, however, is inherently limited to the enzymatic incorporation of a single D-amino acid into the C-terminal position of a peptide.

We have found that upon transition from aqueous solution to anhydrous organic solvents, the discriminating stereoselectivity of several proteases drastically relaxes [24]. Consequently these enzymes have been used to prepare numerous peptides containing D-amino acids in various positions [25]. We measured the ratio for subtilisin of specificity constants $(k_{cat}/K_m)_L/(k_{cat}/K_m)_D$, which reflects enantioselectivity of an enzyme in water and in several organic solvents. The enzymatic reaction studied in water was the hydrolysis of 2-chloroethyl esters of *N*-acetyl-L- and D-amino acids. In organic solvents we examined the subtilisin-catalyzed transesterification reaction between the same esters and propanol. For all nonaqueous solvents tested, this enantioselectivity factor was found to be 10–100-fold lower. Also, there is a linear correlation between the difference in free energy of activation ($\Delta\Delta G^{\#}$) for subtilisin-catalyzed transesterification of L- and D-enantiomers of *N*-Ac-Ala-OCH_2CH_2Cl in 13 different organic solvents and the hydrophobicity of the solvent (Figure 9.1).

These results indicate that an enzyme's enantioselectivity decreases as the hydrophobicity of the solvent increases. These data can be rationalized using the following simple model. When a substrate interacts with the enzyme's active center, water must be excluded between them [26]. In particular, the binding of an *N*-acyl-L-amino acid substrate to the active center of a proteolytic enzyme results in the release of a certain number of water molecules from the hydrophobic binding pocket of the protease; this process is a driving force in the enzyme-substrate interaction [11]. One would expect that the release of water molecules into the reaction medium would become thermodynamically less favorable with increasing hydrophobicity of the solvent. If the D-enantiomer binds to the active center of a protease in the same manner as the *N*-acyl-L-amino acid substrate, then the scissile bond will not face the enzyme's nucleophile, and hence cleavage cannot occur [27]. In order for the D-isomer to be reactive, it must bind to the enzyme "incorrectly" [27]; this binding

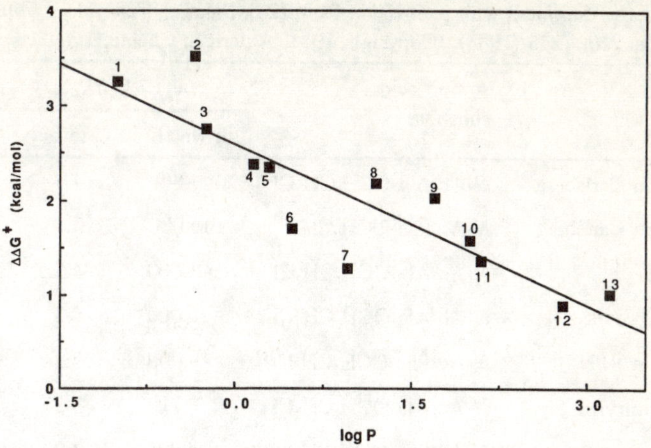

Figure 9.1. The Dependence of the Enantioselectivity of Subtilisin Carlsberg on the Hydrophobicity of the Solvent. Solvents: 1 = dimethylformamide; 2 = acetonitrile; 3 = acetone; 4 = methyl acetate; 5 = butanone; 6 = tetrahydrofuran; 7 = cyclohexanone; 8 = propyl acetate; 9 = butyl acetate; 10 = benzene; 11 = methylene chloride; 12 = butyl ether; and 13 = cyclohexane. Their hydrophobicities, *P*, are expressed as the logarithm of the partition coefficients between octanol and water. The correlation coefficient *R* for the linear free energy relationship shown is 0.88.

will release fewer molecules of water from the hydrophobic binding pocket than that of its L-counterpart [11]. Therefore, the reactivity of the L-enantiomer should be diminished to a greater extent, and consequently the enzyme's enantioselectivity should decrease, when the hydrophobicity of the solvent increases.

To test the generality of the observed phenomenon, we investigated the enantioselectivity of five other serine proteases in water and in butyl ether. As seen in Table 9.1, elastase, α-lytic protease, subtilisin BPN', α-chymotrypsin, and trypsin all exhibit striking enantioselectivity in water, but not in organic solvents: while the $(k_{cat}/K_m)_L/(k_{cat}/K_m)_D$ in water is on the order of 10^3–10^4, in butyl ether it does not exceed even one order of magnitude. These results demonstrate the predictable and rational control of enantioselectivity of enzymes by changing the reaction medium. The phenomenon thus discovered allowed us to solve the problem of enzymatic synthesis of peptides containing D-amino acid residues. Using subtilisin Carlsberg as a catalyst and *tert*-amyl alcohol as a reaction medium, we synthesized several peptides with D-amino acids (Ala, Phe, Trp, Asn) in the N-terminal position (Table 9.2). One can see that various D-amino acids (Ala, Phe, Trp, and Asn) were readily incorporated enzymatically in the N-terminal position. It is worth noting that various amino acid

TABLE 9.1 Enantioselectivity of Various Proteolytic Enzymes in Water and in Butyl Ether[a]. Reprinted with permission from [24], Sakurai, T., *et al.*, *J. Am. Chem. Soc. 110*, 7236 (1988). Copyright 1988, American Chemical Society.

Enzyme	Amino acid substrate	$(k_{cat}/K_m)_L/(k_{cat}/K_m)_D$	
		in water	in butyl ether
subtilisin Carlsberg	*N*-Ac-Ala-OCH_2CH_2Cl	1,800	4.4
subtilisin Carlsberg	*N*-Ac-Phe-OCH_2CH_2Cl	15,000	5.4
elastase	*N*-Ac-Ala-OCH_2CH_2Cl	> 1,000[b]	4.5
α-lytic protease	*N*-Ac-Ala-OCH_2CH_2Cl	> 10,000[b]	8.3
subtilisin BPN'	*N*-Ac-Phe-OCH_2CH_2Cl	16,000	7.3
α-chymotrypsin	*N*-Ac-Ala-OCH_2CH_2Cl	710	3.2[c]
trypsin	*N*-Ac-Phe-OCH_2CH_2Cl	> 4,000[b]	3.2

[a] In water, enzymatic hydrolysis of the esters was kinetically examined at 30 °C and pH 7.0. In butyl ether, enzymatic transesterification of the esters with *n*-propanol (1 M) was kinetically investigated at 30 °C. In all cases no appreciable reaction was detected without enzyme or in the presence of proteases preinactivated with phenylmethanesulfonyl fluoride.

[b] $(k_{cat}/K_m)_D$ was too low to measure accurately. Thus, the value given is the lower limit of enantioselectivity.

[c] In the case of α-chymotrypsin, butyl ether was supplemented with 0.1% distilled water.

TABLE 9.2 Synthesis of Peptides Containing D-Amino Acids Catalyzed by Subtilisin in Anhydrous *tert*-Amyl Alcohol. Reprinted with permission from [25], Margolin, A. L., *et al.*, *J. Am. Chem. Soc. 109*, 7885 (1987). Copyright 1988, American Chemical Society.

Substrates (mmol)		Product	Yield %[a]
Amino acid ester	Nucleophile		
N-Ac-D-Phe-OCH_2CH_2Cl (3.1)	L-Phe-NH_2 (3.1)	*N*-Ac-D-Phe-L-NH_2	67
N-F-D-Ala-OCH_2CH_2Cl (5.1)	L-Phe-NH_2 (15.6)	*N*-F-D-Ala-L-Phe-NH_2	82
N-Ac-D-Asn-OCH_2CH_2Cl (5.0)	L-Le-NH_2 (7.5)	*N*-Ac-D-Asn-L-Leu-NH_2	71
N-Ac-D-Trp-OCH_2CH_2Cl (8.4)	L-Phe-NH_2 (8.4)	*N*-Ac-D-Trp-L-Phe-NH_2	47
N-F-D-Ala-OCH_2CH_2Cl (6.0)	D-Ala-NH_2 (6.0)	*N*-F-D-Ala-D-Ala-NH_2	65
N-F-D-Ala-OCH_2CH_2Cl (16.0)	L-Phe-L-Leu-NH (16.0)	*N*-F-D-Ala-L-Phe-L-Leu-NH_2	61
N-CBZ-L-Tyr-OCH_2CH_2Cl (2.5)	D-Ala-NH-$(CH_2)_3$-Ph (7.8)	*N*-CBZ-L-Tyr-D-Ala-NH-$(CH_2)_3$-Ph	54

[a] Yield of isolated product, %.

derivatives, including both L- and D-isomers and dipeptides, were utilized as nucleophiles. This led us to the first enzymatic synthesis of a dipeptide (*N*-F-D-Ala-D-AlaNH_2) composed of only D-amino acids. Thus, a radically altered stereospecificity of subtilisin in organic solvents affords facile enzymatic preparation of diverse peptides containing D-amino acids.

9.4 KINETICALLY CONTROLLED PEPTIDE SYNTHESIS CATALYZED BY NONPROTEASES

The peptide bond can be formed either by the direct reversal of the hydrolytic process (thermodynamically controlled synthesis) or by aminolysis of N-protected amino acid ester (kinetically controlled synthesis).

Kinetically controlled peptide synthesis is catalyzed by serine and thiol proteases and proceeds via formation of an acyl-enzyme complex. This mechanism implies that other enzymes capable of forming an acyl-enzyme intermediate followed by deacylation with water or other nucleophiles may catalyze peptide synthesis. This concept appears to be very appealing, since nonproteases may not have the same limitations as proteases. Indeed, if a nonprotease such as lipase is used, the reverse reaction (alcoholysis of the peptide) and the secondary hydrolysis of the growing polypeptide chain should be kinetically invisible in the reaction time frame.

These ideas have been implemented by us [28] and others [29] using lipases as catalysts for peptide synthesis. The reaction procedure is very simple. In a representative experiment, 18.5 mmol of *N*-acetyl-L-Phe chloroethyl ester and 16.7 mmol of L-LeuNH_2 are dissolved in 330 ml of anhydrous toluene, porcine pancreatic lipase is added and the suspension is shaken at 45 °C. The precipitated product, *N*-Ac-Phe-LeuNH_2, is extracted with ethanol and recrystallized to give 84% yield of the pure dipeptide. This methodology was used for the synthesis of other dipeptides (Table 9.3). The results obtained indicate that porcine pancreatic lipase catalyzes the reaction with other amino acids such as Tyr, Met, and Ala. A number of both amide and ester derivatives of N-terminal amino acids can be used. Significantly, L- and D-isomers of Leu and Ala serve equally well as nucleophiles. Porcine pancreatic lipase was by no means a unique nonprotease in forming a peptide bond. Lipases from *Chromobacterium viscosum*, *Pseudomonas* sp., *Mucor* sp., and *Candida cylindracea* catalyzed peptide synthesis, although their substrate specificity was different. These enzymes exhibited significant catalytic activity in other dry organic solvents such as xylenes, acetonitrile, *tert*-butyl alcohol, isopropyl and butyl ethers, styrene, and cyclohexanone. West and Wong [25] reported the synthesis of dipeptides using esters of Phe and Gly as acyl donors. These reactions were catalyzed by porcine pancreatic lipase in pyridine containing 1% water, or by lipase from *Candida cylindracea* in ethyl acetate. In the last case the enzyme was adsorbed on Amberlite XAD-8.

TABLE 9.3 Peptide Synthesis Catalyzed by Porcine Pancreatic Lipase in Organic Solvents. Reprinted with permission from [28], Margolin, A. L., Klibanov, A. M., *J. Am. Chem. Soc. 109*, 3802 (1987). Copyright 1987, American Chemical Society.

Substrates, amino acid fragment: Carboxy terminal	Amino terminal	Solvent	Product	Initial rate, nmol/min (reaction half-time, days)	Yield of isolated product, %
N-Ac-L-Phe-OCH_2CH_2Cl	L-Leu-NH_2	toluene	*N*-Ac-L-Phe-L-Leu-NH_2	12.8 (1.4)	83
N-Ac-L-Phe-OEt	L-Leu-NH_2	toluene	*N*-Ac-L-Phe-L-Leu-NH_2	4.2 (4.2)	78
N-Ac-L-Tyr-OCH_2CH_2Cl	L-Leu-NH_2	THF	*N*-Ac-L-Tyr-L-Leu-NH_2	7.5 (2.3)	76
N-Ac-L-Tyr-OEt	L-Leu-NH_2	THF	*N*-Ac-L-Tyr-L-Leu-NH_2	4.2 (4.2)	67
N-Ac-L-Met-OCH_2CH_2Cl	L-Leu-NH_2	toluene	*N*-Ac-L-Met-L-Leu-NH_2	4.3 (4.0)	57
N-Ac-L-Phe-OCH_2CH_2Cl	L-Ala-NH_2	toluene	*N*-Ac-L-Phe-L-Ala-NH_2	11.9 (1.5)	82
N-Ac-L-Phe-OCH_2CH_2Cl	L-Val-NH_2	toluene	*N*-Ac-L-Phe-L-Val-NH_2	11.8 (1.5)	86
N-Ac-L-Phe-OCH_2CH_2Cl	L-Phe-NH_2	toluene	*N*-Ac-(L-Phe)$_2$-NH_2	4.9 (3.5)	48
N-Ac-L-Phe-OCH_2CH_2Cl	D-Leu-NH_2	toluene	*N*-Ac-L-Phe-D-Leu-NH_2	9.8 (2.0)	76
N-Ac-L-Phe-OCH_2CH_2Cl	L-Ala-OMe	toluene	*N*-Ac-L-Phe-L-Ala-OMe	3.8 (4.5)	–
N-Ac-L-Phe-OCH_2CH_2Cl	D-Ala-OMe	toluene	*N*-Ac-L-Phe-D-Ala-OMe	3.7 (4.5)	–
N-Ac-L-Phe-OCH_2CH_2Cl	L-Leu-NHNph	toluene	*N*-Ac-L-Phe-L-Leu-NHNph	2.1 (8.3)	51

9.5 CHEMOSELECTIVITY OF ENZYMATIC PEPTIDE SYNTHESIS

Chemoselectivity is one of the hallmarks of enzymatic peptide synthesis. The ability to exclusively select α-carboxy and α-amino groups without protection and subsequent deprotection of highly reactive side chains makes enzymes valuable practical catalysts.

When thermolysin, suspended in *tert*-amyl alcohol containing 1% water and 9% formamide, was presented with *N*-Ac-Phe and Phe-Lys-*O*-*tert*-Bu as substrates, only the natural Phe-Phe (as opposed to the unnatural Phe-ε-Lys) linkage was formed [30].

We studied a model reaction of *N*-Ac-Phe-OCH_2CH_2Cl with 6-amino-1-hexanol in *tert*-amyl alcohol and have shown that subtilisin Carlsberg catalyzes the acylation of the NH_2 group 50 times faster than that of the OH group [31]. We applied this procedure to the synthesis of peptides containing trifunctional amino acids with a hydroxyl group in a side chain (Tyr, Thr, Ser) (Table 9.4). As an example, we utilized both subtilisin Carlsberg and BPN' to prepare dipeptide fragments of octapeptide T which had been reported to possess antiretroviral activity [32]. The last three entries in Table 9.4 indicate the subtilisins catalyzed formation of only the natural peptide bond. The synthesis of *N*-Ac-D-Ala-Ser-NH-Nph is especially illustrative. It shows that while the enantioselectivity of subtilisin dramatically relaxes in organic solvent, the chemoselectivity of the enzyme remains untouched.

TABLE 9.4 Chemoselective Enzymatic Acylation of Aminoalcohols in *tert*-Amyl Alcohol. Reprinted with permission from [31], Chinsky, N., *et al., J. Am. Chem. Soc. 111*, 386 (1989). Copyright 1989, American Chemical Society.

Acylating agent (mmol)	Aminoalcohol (mmol)	Enzyme	Bond [a]	Product	Yield % [b]
N-Ac-L-Phe-OCH_2CH_2Cl (8.4)	6-amino-1-hexanol (7)	subtilisin Carlsberg	C-N	*N*-Ac-L-Phe-$NH(CH_2)_6$-OH	74
N-CBZ-L-Tyr-OCH_2CH_2Cl (20)	L-Thr-NH_2 (20)	subtilisin Carlsberg	C-N	*N*-CBZ-L-Thr-L-Thr-NH_2	73
N-CBZ-L-Thr-OCH_2CH_2Cl (20)	L-Thr-L-Thr-NH-Ind (20)	subtilisin Carlsberg	C-N	*N*-CBZ-L-Thr-L-Thr-NH-Ind	81
N-Ac-D-Ala-OCH_2CH_2Cl (3.7)	L-Ser-NH-Nph (3.7)	subtilisin BPN'	C-N	*N*-Ac-D-Ala-L-Ser-NH-Nph	66

[a] Bond formed in the acylation reaction.

[b] Yield of isolated product, %.

9.6 ENZYME-CATALYZED SEGMENT CONDENSATION OF PEPTIDES AND THE PROBLEM OF SECONDARY HYDROLYSIS

The construction of large peptides via segment condensation remains a very appealing and challenging target of peptide synthesis. The segments can be easily purified and characterized, reducing the need for laborious purification of the final product, which is often required for large peptides produced by stepwise solid-phase synthesis [33]. The chemical coupling of peptide segments often suffers from racemization, the low solubility of segments, and low yield [33]. Despite the attractiveness of protease-catalyzed peptide synthesis, the use of enzymes as catalysts for peptide segment coupling has been rare, mainly because of secondary hydrolysis of the polypeptide.

The use of organic solvent instead of water as a reaction medium should alleviate this problem. It may also help to increase the yield of kinetically controlled reactions, since it suppresses the hydrolysis of the ester of an N-protected amino acid. Indeed, 7.3 g (88% overall isolated yield) of Phe-LeuNH$_2$ was obtained following deprotection when a reaction between Z-Phe-OCH$_2$CH$_2$Cl and LeuNH$_2$ catalyzed by subtilisin took place in *tert*-amyl alcohol. However, only hydrolysis of Z-Phe-OCH$_2$CH$_2$Cl, but not peptide formation, has been noticed in water (10% or 40% DMF, pH 7 or 10) under identical conditions [25]. In another example Kitaguchi *et al.* [34] demonstrated that subtilisin Carlsberg affords an overwhelming selectivity (> 99%) toward the ε-NH$_2$ group in the reaction between *N*-Ac-Phe-OEt and Lys-*O*-*tert*-Bu both in water and an organic solvent. However, the degree of conversion in this reaction in water never exceeded 30% because of competing enzymatic hydrolysis. Replacement of water with *tert*-amyl alcohol as a reaction medium resulted in 85% isolated yield of *N*-Ac-Phe-(ε)-Lys-*O*-*tert*-Bu. Similar data were obtained when *Pseudomonas* sp. lipoprotein lipase was used as a catalyst in anhydrous toluene. A similar result has been reported by Bobonneau *et al.* [35] for peptide synthesis in a mixture of *tert*-amyl alcohol and trichloroethylene (50/50), catalyzed by poly(ethylene glycol)–modified α-chymotrypsin. These authors performed the kinetically controlled synthesis of several hydrophobic dipeptides and found limited hydrolysis of the starting esters and no hydrolysis of the newly formed peptides. The fragment coupling of an ester of Z-Phe with Phe-LeuNH$_2$ proceeded without secondary hydrolysis. The only product detected was Z-Phe-Phe-Leu-NH$_2$, without any traces of dipeptides Z-Phe-LeuNH$_2$ or Z-Phe-Phe. Coupling of Boc-Tyr-Gly-Gly-Phe-Cam and LeuNH$_2$ has also been performed in organic solvent without any hydrolysis of other sensitive peptide bonds [34].

Recently Kimura *et al.* reported the synthesis of Z-Tyr-Gly-Gly-Phe-LeuOEt from Z-Tyr-Gly-Gly and Phe-LeuOEt catalyzed by thermolysin immobilized on Amberlite XAD-7 in ethyl acetate [36]. The authors have synthesized the resulting pentapeptide in good yield (68%) and without protection of the side chain of Tyr. The same group has developed a continuous synthesis of Z-Gly-Phe-LeuNH$_2$ in ethyl acetate by successive condensation and transesterification catalyzed by immobilized thermolysin and α-chymotrypsin. The series plug-flow reactor was operated for 220 h with a yield of more than 80%. The residual activities of immobilized thermolysin and α-chymotrypsin were 94% and 100%, respectively [37].

A totally new approach to enzymatic segment condensation in organic solvents has been reported by Kitaguchi and Klibanov [30]. These authors attempted thermodynamically controlled segment condensation in *tert*-amyl alcohol catalyzed by thermolysin. Thermolysin turns out to be inactive in anhydrous *tert*-amyl alcohol. The addition of up to 4% water significantly increases enzyme reactivity towards formation of a tetrapeptide from Z-Gly-Gly-Phe and PheNH$_2$. However, this increase in activity is accompanied by substantial secondary hydrolysis. After 6 h, when 82% of Z-Gly-Gly-Phe has reacted, almost one-third of the product is the dipeptide

Z-Gly-Gly. To reconcile the opposing effects of water on the product yield and enzymatic reaction rate, the authors replaced part of the 4% water with dimethylformamide. Since water's role as a molecular lubricant in protein is due to its ability to form multiple hydrogen bonds, other solvents mimicking water in this respect (DMF, ethylene glycol, glycerol) may, at least partially, substitute for it without promoting hydrolytic reactions. When the synthesis of Z-Gly-Gly-Phe-Phe-NH_2 was performed in *tert*-amyl alcohol containing 1% water and 9% formamide, no formation of by-products was detected, in contrast to the situation observed at 4% water content. Three other polypeptides have been prepared via segment condensation under similar conditions (Table 9.5).

TABLE 9.5 Preparative Synthesis of Oligopeptides via Segment Condensation Catalyzed by Thermolysin in *tert*-Amyl Alcohol Containing Water and 9% of a Water Mimic. Reprinted with permission from [30], Kitaguchi, H., Klibanov, A. M., *J. Am. Chem. Soc. 111*, 9273 (1989). Copyright 1989, American Chemical Society.

COOH donor	NH_2 donor	Reaction product	Yield %[a]
Z-Gly-Gly-Phe	Phe-NH_2	Z-Gly-Gly-Phe-Phe-NH_2	76
Z-Gly-Gly-Phe	Phe-Phe-NH_2	Z-Gly-Gly-Phe-Phe-Phe-NH_2	72
Z-Gly-Pro-Phe-Pro-Leu	Leu-NH_2	Z-Gly-Pro-Phe-Pro-Leu-Leu-NH_2	73
Z-Gly-Pro-Gly-Gly-Pro-Ala	Leu-Leu-Phe-NH_2	Z-Gly-Pro-Gly-Gly-Pro-Ala-Leu-Leu-Phe-NH_2	67

[a] Yield of isolated product, %.

9.7 ENZYMATIC SYNTHESIS OF UNUSUAL CONJUGATES OF AMINO ACIDS

The significant change in thermodynamics, along with the broadening of substrate specificity and enantioselectivity of enzymes in organic solvents, allow for new transformations impossible in water [9]. One such transformation is the enzymatic acylation of carbohydrates with amino acids.

Sugars are reasonably soluble in only a few, very hydrophilic organic solvents, such as pyridine and dimethylformamide; unfortunately, most enzymes are inactive in these solvents. Even so, one enzyme, subtilisin Carlsberg, has been shown to exhibit a high level of catalytic activity even under these drastic conditions. This enzyme was used as a catalyst in the reaction of activated esters of several *N*-acetyl amino acids with model monosaccharides in anhydrous DMF [38]. This reaction shows a preference for the primary hydroxyl group of sugars. For instance, when glucose was employed as a nucleophile, the final product *N*-Ac-Phe-glucose consisted of three positional isomers: 6-*O* (80%), 3-*O* (15%) and 2-*O* (5%).

Recently we employed subtilisin-catalyzed acylation in pyridine for the synthesis of conjugates of amino acids with the alkaloid castanospermine [39], which has a number of important biological activities [40]. Castanospermine contains four secondary hydroxyl groups of similar reactivity (Figure 9.2) and therefore represents a challenging target for regioselective modification. The synthesis of individual esters of castanospermine requires 5–6 steps, involving protection and deprotection of neighboring hydroxyl groups. On the other hand, the enzymatic acylation of castanospermine with *N*-Ac-Phe-OCH_2CH_2Cl proceeds in one step to give exclusively 1-*O*-(*N*-acetylphenylalanyl) castanospermine with an isolated yield of 49%. Similar results have been obtained for acylation of castanospermine with activated esters of both *N*-Ac-L-Ala and *N*-Ac-D-Ala (Table 9.6).

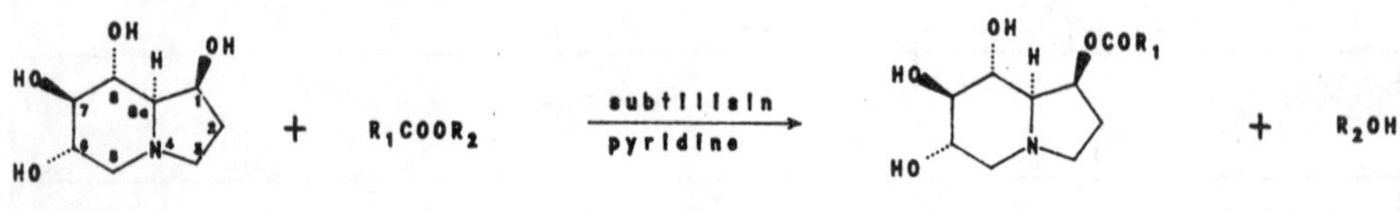

Figure 9.2. Subtilisin-Catalyzed Acylation of Castanospermine. R_1 is α-aminoalkyl; R_2 is chloroethyl.

TABLE 9.6 Preparative Regioselective Subtilisin-Catalyzed Synthesis of 1-*O*-Acylcastanospermine in Anhydrous Pyridine [39].

Acylating agent	Reaction time (h)	Isolated yield (%)
N-Ac-L-Phe-OCH_2CH_2Cl	96	49
N-Ac-L-Ala-OCH_2CH_2Cl	96	44
N-CBZ-L-Ala-OCH_2CH_2Cl	120	64
N-Ac-D-Ala-OCH_2CH_2Cl	120	31

9.8 SUMMARY

The advantages of enzymatic peptide synthesis, such as mild reaction conditions, absence of racemization, minimal protection and activation requirements, are well recognized. Several problems should be solved before enzymatic methods become a routine synthetic procedure. These problems—unfavorable thermodynamics of peptide synthesis in water, narrow substrate specificity and stereoselectivity of enzymes, and undesirable proteolysis of the polypeptide chain—may be overcome by the use of enzymes in nonaqueous media. Not only is the equilibrium of peptide synthesis in organic solvent shifted toward peptide bond formation, but also the substrate specificity of enzymes broadens and their stereoselectivity relaxes. On the other hand enzymes still possess a high level of chemo- and regioselectivity in nonaqueous media. This unique combination makes enzymatic catalysis in organic solvents a powerful tool for peptide synthesis.

9.9 REFERENCES

1. Fruton, J. S., *Adv. Enzymol. 53*, 239 (1982).
2. Oyama, K., Kihara, K., *CHEMTECH 14*, 100 (1984).
3. Jakubke, M.-D., Kuhl, P., Konnecke, A., *Angew. Chem. Int. Ed. Engl. 24*, 85 (1985).
4. Morihara, K., *TIBTECH 5*, 169 (1987).
5. Kullmann, W. *Enzymatic Peptide Synthesis*; CRC Press, Inc., Boca Raton, Florida, 1987.
6. Klibanov, A. M., *CHEMTECH 16*, 354 (1986).
7. Deetz, J. S., Rozzell, J. D., *TIBTECH 6*, 15 (1988).
8. Chen, C.-S., Sih, C. J., *Angew. Chem. Int. Ed. Engl. 28*, 695 (1989).
9. Klibanov, A. M., *Trends Biochem. Sci. 14*, 141 (1989).
10. Dordick, J. S., *Enzyme Microb. Technol. 11*, 194 (1989).
11. Fersht, A. *Enzyme Structure and Mechanism*; W. H. Freeman and Co., New York, 1985.
12. Zaks, A., Klibanov. A. M., *J. Am. Chem. Soc. 108*, 2767 (1986).
13. Gaertner, H., Puigserver, A., *Eur. J. Biochem. 181*, 207 (1989).
14. Ferjancic, A., Puigserver, A., Gaertner, H., *Biotechnol. Lett. 10*, 101 (1988).
15. Shoji, J., *Adv. Appl. Microbiol. 24*, 187 (1978).
16. Hansen, P. E., Morgan, B. A. In *The Peptides: Analysis, Synthesis, Biology*; Udenfriend, S., Meienhofer, J., Eds.; Academic Press, Orlando, 1984; Vol. 6, p 269.
17. Wolfe, S.' Demain, A. L., Jensen, S. E., Westlake, D. W. S., *Science 226*, 1386 (1984).
18. Geysen, H. M., Rodda, S. J., Mason, T. Y. In *Synthetic Peptides as Antigens*; Ciba Foundation Symposium No. 119; Wiley, Chichester, 1986; p 130.
19. See [11], p 405.
20. Petkow, D., Stoineva, I. B., *Tetrahedron Lett. 25*, 3751 (1984).
21. West, J. B., Wong, C.-H., *J. Org. Chem. 51*, 2728 (1986).
22. Fortier, G., MacKenzie, S. L., *Biotechnol. Lett. 8*, 873 (1986).
23. Barbas, C. F., Wong, C.-H., *J. Chem. Soc., Chem Commun.*, 533 (1987).
24. Sakurai, T., Margolin, A. L., Russell, A. J., Klibanov, A. M., *J. Am. Chem. Soc. 110*, 7236 (1988).
25. Margolin, A. L., Tai, D.-F., Klibanov, A. M., *J. Am. Chem. Soc. 109*, 7885 (1987).
26. Dewar, M. J. S., *Enzyme 36*, 8 (1986).
27. Blow, D. M., *The Enzymes 3*, 185 (1971).
28. Margolin, A. L., Klibanov, A. M., *J. Am. Chem. Soc. 109*, 3802 (1987).
29. West, J. B., Wong, C.-H., *Tetrahedron Lett. 28*, 1629 (1987).

30. Kitaguchi, H., Klibanov, A. M., *J. Am. Chem. Soc. 111*, 9273 (1989).
31. Chinsky, N., Margolin, A. L., Klibanov, A. M., *J. Am. Chem. Soc. 111*, 386 (1989).
32. Ruff, M. R., Martin, B. M., Ginns, E. I., Farrar, W. L., Pert, C. B., *FEBS Lett. 211*, 17 (1987).
33. Nakatsuka, T., Sasaki, T., Kaiser, E. J., *J. Am. Chem. Soc. 109*, 3808 (1987).
34. Kitaguchi, H., Tai, D.-F., Klibanov, A. M., *Tetrahedron Lett. 29*, 5487 (1988).
35. Babonneau, M.-T., Jacquier, R., Lazaro, R., Viallffont, P., *Tetrahedron Lett. 30*, 2787 (1989).
36. Kimura, Y., Nakanishi, K., Matsuno, R., *Enzyme. Microb. Technol. 12*, 272 (1990).
37. Kimura, Y., Yoshida, T., Muraga, K., Nakanishi, K., Matsuno, R., *Agr. Biol. Chem. 54*, 1433 (1990).
38. Riva, S., Chapineau, J., Kieboom, A. P. G., Klibanov, A. M., *J. Am. Chem. Soc. 110*, 584 (1988).
39. Margolin, A. L., Delinck, D. L., Whalon, M. R., *J. Am. Chem. Soc. 112*, 2849 (1990).
40. Fleet, G. W. J., Karpas, A., Dwek, R. A., Fellows, L. E., Tyms, A. S., Petursson, S., Hamgoong, S. K., Ransden, N. G., Smith, P. W., Son, J. C., Wilson, F., Witty, D. R., Jacob, G. S., Rademacher, T., *FEBS Lett. 237*, 128 (1988).

ASPARTAME SYNTHESIS IN A MISCIBLE ORGANIC SOLVENT

C. Bull, D. R. Durham, A. Gross, R. J. Kupper, and J. F. Walter

CONTENTS

10.1 INTRODUCTION: THE ASPARTAME MARKET

Aspartame is a synthetic sweetener marketed by The NutraSweet Co. [1]. Its structure, shown in Figure 10.1, contains the natural form of two amino acids, aspartic acid and phenylalanine, as a dipeptide methyl ester. Aspartame has some particularly desirable properties as a sweetener, being 200-fold sweeter than sucrose. In addition, its safety has been thoroughly investigated [2]. It has been formulated into a wide variety of products. Worldwide, sales of aspartame are estimated to exceed a billion dollars in 1990 [3]. The basic use patent [4] will expire in 1992 in the U.S., and has already expired for the Canadian and European markets.

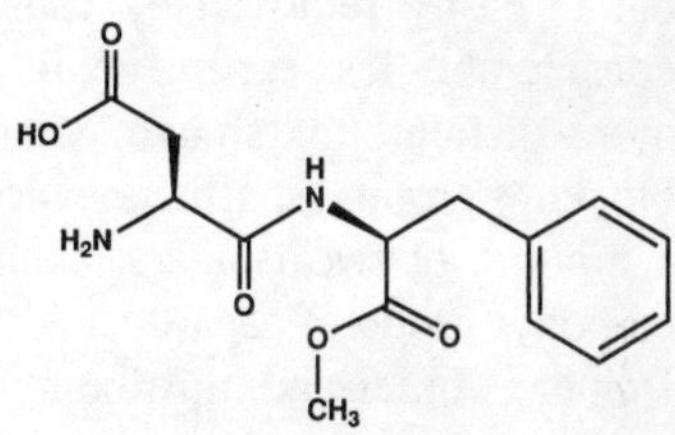

Figure 10.1. Aspartame Is the Methyl Ester of the Dipeptide L-Aspartyl-L-phenylalanine. If the phenylalanine is attached to the β carboxyl group instead, the bitter β isomer results.

10.2 ROUTES TO ASPARTAME SYNTHESIS

We will review here five general approaches to a cost-effective synthesis. The first is a chemical route wherein N-blocked aspartic anhydride is coupled with phenylalanine methyl ester. The second is similar except that coupling the anhydride is done with phenylalanine directly and the methyl ester is added later. The third general approach approach uses an activated, unblocked aspartic acid and phenylalanine methyl ester. The last two routes use an enzymatic coupling of blocked aspartic acid and phenylalanine methyl ester, and differ in how the reaction is driven in the direction of synthesis. For each of these general routes there may be several variations, for example, the nature of the blocking group, or whether the reaction can use racemic or chiral phenylalanine. After this review of the existing technology, we present our results using a new protease in a new solvent system, and an alternative hydroxy-aspartame route.

10.2.1 Blocked Aspartic Anhydride

The structures of two common blocking groups, formyl (f) and carbobenzyloxy (Z), on aspartic anhydride are shown in Figure 10.2. The Z group is more expensive

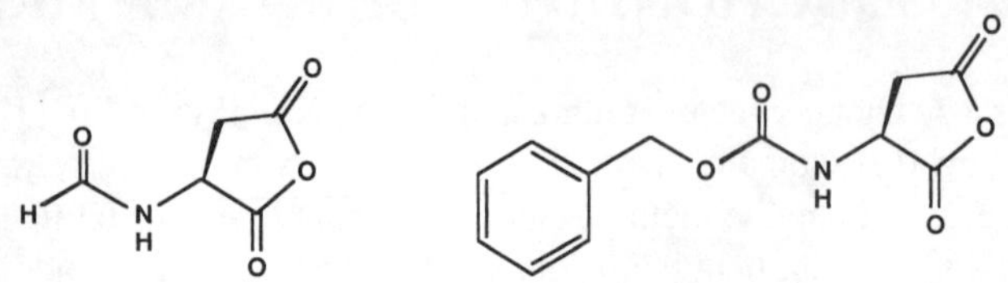

Figure 10.2. Structures of Formyl- and Z-Blocked L-Aspartic Anhydride. Both react with L-phenylalanine methyl ester to give a 4:1 mixture of the correspondingly blocked α:β aspartame.

but is easily removed by catalytic hydrogenation. Although the formyl group is considerably less expensive, its removal requires acid hydrolysis and so can be much more difficult to remove without cleaving off the methyl ester [5]. With either blocking group, aspartic anhydride couples with L-phenylalanine methyl ester to give a mixture of two isomers resulting from substitution at either the alpha or the beta carboxyl group of aspartic acid. The alpha isomer is the sweet isomer, while the beta isomer is bitter and must be removed. The anhydride coupling method gives at best a 4:1 ratio of the alpha to beta isomer. Due to the cost of phenylalanine, the beta isomer must be recovered and recycled for its phenylalanine value.

10.2.2 Blocked Aspartic Anhydride with Phenylalanine

This route couples the anhydride with L-phenylalanine directly. Again two isomers are formed, but these can be separated more easily than with the methyl ester. For the Z system in which the blocking group is easily removed by hydrogenation, there is no strong advantage to a process that adds the methyl ester to Z-asp-phe after coupling, compared to a process that couples Z-asp with L-phenylalanine methyl ester. However there is an advantage for the formyl system since the formyl group can be removed in the same step that forms the methyl ester. Normally an acidic methanolic esterification will give a mixture including some of the undesired ester of the beta carboxyl group of aspartic acid. A selective crystallization process has been described to continuously remove only the desired product [6]. This process results in the conversion a large fraction of the dipeptide f-asp-phe into aspartame and is believed to be the current industrial process.

10.2.3 Unblocked *N*-Thiocarboxy Anhydride

The structure of *N*-thiocarboxy aspartic anhydride (NTA) is shown in Figure 10.3. This compound reacts directly with L-phenylalanine methyl ester to give only the

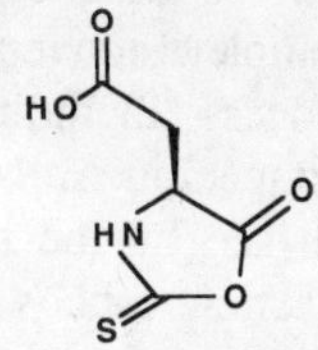

Figure 10.3. Structure of *N*-Thiocarboxy Aspartic Anhydride. Reaction with phenylalanine methyl ester gives only the α isomer of aspartame.

desired alpha isomer and no further deblocking step is necessary. The NTA route leaves traces of sulfur-containing compounds that must be removed by a controlled oxidation [7]. While the coupling step is particularly easy for this route, the synthesis of NTA in good yield is not completely simple [8]. The corresponding route using *N*-carboxy aspartic anhydride is possible [9] but suffers difficulties in forming the required anhydride and in the basic coupling conditions.

10.2.4 Enzymatic Coupling in Aqueous Solutions

Oyama and colleagues at Toyo-Soda showed that several neutral proteases could couple Z-aspartic acid with L-phenylalanine methyl ester [10]. Normally these enzymes catalyze complete peptide cleavage. However, the presence of the blocking group on the aspartic acid, and the methyl ester on the phenylalanine, change the the otherwise unfavorable thermodynamics so that some synthesis occurs in aqueous solution [11]. In the case of the Z blocking group, Z-aspartame forms a complex of limited solubility with excess phenylalanine methyl ester. By starting with a high concentration of substrates, the resulting product complex will precipitate and drive the overall reaction near completion [12]. *Thermolysin* (known as *thermoase* in its crude form) is reported to be a particularly favorable protease.

This enzyme is specific for L-phenylalanine methyl ester, allowing a process to be run with DL-phenylalanine methyl ester. Only the desired product, Z-L-aspartyl-L-phenylalanine methyl ester, is formed and precipitates as the salt with D-phenylalanine methyl ester. The salt can be split and the D-phenylalanine methyl ester recovered, racemized to DL-phenylalanine and recycled to methyl ester synthesis. This process does not work with f-aspartic acid for two reasons. First, the enzyme binds this material very poorly so that rates are extremely low. Second, the resulting complex is more soluble so the overall yield is lowered [13].

10.2.5 Enzymatic Coupling in Biphasic Systems

Oyama also found that blocked aspartame will partition into organic solvents much more readily than the substrates [14]. Thus adding an immiscible organic solvent increases the overall production of blocked aspartame. This effect increases as the ratio of organic phase to aqueous phase is increased and can be used to force nearly complete conversion of substrates into product. There are several variations on this process, including one wherein a thin aqueous layer of enzyme is prepared by dampening Amberlite™ beads with enzyme solution. The enzyme is trapped on the surface by the addition of an organic phase containing the substrates. Once the reaction reaches completion, the organic solution is removed and replaced with a fresh substrate solution. If the enzymatic activity is lost, the beads may be simply washed and coated with new enzyme. Repeated batchwise operation of these "immobilized" enzyme beads performed better than packing them into a continuous column reactor [15].

Although a buffer is used in the enzyme solution, it is generally present at much lower levels than the two substrates. Thus the aqueous solution pH is controlled by the ratio of these substrates in the organic phase. Z-aspartic acid and L-phenylalanine methyl ester are not drastically different in partitioning between the aqueous and organic phases, so that the aqueous pH values near 6 are easily obtained. The f-aspartic acid partitions strongly into an aqueous phase, and cannot be used in this mode because it would make the aqueous phase very acidic and compromise enzyme stability.

10.3 ENZYMATIC COUPLING IN WATER-MISCIBLE SYSTEMS

An enzymatic synthesis was attractive in that it avoided the isomer separation, and gave high yields. In addition, the use of enzymes in organic solvents was relatively new and provided hitherto unsuspected opportunities. While enzymatic coupling of aspartame precursors had been studied in miscible organic solvents using thermolysin, the results had not looked promising [16]. Furthermore the cost of thermolysin (then $300/lb in quantity for a crude material containing only 1/5 active protease) required that the enzyme have an extended lifetime in whatever reactor configuration was used.

10.3.1 The Neutral Protease Vibriolysin

Therefore, we started screening for other proteases which might be synthetically active. The most promising candidate to emerge from this program was a neutral protease designated *vibriolysin* from *Vibrio proteolyticus.* This microorganism was first described as *Aeromonas proteolytica* in 1964 [17]. Both a neutral protease and an aminopeptidase have been isolated from fermentations of this organism [18]. Fermentation titers have been modest, in the range of 0.5 mg/ml [19].

Improvements here in fermentation media and the isolation of a mutant strain resulted in a considerable increase in titer and a simplification of the recovery [20].

The gene for this enzyme has been cloned and sequenced [21]. The enzyme was recovered in two forms, a smaller mature form (34,760 daltons), as described by Prescott, and larger precursor form (44,800 daltons), which appears more hydrophobic. We found conditions which maximized production of the unprocessed form and the work described here used crude preparations containing over 80% this unprocessed form.

This enzyme differs from thermolysin in having lower thermostability and in requiring much less Ca^{2+} for stability. It has none of the unusual solubility characteristics of thermolysin, which in pure form requires high salt concentrations for easy dissolution. Vibriolysin was prepared and stored at 4 °C for extended periods at high concentrations without substantial loss due to autoproteolysis. Furthermore the more hydrophobic higher molecular weight form had some interesting properties in organic solvents.

10.3.2 The Acetonitrile Solvent System

Initially the synthesis of Z-aspartame was demonstrated in a two-phase organic solvent system using vibriolysin trapped on Amberlite™ beads. However the enzyme did not appear as stable on these beads as thermolysin. Although the initial cycle of synthesis proceeded at a reasonable rate, in subsequent cycles the activity diminished greatly. Protein analysis of spent beads showed that most of the enzyme was no longer on the beads. This suggested that the suspended enzyme might be contributing to synthesis. Therefore, a series of synthetic experiments was done with lyophilized enzyme added directly to a variety of organic solvents, including water-miscible solvents such as methanol, dimethylsulfoxide, and acetonitrile. Some solvents such as methanol and DMSO rapidly inactivated the enzyme, but acetonitrile proved not only to preserve activity but also to permit some synthetic activity. Some water (2% to 30%) had to be present in order to get a reasonable synthetic rate.

Lyophilized protease powders are tedious to prepare on a larger scale and present a potential health hazard due to electrostatically charged dusts. Therefore we tried preparing enzyme precipitated using acetonitrile, in the same manner as acetone powders are prepared. The resulting powders proved quite active when suspended in acetonitrile with substrates. Indeed, merely adding acetonitrile to a concentrated solution of enzyme could produce a finely flocculated material with particularly desirable properties. For example, flocculated enzyme sedimented well, could be resuspended easily and appeared indefinitely stable.

The flocculated enzyme was easy to use in kinetic studies. Activity of an enzyme slurry could be tested by simply diluting an aliquot into a tenfold larger volume of aqueous buffer. The enzyme activity remaining was assayed using FAGLA [22]. Since the presence of substrates interfered with this assay, samples containing substrates were washed by centrifuging the slurry and resuspending the pellet in pure acetonitrile. Typical recovery of activity was over 90% of the initial activity present before flocculation. All these slurries were so very finely divided that pipetting operations were convenient and quantitative as long as the pipette orifice was greater than a millimeter. Digital positive displacement pipettes were required for kinetic studies. Air displacement pipettes were avoided due to the higher vapor pressure of acetonitrile.

10.3.3 Influence of Water Content

The synthetic activity of the flocculated enzyme was dependent on the water content of the solvent. Initial rate studies showed a direct dependence on water content in the range of 2% to 30% water. However the increased rate noted when the solvent had the highest water content did not prove useful because the long-term enzyme stability was greatly lowered. Phase separation occurred after several days, with formation of gummy lumps of inactive protein. This did not happen with solvents containing less than 20% water.

The equilibrium constant was also dependent on water content. Since the coupling reaction releases one mole of water per mole of product, the equilibrium level of product synthesized is expected to be inversely dependent on water content. Therefore one must balance synthetic rate and overall yield. We were able to obtain 95% conversion of blocked aspartic acid into product in 10% water and 90% acetonitrile and still retain reasonable synthetic rates.

10.3.4 Kinetic Studies

Several important features of vibriolysin as a coupling enzyme emerged from early kinetic studies in 80% acetonitrile/20% water. Two reactors were started, with the same amount of slurried enzyme and with the same 1:3 ratio of two substrates, formyl-aspartic acid and L-phenylalanine methyl ester. The two reactors differed fourfold in the concentration of substrates: the initial concentrations of formyl-aspartic acid were 80 mM and 20 mM, respectively, and the corresponding concentrations of L-phenylalanine methyl ester were 240 mM and 60 mM. HPLC was used to monitor synthesis of formyl-aspartame (results are shown in Figure 10.4). The most important conclusion was that vibriolysin, unlike thermolysin, freely accepted formyl-aspartic acid as a substrate. The second conclusion was that the reaction proceeded nearly to completion, with conversion of over 90% of the 80 mM formyl-aspartic acid into formyl-aspartame. The enzyme was found to retain full activity at the end of these reactions. Thus this system showed considerable potential utility in aspartame synthesis.

Two other observations from Figure 10.4 were that the first half of each reaction was reasonably linear with time and that this rate differed fourfold between the

Figure 10.4. Synthesis of Formyl Aspartame by Vibriolysin in 80% Acetonitrile. Two reactors, containing 60 mM and 240 mM phenylalanine methyl ester, were started with 20 mM and 80 mM formyl aspartic acid, respectively.

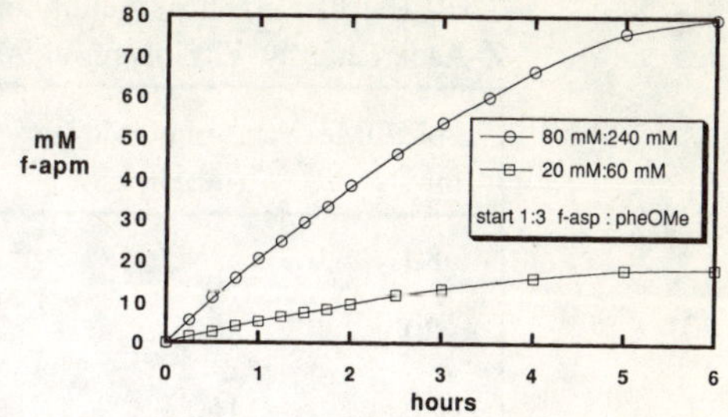

two reactors. Had both substrates been above their respective K_m values, the rates of the two reactors would have been the same. Had both substrates started below their respective K_m values, the two reactors would have differed sixteenfold in rate. Only if one substrate was below its K_m and the other were well above its K_m would there be a fourfold difference with a reasonably linear progress curve. The limiting substrate, formyl-aspartic acid, would also have to be the tighter binding in order to get the linear progress curve. These observations suggest that the reactions started with substrate concentrations above the K_m for formyl-aspartic acid ($K_{m\text{-FA}}$) and below the K_m for phenylalanine methyl ester ($K_{m\text{-PheOMe}}$).

However, it was clearly desirable to run an industrial process at the highest practical concentration without added buffers. In this case, where one substrate is a diacid and the other a base, the ratio of substrates controls the acidity of the solution. Holding the concentration of one substrate fixed while systematically varying the other will change the apparent pH of the solvent system and may alter the kinetics. Attempts at pH control through addition of ammonia or phenylalanine methyl ester hydrochloride both resulted in severe clumping of the flocculated enzyme, compromising any rate studies. Therefore an indirect method had to be used to study the enzymatic affinity towards its substrates.

10.3.5 Use of D-Phenylalanine as Buffer

D-Phenylalanine methyl ester has been shown to bind weakly or not at all to thermolysin in the biphasic system. If this were also true for vibriolysin in acetonitrile, D-phenylalanine methyl ester could be used to adjust the apparent pH, independent of changes in the L-phenylalanine methyl ester concentration. The initial rate of formation of Z-aspartame in the presence of varying amounts of D- and L-phenylalanine methyl ester is shown in Table 10.1. Z-aspartic acid was held constant in this experiment at 80 mM. The apparent acidity of the solution was then controlled by the total of the concentrations of D- and L-phenylalanine methyl ester shown in the first two columns of Table 10.1. In this table, data taken at equivalent acidity have been arranged into groups, and the groups ordered by decreasing acidity.

TABLE 10.1 Effect of Acidity on Synthesis of Z-Aspartame by Vibriolysin in 90% Acetonitrile*

L-pheOMe (mM)	D-pheOMe (mM)	Initial rate (g/h·g enz)	Initial rate/ L-pheOMe
80	–	0.4	0.005
80	80	0.7	0.009
80	160	1.0	0.013
240	–	3.5	0.015
240	80	3.9	0.016
320	–	5.5	0.017

*Z-aspartic acid was 80 mM initially for each experiment. The total amount of added phenylalanine methyl ester controlled the acidity, resulting in four different acidities. The last column shows the effect of the different acidities on the enzymatic activity after correction for the varying concentrations of L-phenylalanine.

The third and fourth columns in Table 10.1 show the results expressed in two ways. The third column shows the initial rate of Z-aspartame synthesis as grams of Z-aspartame produced per hour per gram of flocculated enzyme. These data were then divided by the L-phenylalanine methyl ester concentration to obtain the fourth column. This procedure should give a rate constant for binding L-phenylalanine methyl ester to the enzyme at each acidity. These data show that the reaction rate decreases as acidity increases, and that D-phenylalanine has little effect on the rate.

10.3.6 Product Inhibition with the Z Protecting Group

Use of the Z protecting group allowed operation at higher concentrations than reactions run using the formyl protecting group. The limiting feature in both cases is precipitation of the blocked aspartame-phenylalanine methyl ester salt from the acetonitrile solvent onto the slurried enzyme. This precipitation occurred at much higher concentrations than in a purely aqueous solution (in the Z case at 30 °C in 90% acetonitrile, precipitation occurred only above 0.2 M product). This was over 8% (by weight) product in solution, desirable from an engineering perspective.

However, product inhibition became important at these higher concentrations (Figure 10.5). The progress curves were no longer linear in the earlier stages of the reaction. Two reactors were started with identical enzyme content and 400 mM

Figure 10.5. Use of Z-Aspartic Acid during Z-Aspartame Synthesis by Vibriolysin. Two reactors were started with 400 mM phenylalanine methyl ester. One also contained 200 mM Z-aspartic acid while the other contained both Z-aspartic acid and Z-aspartame to mimic 75% completion of synthesis. Both reactors showed an exponential approach to equilibrium with the same half-life.

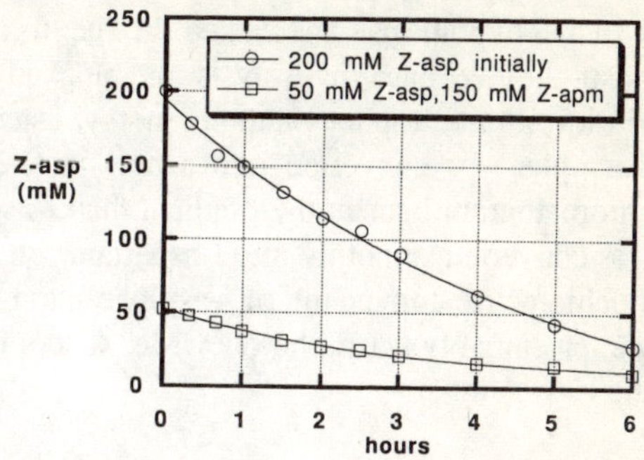

L-phenylalanine methyl ester. One reactor used 200 mM Z-aspartic acid and the other 50 mM Z-aspartic acid and 150 mM Z-aspartame (that is, it was 3/4 reacted initially). Both showed an exponential approach to equilibrium with the identical rate constant (a two-hour half-life for unreacted Z-aspartic acid). Normally such product inhibition would limit the concentrations which may be used practically, but in this system the overall productivity is not adversely affected. Basically, increasing the total concentration also increases the rate due to the higher concentration of L-phenylalanine methyl ester, offsetting the effect of product inhibition.

10.3.7 Control of By-Products

All of the above results depended on our obtaining enzyme that gave minimal by-product formation. Crude vibriolysin contains low levels of an esterase activity that limited yield. The limited yield was caused by accumulations of phenylalanine and of the cyclized phenylalanine dipeptide, dibenzyldiketopiperazine (DBDKP). No loss of blocked aspartic acid occurred. Phenylalanine was formed by direct action of the esterase, but large amounts of DBDKP were formed in the absence of a blocked aspartic substrate or product. DBDKP was particularly troublesome because it was poorly soluble in the acetonitrile solvent and accumulated in continuous reactors. The following hypothesis can account for the formation of DBDKP.

It appears that the esterase is more active on the blocked aspartame product than on L-phenylalanine methyl ester. For the Z system, cleavage of the product methyl ester would lead to production of the dipeptide Z-asp-phe. This would also be a substrate for vibriolysin (probably a better substrate than Z-aspartic acid) and lead to synthesis of the tripeptide Z-asp-phe-pheOMe. Such a tripeptide would also serve as a substrate for vibriolysin with cleavage at the adjacent peptide bond, leading to Z-aspartic acid and the dipeptide phe-pheOMe. The latter will spontaneously cyclize to form DBDKP. The former would be converted into product normally. Thus, formation of DBDKP is an indirect result of product methyl ester cleavage resulting from the action of both enzymes.

This hypothesis was tested in the formyl system by preparing the dipeptide f-asp-phe by base hydrolysis of blocked product and reacting that with crude vibriolysin and L-phenylalanine methyl ester in acetonitrile. Within one hour, all the f-asp-phe was converted into a new material eluting from a reversed-phase HPLC chromatogram later in the gradient than f-aspartame. Subsequently this new material was converted smoothly into f-aspartame and DBDKP. This experiment demonstrated stochiometric conversion of f-asp-phe into f-aspartame and DBDKP via an intermediate, presumably f-asp-phe-pheOMe. There is every reason to expect similar events in the Z system.

10.3.7.1 Elimination of Esterase Activity

The esterase activity in crude vibriolysin was correlated with the aminopeptidase activity described in this organism by Prescott [23]. This enzyme and vibriolysin were difficult to separate using common column chromatographic methods, being similar in size, shape, and hydrophobicity. A study of the pH stability of the two enzymes led to a much simpler solution to the problem [24] as shown in Figure 10.6. At low pH, both the aminopeptidase and vibriolysin were prone to lose their cofactor, zinc. This effect was reversible and did not provide a long-term solution. At high pH, however, the loss of aminopeptidase activity was irreversible. Treatment of concentrated vibriolysin at pH 11.6 in 0.5% sodium carbonate solution for six hours could completely eliminate formation of DBDKP in the acetonitrile coupling reactors.

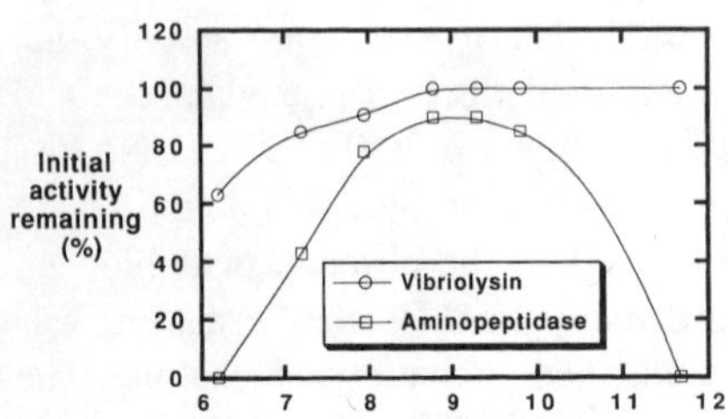

Figure 10.6. Stability of Vibriolysin and Aminopeptidase as a Function of pH. Samples were incubated overnight in various buffers at room temperature.

10.3.7.2 Other Minor By-Products

Another series of reactions involving a tripeptide may account for the presence of trace amounts of unblocked aspartame in the Z reaction system (generally <1%). Typical preparation of Z-aspartic acid leads to concomitant production of an impurity, the dipeptide Z-asp-asp. This should also be a substrate for the enzyme, leading to formation of Z-asp-asp-pheOMe. Again, this tripeptide could be split enzymatically into Z-asp and asp-pheOMe and the Z-asp will eventually appear as product. While these conversions have not been tested directly, they are plausible and would explain why crude Z-asp can be used in enzyme coupling reactions. [25] The production of unblocked aspartame is not directly a problem. However, if it cyclizes during subsequent treatment, then a small amount of benzylcarboxymethyldiketopiperazine will be found. Such by-products were generally produced in only trace quantities.

Using the formyl blocking group, there will be a by-product resulting from chemical transformylation during the reaction. In this reaction, pheOMe displaces aspartic acid from formyl-aspartic acid. Aspartic acid is a poor substrate while formyl-phenylalanine methyl ester is fully blocked, so neither reacts further. Similarly, transformylation of product can produce asp-pheOMe and hence benzylcarboxymethyl-diketopiperazine. Both processes were slow compared to the enzymatically catalyzed reaction and hence contributed only trace levels of by-product.

10.4 DESIGN OF CONTINUOUS REACTORS

The flocculated enzyme and acetonitrile solvent system proved very amenable to the design of continuous reactors. On a small scale (15 to 500 ml), standard ultrafiltration cells, with modifications, from Spectrum Medical Industries worked well. The polycarbonate barrels of these cells must be replaced by equivalent glass barrels that easily handle the low operating pressures and solvents used. Teflon membranes ranging from 100,000-MW cutoff to 0.1-μm pores were held in place with silicone o-rings from Apple Rubber Products. These membranes provided remarkably high flow rates with acetonitrile, and showed little tendency to become fouled by the flocculated enzyme. A typical 90-ml CSTR reactor with a 43-mm membrane containing 20 mg/ml enzyme could be run at 50 ml/hr flow rate with only 20 cm of acetonitrile head pressure. Gravity feed from a Mariott flask was typically used, with the substrate solution replenished every few days. Continuous operation at room temperature for over thirty days was demonstrated.

For larger scale tests, a mini-pilot plant was assembled. In this process, when a 2.5-l batch reactor containing 20 mg/ml flocculated enzyme reached the desired conversion, the slurry was concentrated using a ceramic microporous filter (Norton) and a standard magnetically coupled gear pump. The slurry could be concentrated to about 100 mg/ml before becoming too sluggish to pump. The collected permeate was sent on through acidic ion exchange resins which retained excess L-phenylalanine methyl ester from the acetonitrile solution containing product. Fresh substrate solutions were charged at that point through the ceramic filter. L-Phenylalanine methyl ester was recovered from the ion exchange columns with a solution of ammonia in acetonitrile, stripped of solvent, and recycled to the process. A schematic of the mini-pilot plant is shown in Figure 10.7 and the results for the first three operating cycles at two enzyme loads are given in Table 10.2.

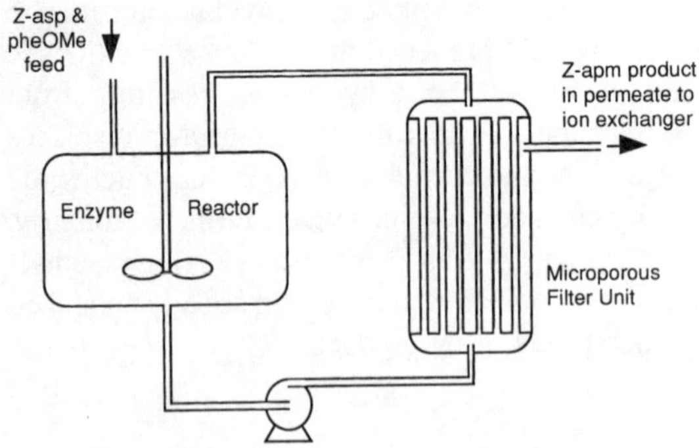

Figure 10.7. Mini-Pilot Plant for Synthesis of Z-Aspartame using Vibriolysin in 90% Acetonitrile. Once the reaction was complete, the enzyme slurry was concentrated fivefold with the microporous filter; 80% of the product was recovered. Fresh substrates were then added to start the next cycle.

TABLE 10.2 Performance of Mini-Pilot Plant Using Slurried Vibriolysin to Produce Z-Aspartame at 30 °C

Cycle	Final Z-apm concentration (g/l)	Conversion of Z-asp (%)	Selectivity based on Z-asp used (%)	Selectivity based on pheOMe used (%)
20 g/l vibriolysin, 5.3 h reaction time				
1	77	98	98	96
2	76	95	98	94
3	80	96	97	97
6 g/l vibriolysin, 24 h reaction time				
1	73	88	98	94
2	74	93	97	92
3	79	92	96	89

10.5 HYDROXYASPARTAME

An interesting variant on this process involved the use of phenylserine methyl ester instead of phenylalanine methyl ester [26]. Vibriolysin was specific for only one of the four possible phenylserine stereoisomers: L-*erythro*-phenylserine. The product was an N-blocked hydroxyaspartame. Hydroxyaspartame has a similar sweetness to aspartame. It also can be hydrogenated conveniently to remove the benzylic hydroxyl group, giving aspartame as the final product.

Enzymatic coupling using L-*erythro*-phenylserine methyl ester was slightly faster than using L-phenylalanine methyl ester, with similarly high yields. L-*erythro*-Phenylserine methyl ester could be prepared by a process described below. Neat L-*erythro*-phenylserine methyl ester was stable for weeks if kept cold. Prepared substrate solutions also containing blocked aspartic acid in acetonitrile showed no tendency to undergo retroaldol cleavage to glycine methyl ester and benzaldehyde when stored for several days at room temperature. The overall attractiveness of this route depended primarily on the relative expense of L-*erythro*-phenylserine methyl ester versus L-phenylalanine methyl ester.

10.5.1 Serine Hydroxymethyltransferase

Synthesis of both L-*erythro*-phenylserine and L-*threo*-phenylserine can be catalyzed by the enzyme *serine hydroxymethyltransferase* (*SHMT*) [27]. This enzyme condenses glycine with benzaldehye to give only the L configuration at the alpha carbon, but shows less stereochemical preference at the beta carbon, as shown in Figure 10.8. L-*erythro*-Phenylserine is less stable than L-*threo*-phenylserine by a factor of three, but is the kinetically favored isomer. This enzyme is ubiquitous in nature, and has been cloned [28]. It is highly active in synthesizing L-*erythro*-phenylserine.

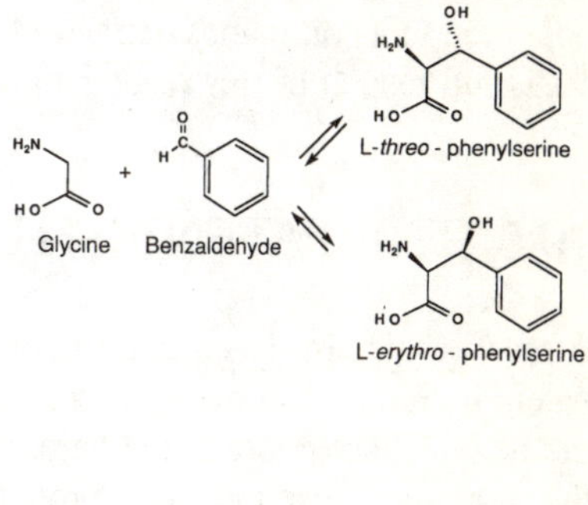

Figure 10.8. Serine Hydroxymethyltransferase Can Produce Two Phenylserine Isomers. While the L-*erythro* isomer is kinetically favored, the L-*threo* isomer is thermodynamically more stable by threefold.

A convenient method for condensing glycine and benzaldehyde to provide the phenylalanine portion of aspartame could be a distinct advantage economically. Unfortunately, the equilibrium constant for L-*erythro*-phenylserine formation does not favor extensive synthesis in water. At 130 g/l glycine and saturating benzaldehyde, only 16 g/l of L-*erythro*-phenylserine was formed at equilibrium. Since the final solution at equilibrium also contains about 100 g/l glycine and 50 g/l L-*threo*-phenylserine, separation and purification of the desired product is difficult.

This problem was solved by maintaining kinetic control of the product, rather than equilibrium control. Raising the pH had a distinct effect on the rates at which the two isomers are synthesized by SHMT, slowing that for L-*threo*-phenylserine more than that for L-*erythro*-phenylserine. Also helpful were lowering the benzaldehyde concentration and limiting the extent of the reaction. Our studies showed that the best ratio of L-*erythro*-phenylserine to L-*threo*-phenylserine that could be achieved was 30:1 by using all three factors. However, running a reaction at low concentrations and rates to a partial yield has never appealed to industrial engineers.

10.5.2 Extractor Reactor for L-*erythro*-Phenylserine

A process for continuously extracting product removed these severe limitations to purity, rate, and yield. In this process, an immiscible organic phase containing benzaldehyde continuously extracted product from a high-pH reactor as the Schiff base of L-*erythro*-phenylserine. The organic phase was separated and contacted to an acidic aqueous phase. The acid reversed the Schiff base and re-extracted the product. The benzaldehyde liberated from the Schiff base partitioned back into the organic phase and was returned to the reactor. L-*erythro*-Phenylserine would accumulate in the acidic aqueous phase to high concentration. The success of this scheme depended on finding a solvent which would not extract the substrate glycine. A 200-fold better extraction of L-*erythro*-phenylserine than glycine was obtained using 40% propyl acetate, 40% *n*-butanol and 20% benzaldehyde. This was sufficient to give reasonable operation to a continuous extractor [29]. A schematic of the reactor is given in Figure 10.9, and results of this reactor in operation are given in Table 10.3.

10.6 THE PATENT SITUATION

A market this size has attracted considerable attention, reflected in the large body of patent literature dealing with aspartame synthesis. Despite this extensive work, all of the new technology reported here, involving the use of vibriolysin in acetonitrile for the coupling reaction, the production of hydroxyaspartame, and the selective production of L-*erythro*-phenylserine, proved patentable. Although we have not

Figure 10.9. Extractor Reactor for the Continuous Production of L-*erythro*-Phenylserine Using Serine Hydroxymethyltransferase. Product is separated from the enzyme in the dialysis cartridge and contacts the organic phase containing benzaldehyde. The Schiff base of the product is preferentially extracted, while unreacted glycine and fresh benzaldehyde in solution are returned to the enzymatic reactor. The organic phase containing the product Schiff base is contacted to an acidic aqueous phase which reverses the Schiff base. The benzaldehyde rejoins the organic phase while the product accumulates in the acidic aqueous phase.

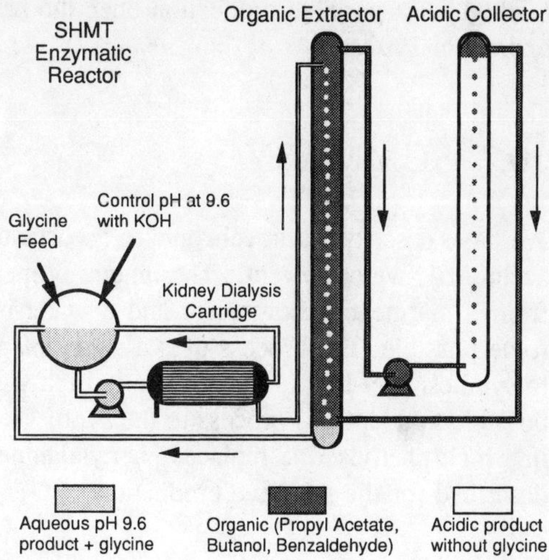

TABLE 10.3 Production of L-*erythro*-Phenylserine Using Serine Hydroxymethyltransferase in the Extractor Reactor*

Time of operation (h)	L-*erythro*-phenylserine (g/l)	L-*threo*-phenylserine (g/l)	Benzaldehyde (g/l)	Glycine (g/l)
1	0	0	4	< 4
3	0.5	0.1	4	< 4
6	3.3	0.3	4	< 4
9	5.2	0.3		< 4
12	8.3	0.6		< 4
18	10	1.0		< 4
27	16	1.3	4	< 4
36	21	1.3	3	< 4
Comparative production without the extractor reactor:				
Equilibrium	6	20	4	144

*Samples of the acidic collecting phase were taken at the indicated times and analyzed by HPLC.

discussed pricing explicitly in this chapter, these routes can provide a cost-competitive method for aspartame production once the basic patent on aspartame as a sweetener expires in 1992.

10.7 SUMMARY

We have described a novel route to aspartame using enzymatic coupling in a mixed acetonitrile-water solvent. The unique properties of the neutral protease vibriolysin from a marine microorganism and a water-miscible solvent system made this new route possible. The process uses a slurry of precipitated enzyme in 90% acetonitrile to synthesize N-blocked aspartame in high yield at high concentration. It also proved possible to adapt this process to the synthesis of a new sweetener hydroxyaspartame, in which phenylserine replaces phenylalanine. A second enzymatic process is also described for the selective production of L-*erythro*-phenylserine.

10.8 REFERENCES

1. Mazur, R. H., Schlatter, J. M., Goldkamp, A. H., *J. Am. Chem. Soc. 97*, 2684 (1969).
2. *Aspartame: Physiology and Biochemistry*; Stegink, L. D., Filer, L. J., Eds.; Marcel Dekker, New York, 1984.
3. *Sweeteners: Discovery, Molecular Design, and Chemorecognition*; Walters, D. E., Orthofer, F. T., DuBois, G. E., Eds.; American Chemical Society, Washington, 1991; p 313.
4. Schlatter, J. M., U.S. Pats. 3,492,131 (1970) and 3,642,491 (1972).
5. Takemoto, T., Kakizaki, F., Ariyoshi, Y., U.S. Pat. 4,071,511 (1978).
6. Bachman, G. L., Vineyard, B. D., U.S. Pat. 4,173,562 (1979).
7. Vinick, F. J., Jung, S., *Tetrahedron Lett. 23*, 1315 (1982); Vinick, F. J., U.S. Pat. 4,238,392 (1980).
8. Vinick, F. J., Jung, S., *J. Org. Chem. 47*, 2199 (1982); Vinick, F. J., U.S. Pats. 4,256,897 (1981) and 4,321,391 (1982).
9. Blacklock, T. J., Shuman, R. F., U.S. Pat. 4,510,083 (1985).
10. Isowa, Y., Ohmori, M., Ichikawa, T., Mori, K., Nonaka, Y., Kihara, K., Oyama, K., Satoh, H., Nishimura, S., *Tetrahedron Lett. 28*, 2611 (1979).
11. Kuhlmann, W. *Enzymatic Peptide Synthesis*; CRC Press, Boca Raton, FL, 1987.
12. Ichikawa, T., Isowa, Y., Kurita, H., Mori, K., Ohmori, M., Oyama, K., Sato, M., U.S. Pat. 4,116,768 (1978); Ichikawa, T., Isowa, Y., Kurita, H., Mori, K., Ohmori, M., Sato, M., U.S. Pat. 4,119,493 (1978); Ichikawa, T., Isowa, Y., Kihara, K., Mori, K., Nishimura, S., Nonaka, Y., Ohmori, M., Oyama, K., Satoh, H., U.S. Pat. 4,165,311 (1979); Ichikawa, T., Isowa, Y., Kihara, K., Mori, K., Nishimura, S., Nonaka, Y., Ohmori, M., Oyama, K., Satoh, H., U.S. Pat. 4,256,836 (1981); and Ichikawa, T., Isowa, Y., Kihara, K., Mori, K., Nishimura, S., Nonaka, Y., Ohmori, M., Oyama, K., Satoh, H., U.S. Pat. 4,436,925 (1984).

13. Likos, J. J., Glover, G. I., Eur. Pat. Applic. 0 149 594 and Internat. Pat. Applic. WO 85/04924 (1985); (Stepanov, V. M., Lyublinskaya, L. A., Chikindas, S. E., Russian Applic. SU86/00010 {1986}).
14. Oyama, K., Nishimura, S., Nonaka, Y., Kihara, K., Hashimoto, T., *J. Org. Chem. 46*, 5241 (1981); Oyama, K., Nishimura, S., Nonaka, Y., Hashimoto, T., Kihara, K., U.S. Pat. 4,284,721 (1981).
15. Nakanishi, K., Kamikubo, T., Matsuno, R., *Biotechnology 3*, 459 (1985).
16. Rodgers, P. D., Durrant, I., Benyon, R. J. In *Biocatalysis in Organic Media*; Laane, C., Tramper, J., Lilly, M. D., Eds.; Elsevier, Amsterdam, 1986; pp 405–409.
17. Merkel, J. R., Traganza, E. D., Mukherjee, B. B., Griffin, T. B., Prescott, J. M., *J. Bact. 87*, 1227 (1964).
18. Griffin, T. B., Prescott, J. M., *J. Biol. Chem. 245*, 1348 (1970).
19. Wilkes, S. H., Prescott, J. M., *Methods Enzymol. XIV*, 404 (1976).
20. Fortney, D. Z., Durham, D. R., unpublished.
21. David, V. A., Deutch, A. H., Sloma, A., Pawlyk, D., Ally, A., Durham, D. R., *Gene*, in press; Deutch, A. H., David, V. A., U.S. Pat. 4,966,846 (1990).
22. FAGLA is a synthetic peptide which allows spectroscopic monitoring of activity of a variety of proteases: Feder, J., *Biochemistry 6*, 2088 (1967).
23. Prescott, J. M., Wilkes, S. H., *Methods Enzymol. 45B*, 530 (1976).
24. Durham, D. R., *Appl. Environ. Microbiol. 56*, 2277 (1990).
25. Harada, T., Irino, S., Nakamura, M., Oyama, K., U.S. Pat. 4,521,514 (1985).
26. Chmurny, A. B., Gross, A. T., Kupper, R. J., Roberts, R. L., U.S. Pats. 4,710,583 (1987); 4,810,817 (1989); and 4,873,359 (1989).
27. Schirch, L., Diller, A., *J. Biol. Chem. 246*, 3961 (1971); Nakazawa, H., Enei, H., Kubota, K., Okumura, S., U.S. Pat. 3,871,958 (1975); and Ulevich, R. J., Kallen, R. G., *Biochemistry 16*, 5355 (1977).
28. Plamann, M. D., Stauffer, L. T., Urbanowski, M. L., Stauffer, G. V., *Nucleic Acids Res. 11*, 2065 (1983); and Stauffer, G. V., Plamann, M. D., Stauffer, L. T., *Gene 14*, 63 (1981).
29. Walter, J. F., Bull, C., U.S. Pat. 5,102,792 (1992).

ENZYMATIC PRODUCTION OF SULFUR- AND SELENIUM-CONTAINING AMINO ACID ANALOGS

Nobuyoshi Esaki and Kenji Soda

CONTENTS

11.1 INTRODUCTION

Amino acids have been prepared by isolation from natural materials, by microbial and enzymatic procedures, or by chemical synthesis. The first two procedures give optically active (usually L) amino acids. However, the chemical methods generally produce the racemates: an additional optical resolution step is required to obtain optically active amino acids. Auxotrophs and regulatory mutants have been developed to produce optically active amino acids on an industrial scale. Amino acids can be obtained by fermentation of simple and cheap raw materials, but the production is time-consuming and requires separation of the amino acids from the other amino acids and various contaminants including microbial cells. The waste may cause water pollution, in particular when molasses is used as a carbon source. The enzymatic procedure replaces the microbial one. It is less time-consuming and usually more efficient, and does not produce complex impurities and wastes. However, substrates are more expensive than the materials used in the microbial methods. Although the enzymes are neither cheap nor stable, various procedures such as immobilization of enzymes have been developed to diminish these drawbacks of the enzymatic methods.

Various sulfur-containing amino acids occur in nature, and many of them are physiologically active. Selenium is a homologue of sulfur and an essential micronutrient for mammals, birds, fish, and some bacteria: it occurs as a constituent of amino acids, nucleotides, and other organic compounds, as well as in a free state. Selenocysteine and selenomethionine have been found in the polypeptide chains of several selenium-containing enzymes such as glutathione peroxidase and bacterial glycine reductase. The biological role of optically active sulfur and selenium amino acids has received considerable attention, but studies of their functions have been hampered by difficulty in their synthesis. Recently, Soda and his co-workers established facile enzymatic procedures to produce optically active sulfur- and selenium-containing amino acids. A number of pyridoxal 5'-phosphate–dependent enzymes catalyzing β- and γ-replacement reactions, and γ-addition reactions, have been used to synthesize them. Aminotransferases and NAD(P)-dependent amino acid dehydrogenases also have been utilized to produce them from their α-keto analogs.

We describe here enzymatic methods for the synthesis of L-selenocystine, L-selenohomocystine, *S*-substituted L-cysteines, *S*-substituted L-homocysteines, *Se*-substituted L-selenocysteines, and *Se*-substituted L-selenohomocysteines with pyridoxal 5'-phosphate–dependent enzymes such as L-methionine γ-lyase, *S*-alkyl-L-cysteine α,β-lyase, tryptophan synthase, *O*-acetylhomoserine sulfhydrylase, and NAD-dependent L-leucine dehydrogenase.

11.2 L-METHIONINE γ-LYASE

L-Methionine γ-lyase (EC 4.4.1.11) is a pyridoxal-P enzyme catalyzing the conversion of L-methionine into α-ketobutyrate, methanethiol, and ammonia; it plays an important role in the bacterial metabolism of methionine. Screening was carried out in order to find bacterial strains that would produce a high activity of L-methionine γ-lyase under aerobic and anaerobic conditions [1]. The enzyme is widely distributed in pseudomonads. The enzyme is inducibly produced by addition of L-methionine to the medium, and was purified to homogeneity from the crude extract of *Pseudomonas putida* ICR 3460, the best producer of the enzyme, by a simple and convenient purification procedure (average yield, 75%) [2]. Recently, *Aeromonas* sp. isolated from a lake was also found to produce L-methionine γ-lyase abundantly, and the *Aeromonas* enzyme was purified to homogeneity by a similar procedure to that for the *Pseudomonas* enzyme by Nakayama *et al.* [3]. Table 11.1 summarizes the physicochemical properties of the enzyme purified from both strains.

TABLE 11.1 Properties of Methionine γ-Lyase

Properties	Source	
	Pseudomonas putida	*Aeromonas* sp.
$S^{\circ}_{20,w}$	8.3S	—
Molecular weight:		
Sedimentation equilibrium	165,000	—
Gel permeation chromatography	—	149,000
Low-angle light-scattering	174,000	159,000
Absorption maxima	278 nm (ε = 134,000) 420 nm (ε = 38,900)	278 nm (ε = 159,000) 423 nm (ε = 31,300)
Number of subunits	4	4
Molecular weight of subunits	43,000	41,000
Pyridoxal-P content (mol/mol of enzyme)	4	4

In addition to methionine, several derivatives and analogs of L-methionine and L-cysteine, such as homocysteine, ethionine, *O*-ethylhomoserine, *O*-acetylhomoserine, *S*-methylcysteine, and *O*-acetylserine are effective substrates, whereas D-methionine, D-cysteine, L-cystathionine, and L-norleucine do not react (Table 11.2) [2, 3]. These results show that the enzyme catalyzes both α,γ- and α,β-replacement reactions. The enzyme also catalyzes γ-replacement reactions. Ethionine is synthesized from L-methionine and ethanethiol by means of a γ-replacement reaction [1]. The synthetic reaction proceeds as a function of enzyme concentration and incubation time. Successive additions of enzyme and ethanethiol are very effective, giving a yield of more than 95% [4]. L-Ethionine is also produced by a γ-replacement reaction between the derivatives of methionine which are substrates for the α,γ-elimination (for example, homocysteine, methionine sulfone, or *O*-acetylhomoserine) and ethanethiol (Table 11.2). When ethanethiol is replaced by various alkanethiols (C_3–C_7) and arylthioalcohols (for example, benzenethiol and β-naphthalenethiol) in the reaction system, new sulfur amino acids corresponding to the thiols are synthesized [4]. Longer-chain alkanethiols (C_8–C_{10}) and alcohols, however, are inert substituent donors. The derivatives of ethanethiol (for example, β-mercaptoethanol and cysteamine) also serve as substituent donors to methionine [4]. In addition, the enzyme catalyzes the β-replacement reactions (Table 11.2). For example, *S*-methyl-L-cysteine reacts with ethanethiol to produce *S*-ethyl-L-cysteine and methanethiol [1]. Thus, one can synthesize various sulfur amino acids easily from thiols and methionine, or some cheaply available sulfur amino acids, by replacement reactions.

Vinylglycine has been reported to inactivate several transaminases as a suicide substrate. This is not the case with methionine γ-lyase. L-Methionine γ-lyase catalyzes the deamination of vinylglycine to produce α-ketobutyrate and ammonia [5]. The enzyme catalyzes the γ-addition reaction of various alkanethiols to vinylglycine to yield the corresponding *S*-substituted homocysteines [5]. The relative activity of the enzyme for various alkanethiols in the γ-addition reaction of vinylglycine is close to that obtained for the γ-replacement reaction of methionine.

L-Methionine γ-lyase catalyzes the α-γ-elimination of selenomethionine to yield α-ketobutyrate, ammonia and methaneselenol, and also the γ-replacement reaction with various thiols to produce *S*-substituted homocysteines [6]. Selenomethionine is a better substrate than methionine for α-γ-elimination based on their V_{max} and K_m values, but is less effective for γ-replacement. In addition, L-methionine and its derivatives, which are substrates for the α-γ-elimination, react with selenols to form the corresponding *Se*-substituted selenohomocysteines, although selenols are less efficient substituent donors than thiols [6]. The enzymatic β-replacement reaction also occurs between *S*-substituted cysteines of *O*-substituted serines and selenols. Vinylglycine also serves as a substrate in the γ-addition reaction with selenols to form the corresponding selenohomocysteines [6]. These enzymatic replacement and γ-addition reactions facilitate the synthesis of various selenium amino acids labeled with ^{14}C, ^{75}Se and ^{3}H.

TABLE 11.2 Substrate Specificity of L-Methionine γ-Lyase

Substrate	Relative activity (%)	
	Elimination	Replacement
L-Methionine	100	100
D-Methionine	0	0
L-Ethionine	55	57
DL-Methionine sulfone	55	39
DL-Methionine DL-sulfoxide	21	43
L-Methionine DL-sulfoximine	1	5
S-Methyl-L-methionine	4	0
L-Homocysteine	180	176
O-Acetyl-L-homoserine	140	202
O-Ethyl-L-homoserine	99	nd
DL-Selenomethionine	282	114
L-Norleucine	0	0
L-Norvaline	0	0
L-Cysteine	10	59
D-Cysteine	0	0
L-Cystathionine	0	nd
S-Methyl-L-cysteine	9	125
O-Acetyl-L-serine	15	nd

nd = not determined.

According to the general concept of pyridoxal-P–dependent enzyme reactions, a Schiff base is formed between a sulfur-, selenium-, or *O*-acyl amino acid and the enzyme-bound pyridoxal-P, and is converted to a ketimine quinoid intermediate, through the release of the α-hydrogen, followed by loss of a β-hydrogen and a γ-substituent. The resulting β,γ-unsaturated intermediate is a ketimine intermediate of pyridoxal-P and vinylglycine. The intermediate can undergo the γ-addition of a thiol or selenol to form a new sulfur or selenium amino acid. Alternatively, in the absence of thiols or selenols, the vinylglycine intermediate undergoes tautomerization, followed by hydrolysis to form α-ketobutyrate and ammonia. Thus, the α,γ-elimination and γ-replacement reactions of sulfur-, selenium-, or *O*-acyl amino acids are accomplished. The α,β-elimination and β-replacement reactions of amino acids proceed through an aldimine Schiff base between the substrates and enzyme-bound pyridoxal-P, and an aminoacrylate intermediate [7]. Incubation of the enzyme with vinylglycine results in the appearance of a new absorption band at 480 nm, which is also observed with the substrates such as methionine, *O*-acetylhomoserine and selenomethionine for α,γ-elimination, but not with the α,β-elimination substrates. These findings support the mechanism through a vinylglycine-pyridoxal-P quinoid intermediate as shown in Figure 11.1 [7].

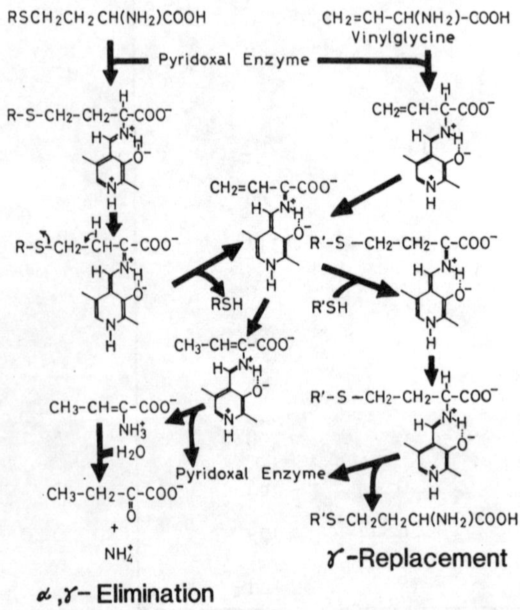

Figure 11.1. Mechanism of α,γ-Elimination and γ-Replacement Reactions Catalyzed by L-Methionine γ-Lyase.

L-Methionine γ-lyase catalyzes the exchange of the α-H and β-H of L-methionine and *S*-methyl-L-cysteine with ^2H in ^2H_2O; the rate of α-H exchange is about 40 times higher than that of the elimination reaction [7]. The enzyme can also exchange the α- and β-H of the following straight-chain amino acids that are not substrates for the elimination reaction: alanine, α-aminobutyrate, norvaline, and norleucine. The exchange of the two prochiral β-hydrogens is not stereospecific. No exchange is observed for D-isomers, L-valine, L-leucine, L-glutamate, L-aspartate, L-lysine, and L-arginine, although glycine, L-tryptophan and L-phenylalanine slowly undergo the α-H exchange. This enzymatic hydrogen exchange reaction facilitates the synthesis of various amino acids labeled with ^{2}H and ^{3}H.

11.3 *S*-ALKYLCYSTEINE α,β-LYASE

Various *S*-substituted cysteines and their sulfoxide derivatives occur in higher plants and bacteria. Alliin lyase (EC 4.4.1.4) catalyzes α,β-elimination of *S*-alkyl-L-cysteine sulfoxides to produce pyruvate, ammonia, and alkylsulfenate. The enzyme has been purified to homogeneity from *Allium cepa* [8], *A. sativum* [9], and *Lentinus edodes* [10], and characterized extensively. *S*-Alkylcysteine α,β-lyase (EC 4.4.1.6) differs from alliin lyase in that it acts on both *S*-alkyl-L-cysteines and their sulfoxide derivatives. *S*-Alkylcysteine α,β-lyase has been partially purified from *Pseudomonas cruciviae* [11] and purified to homogeneity from etiolated seedlings of *Acacia farnesiana* [12].

Tanaka *et al.* also found that *Pseudomonas putida* ICR 3640 produces *S*-alkylcysteine α,β-lyase abundantly by addition of *S*-methyl-L-cysteine to the medium [13]. However, the addition of L-cysteine increases the enzyme activity only slightly. The enzyme has been purified about 370-fold to near homogeneity from the crude extract, with an overall yield of 23% [13]. The enzyme is composed of six subunits identical in molecular weight (about 37,000). In addition to *S*-methyl-L-cysteine, *O*-methyl-L-serine and *Se*-methyl-DL-selenocysteine are also effective substrates. In particular, *Se*-methyl-L-selenocysteine is better as a substrate than *S*-methyl-L-cysteine. Homologues of *S*-methyl-L-cysteine, such as *S*-ethyl-L-cysteine, *S*-benzyl-L-cysteine, and L-djenkolate also serve as effective substrates. Conversion of the thioether of *S*-methyl-L-cysteine into sulfoxide does not affect the ability of the enzyme to utilize it as the substrate. *S*-Carboxymethyl-L-cysteine, *O*-acetyl-L-serine, L-cystathionine, L-cysteine, and L-cystine also serve as substrates, though slowly. The following amino acids are inert: *S*-methyl-D-cysteine, L-methionine, L-serine, L-alanine, L-tyrosine, L-tryptophan, L-phenylalanine, L-norvaline, and L-methionine sulfoxide.

The enzyme catalyzes the β-replacement reaction between *S*-methyl-L-cysteine and various thiols to yield the corresponding *S*-substituted L-cysteines [13]. In addition to *S*-methyl-L-cysteine, *Se*-methyl-DL-selenocysteine and *O*-methyl-L-serine also serve as substrates for the replacement reaction. The β-replacement reactivities between amino acids and various thiols are summarized in Table 11.3. Various sulfur amino acids can be prepared by the β-replacement reaction in the same manner as L-methionine γ-lyase.

TABLE 11.3 Substrate Specificity of *S*-Alkylcysteine α,β-Lyase

Thiol (RSH)	Relative activity on amino acid substrates*		
R=	SeMC	SMC	OMS
Ethyl	100	121	91
n-Propyl	34	12	35
Phenyl	7	0	0
Benzyl	12	9	8

* SeMC = *Se*-methyl-L-selenocysteine; SMC = *S*-methyl-L-cysteine; OMS = *O*-methyl-L-serine.

11.4 D-SELENOCYSTINE α,β-LYASE

Esaki *et al.* have found a novel enzyme that catalyzes α,β-elimination of D-selenocystine to produce H_2Se_2 and a double equivalent amount of pyruvate and ammonia [14]. The enzyme occurs in *Clostridium sticklandii* and *C. sporogenes*, and was named D-selenocystine α,β-lyase. Clostridial strains are classified into four groups according to nutritional characteristics: proteolytic and saccharolytic (group A), only proteolytic (group B), only saccharolytic (group C), and neither proteolytic nor saccharolytic (group D) [15]. *C. sporogenes* and *C. sticklandii* are classified as members of group A and D, respectively. Therefore, the enzyme distribution cannot be correlated with the nutritional characteristics of the strains.

D-Selenocystine α,β-lyase was purified to homogeneity from the cell extract of *C. sticklandii* [14]. The enzyme is composed of two subunits with identical molecular weight of about 35,000. The isoelectric point of the enzyme was determined to be 4.5. The enzyme contains pyridoxal-P as a coenzyme.

D-Selenocystine serves as the best substrate of D-selenocystine α,β-lyase [14]. D-Cystine, D-lanthionine, and β-chloro-D-alanine are also effective substrates. Although D-cysteine undergoes α,β-elimination slowly, it shows the highest affinity for the enzymes. However, D-selenocysteine is inert. DL-Selenohomocystine, DL-selenohomocysteine, cystine, and cysteine are not substrates. The enzyme catalyzes the β-replacement reaction between D-selenocystine and various thiols to yield the corresponding *S*-substituted D-cysteines [14]. Selenols also can serve as a substrate in the β-replacement reaction. Thus, D-selenocystine α,β-lyase can be used to prepare D-enantiomers of *S*-substituted cysteines and *Se*-substituted selenocysteines [14].

11.5 TRYPTOPHAN SYNTHASE

Tryptophan synthase (EC 4.2.1.20) is a unique pyridoxal-P enzyme with a multicatalytic function (Table 11.4). It has been purified to homogeneity and crystallized from *E. coli*, and characterized extensively. The enzyme catalyzes not only reaction 1 (that is, the physiological one) but also reactions 2 and 3, which are regarded as partial reactions of reaction 1 (Table 11.4).

TABLE 11.4 Reactions Catalyzed by Tryptophan Synthase

Reaction	Catalyzed by
1. Indole-3-glycerol phosphate + L-serine → L-tryptophan + D-glyceraldehyde 3-phosphate + H_2O	$\alpha_2\beta_2$
2. Indole-3-glycerol phosphate → indole + D-glyceraldehyde 3-phosphate	α; $\alpha_2\beta_2$
3. Indole + L-serine → L-tryptophan + H_2O	β_2; $\alpha_2\alpha_2$
4. L-serine → pyruvate + ammonia	β_2
5. β-Mercaptoethanol + L-serine → *S*-(β-hydroxyethyl)-L-cysteine + H_2O	β_2; $\alpha_2\beta_2$
6. β-Mercaptoethanol + L-serine + pyridoxal-P → *S*-pyruvylmercaptoethanol + pyridoxamine-P + H_2O	β_2

Tryptophan synthase is found widely in various bacteria, yeasts, molds, and plants. The enzyme of *E. coli* is composed of two kinds of proteins, α (M_r, 29,000) and β (M_r, 44,200). Pyridoxal-P is bound to the β-subunit through a Schiff base. Two α-subunits combine with one β_2-dimer to form an $\alpha_2\beta_2$-complex (M_r, 147,000) that catalyzes the physiological reaction. Each of the subunits catalyzes its own specific reaction as well. The crystalline $\alpha_2\beta_2$-complex is obtained after a sixfold purification from *E. coli* trp R$^-$ ΔtrpED102/F'ΔtrpED102. About 16% of the intracellular soluble protein of this mutant is the tryptophan synthase complex. Goldberg and Baldwin [16] and Miles and her co-workers [17] have shown that methanethiol and β-mercaptoethanol serve as *S*-substituent donors to serine to yield *S*-methyl-L-cysteine and *S*-(β-hydroxyethyl)-L-cysteine, respectively, by the $\alpha_2\beta_2$ and β_2-complexes. Esaki *et al.* [18] have studied the enzymatic synthesis of various *S*-substituted-L-cysteines from L-serine and its derivatives (for example, β-chloro-L-alanine and *O*-methyl-L-serine) with the $\alpha_2\beta_2$-complex (Table 11.5). Thiols such as α-toluenethiol, 1-propanethiol, and 1-butanethiol are much more efficient as *S*-substituent donors than indole (Table 11.6). When L-threonine and L-vinylglycine are used as *S*-substituent

TABLE 11.5 Relative Reactivities of *S*-Substituent Acceptors in β-Replacement Reactions with α-Toluenethiol by Tryptophan Synthase

S-Substituent acceptors	Relative rate (%)
L-Serine	100
O-Methyl-DL-serine	41
β-Chloro-L-alanine	35
O-Acetyl-L-serine	13
S-Methyl-L-cysteine	24
S-Ethyl-L-cysteine	12
Se-Methyl-DL-selenocysteine	84
L-Cysteine	0
L-Threonine	1
L-Allothreonine	0
L-Vinylglycine	0.5

acceptors of thiols, the corresponding *S*-substituted β-methyl-L-cysteines are synthesized [18]. The enzyme also catalyzes the β-replacement reactions of L-serine with selenols to produce the corresponding *Se*-substituted L-selenocysteines [19]. *Se*-Benzyl-L-selenocysteine and *Se*-methyl-L-selenocysteine are synthesized from L-serine and α-tolueneselenol and methaneselenol in a similar way with yields of 44% and 16%, respectively, based on L-serine. The relative activity of indole to methanethiol (100) is approximately 150. Those of methaneselenol and α-tolueneselenol are 24% and 30%, respectively. Therefore, the production of *Se*-methyl and *Se*-benzyl-L-selenocysteines proceeds much more rapidly than tryptophan synthesis, the inherent reaction of the enzyme. L-Serine can be replaced by a variety of β-substituted L-alanines, such as β-chloroalanine and *O*-acetylserine, in the reaction system containing these selenols.

According to the general mechanism for the β-replacement reaction catalyzed by pyridoxal-P enzymes, nucleophilic addition of selenols occurs in an intermediate derived from the substrate. Although selenols are more nucleophilic than thiols, selenols are less reactive substituent donors than thiols in the enzymatic β-replacement reactions catalyzed by tryptophan synthase. This is compatible with the reactivities

TABLE 11.6 Relative Reactivities of Thiols in Synthesis of *S*-Substituted Cysteines from L-Serine by Tryptophan Synthase

Thiol	Relative rate (%)
α-Toluenethiol	100
Methanethiol	19
Ethanethiol	21
1-Propanethiol	152
1-Butanethiol	113
2-Butanethiol	24
2-Methylpropane-1-thiol	40
2,2-Dimethylpropane-1-thiol	0
1-Pentanethiol	37
1-Hexanethiol	19
1-Heptanethiol	12
1-Octanethiol	3
1-Nonanethiol	1
1-Decanethiol	0
Benzenethiol	8

of α-tolueneselenol and α-toluenethiol, although methaneselenol is a slightly more efficient substituent donor than methanethiol. Some physicochemical properties of methaneselenol and methanethiol, such as volatility and solubility, may affect their reactivity in the enzyme reaction.

11.6 *O*-ACETYLHOMOSERINE SULFHYDRYLASE

O-Acetylhomoserine sulfhydrylase (*O*-acetylhomoserine (thiol)-lyase, EC 4.2.99.10) is a pyridoxal-P enzyme that catalyzes the synthesis of cysteine and homocysteine from H_2S with *O*-acetyl-L-serine (OAS) and *O*-acetyl-L-homoserine (OAH), respectively. The enzyme from bakers' yeast has been purified and characterized. It is involved in the synthesis *in vivo* of cysteine.

Chocat *et al.* [20] have shown that OAH sulfhydrylase catalyzes the β- and γ-replacement reactions between the *O*-acetyl groups of OAS and OAH, and Na_2Se_2. L-Cysteine *O*-sulfate (SOS) also serves as a substrate of the β-replacement reaction, although its reactivity is lower than that of OAS (Table 11.7). The selenium amino acids produced have been isolated and identified as L-selenocystine and L-selenohomocystine, but this does not necessarily mean that the primary enzymatic product is the diselenide. The initial products of the enzyme reaction are probably *Se*-(selenohydryl) derivatives, that is, $-SeSeCH_2CH(NH_2)COOH$ and $-SeSeCH_2CH_2CH(NH_2)COOH$, which are nonenzymatically oxidized to the corresponding diselenides.

TABLE 11.7 Kinetic Parameters of Reactions Catalyzed by *O*-Acetylhomoserine Sulfhydrylase

Substituent acceptor	Substituent donor	Relative V_{max}	K_m (mM) of substituent	
			Acceptor	Donor
O-Acetylhomoserine	NaHS	100	4.1	0.52
O-Acetylhomoserine	Na_2Se_2	17	5.3	8.9
O-Acetylserine	NaHS	14	2.5	0.70
O-Acetylserine	Na_2Se_2	8.4	5.0	nd
Serine *O*-sulfate	NaHS	5.1	4.0	0.70
Serine *O*-sulfate	Na_2Se_2	1.4	4.0	10
Serine *O*-sulfate	NaHSe	1.3	nd	1.2

nd = not determined.

11.7 COMPARISONS AMONG PYRIDOXAL-P ENZYMES CATALYZING SULFUR AND SELENIUM AMINO ACID SYNTHESIS

L-Methionine γ-lyase is useful for the production of various sulfur and selenium amino acids. However, the enzyme catalyzes not only the replacement reactions, but also the elimination reactions; the yield depends mainly on the concentrations of the substrates, thiols or selenols. The amino acids produced may be decomposed by the elimination reactions. The elimination catalyzed by the enzyme is regarded as a weak point of the enzyme from the point of view of amino acid production. Therefore, an excess of thiols or selenols has to be added to prevent the elimination reaction.

Tryptophan synthase surpasses methionine γ-lyase with regard to the reaction specificity: it does not catalyze the elimination reaction that leads to a decrease in the yield of sulfur and selenium amino acids. However, tryptophan synthase is inferior to methionine γ-lyase in the substrate specificity: it does not catalyze the synthesis of *S*-substituted homocysteines and *Se*-substituted selenohomocysteines by the γ-replacement reactions.

O-Acetylhomoserine (thiol)-lyase catalyzes both the β- and γ-replacement reactions between *O*-acetylhomoserine or *O*-acetylserine and various thiols or selenols, but never catalyzes the elimination reactions. Therefore, the enzyme is also useful for the synthesis of optically active sulfur and selenium amino acids. However, *O*-acetylhomoserine(thiol)-lyase is poorly produced by yeasts, and shows a high substrate specificity with respect to substituent acceptors. Therefore, we have to use expensive substrates, such as *O*-acetylhomoserine and *O*-acetylserine. We cannot generally state which is the best, but we can choose one of them for the desired amino acid synthesis.

D-Selenocystine α,β-lyase gives exclusively D-enantiomers of sulfur and selenium amino acids by the β-replacement reactions. The enzyme catalyzes also α,β-elimination, and the yield of amino acids synthesized is affected by the concentrations of thiols or selenols used as substituent donors in the same manner as L-methionine γ-lyase. Therefore, one needs to add an excess of thiols or selenols to depress the elimination reaction.

11.8 D-AMINO ACID AMINOTRANSFERASE

D-Amino acid aminotransferase (EC 2.6.1.21) is regarded as a key enzyme in bacterial D-amino acid metabolism. The enzyme has been purified from *Bacillus sphaericus* to homogeneity [21]. Recently, Tanizawa *et al.* isolated thermophiles, *Bacillus* sp. YM-1 and YM-2, which produce the enzyme abundantly [22]. D-Amino acid aminotransferase purified to homogeneity from *Bacillus* sp. YM-1 is stable at high temperature: the enzyme activity was not lost by heating at 55 °C for more than 30 minutes.

The gene of *Bacillus* sp. YM-1 was cloned and expressed in *E. coli* C600 with the plasmid vector pBR322 [22]. The amount of enzyme produced is equivalent about 9% of total protein in extracts of the cells into which the gene has been cloned. The enzyme can be purified efficiently by heat treatment.

The enzyme has a low substrate specificity: acidic, basic, and neutral aliphatic and aromatic D-amino acids are transaminated with various α-keto acids [21, 22].

Therefore, the enzyme can be used for the production of D-amino acids. The enzyme was found to act on D-selenomethionine and its α-keto analog, α-keto-γ-methylselenobutyrate.

Esaki *et al.* [23] have established an efficient method for D-selenomethionine synthesis with the enzyme from α-keto-γ-methylselenobutyrate. In order to synthesize D-selenomethionine efficiently, a D-alanine regeneration system was used, as shown in Figure 11.2. Pyruvate produced from D-alanine was reduced with L-alanine dehydrogenase, and L-alanine thus produced was racemized with alanine racemase to regenerate D-alanine. NADH consumed by the pyruvate reduction was also regenerated by coupling with the formate dehydrogenase reaction: the enzyme catalyzes an irreversible oxidation of formate to CO_2 with a concomitant reduction of NAD to NADH. Therefore, D-alanine and NADH are regenerated continuously in the reaction system by consumption of ammonium formate. Ohshima *et al.* [24] reported a continuous method for production of L-leucine from its α-keto analog by means of a membrane reactor system using NAD immobilized on polyethylene glycol. D-Selenomethionine can be produced continuously in the same manner. Pyruvate formed in the reaction system was continuously converted to D-alanine by means of alanine racemase (EC 5.1.1.1), L-alanine dehydrogenase (EC 1.4.1.1), formate dehydrogenase (EC 1.2.1.2), and NADH. Thus, D-selenomethionine was produced in an 80% yield based on α-keto-γ-methylselenobutyrate.

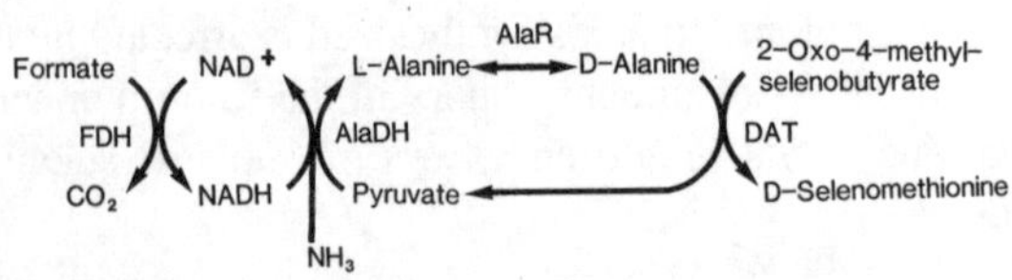

Figure 11.2. Synthesis of D-Selenomethionine from α-Keto-γ-methylselenobutyrate (2-Oxo-4-Methylselenobutyrate), Formate, NAD^+, D-Alanine, and Ammonia by Coupling of Alanine Dehydrogenase (AlaDH), Formate Dehydrogenase (FDH), Alanine Racemase (AlaR), and D-Amino Acid Aminotransferase (DAT).

11.9 LEUCINE DEHYDROGENASE

Leucine dehydrogenase (EC 1.4.1.9) is an NAD-dependent oxidoreductase that catalyzes the reversible deamination of various straight- and branched-chain L-amino acids. Ohshima *et al.* have crystallized the enzyme from *B. sphaericus*, which produces the enzyme most abundantly [25]. Recently, Nagata *et al.* cloned the leucine dehydrogenase gene of *B. stearothermophilus* into *E. coli* with pBR322 to produce the thermostable enzyme abundantly [26]. In addition to branched-chain L-amino acids, straight-chain aliphatic L-amino acids are effectively deaminated by both the *B. sphaericus* and *B. stearothermophilus* enzymes. All the α-keto acids produced by oxidative deamination serve as good substrates for the amination reaction [25, 26]. L-Selenomethionine is a rather poor substrate of the enzyme, but L-selenomethionine was produced from its α-keto analog with a large amount of enzyme. Esaki *et al.* [27] have established an efficient method for L-selenomethionine production with the enzyme: L-selenomethionine can be produced in a 95% yield from α-keto-γ-methylselenobutyrate and ammonia by means of an NADH regeneration system with formate dehydrogenase (Figure 11.3).

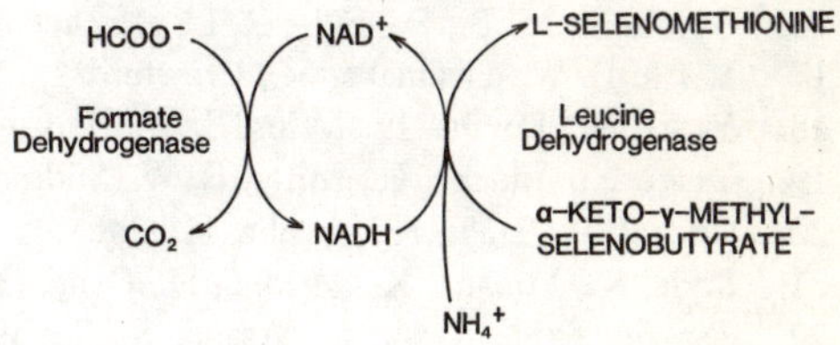

Figure 11.3. Synthesis of L-Selenomethionine from α-Keto-γ-methylselenobutyrate and NH_4^+ by Coupling of Leucine Dehydrogenase and Formate Dehydrogenase.

11.10 SUMMARY

Although various sulfur- and selenium-containing amino acids occur in nature and play important physiological roles, their facile chemical synthesis is not available. We have described the mechanism of these enzymatic procedures, and discuss their merits and drawbacks. We also described enzymatic methods for the synthesis of L-selenocystine, L-selenohomocystine, *S*-substituted L-cysteines, *S*-substituted L-selenohomocysteines with pyridoxal 5'-phosphate–dependent enzymes such as L-methionine γ-lyase and cysteine desulfhydrase, and NAD-dependent L-leucine dehydrogenase.

11.11 REFERENCES

1. Tanaka, H., Esaki, N., Soda, K., *Biochemistry 16*, 100 (1977).
2. Nakayama, T., Esaki, N., Sugie, K., Berezov, T. T., Tanaka, H., Soda, K., *Anal. Biochem. 138*, 421 (1984).
3. Nakayama, T., Esaki, N., Lee, W.-J., Tanaka, I., Tanaka, H., Soda, K., *Agric. Biol. Chem. 48*, 1991 (1984).
4. Esaki, N., Tanaka, H., Yamamoto, T., Soda, K., *Bull. Inst. Chem. Res. Kyoto. Univ. 59*, 321 (1981).
5. Esaki, N., Suzuki, T., Tanaka, H., Soda, K., Rando, R. R., *FEBS Lett. 84*, 309 (1977).
6. Esaki, N., Tanaka, H., Uemura, S., Suzuki, T., Soda, K., *Biochemistry 18*, 407 (1979).

7. Esaki, N., Nakayama, T., Sawada, S., Tanaka, H., Soda, K., *Biochemistry 24*, 3857 (1985).
8. Schwimmer, S., Kjaer, A., *Biochim. Biophys. Acta. 42*, 316 (1960).
9. Mazelis, M., Fowden, L., *Phytochemistry 12*, 1287 (1973).
10. Murakami, F., *Vitamins 20*, 126 (1960).
11. Nomura, J., Nishizuka, Y., Hayaishi, O., *J. Biol. Chem. 238*, 1441 (1963).
12. Mazelis, M., Creveling, R. K., *Biochem. J. 147*, 485 (1975).
13. Tanaka, H., Imahara, H., Esaki, N., Soda, K., *Sulfur Amino Acids 3*, 275 (1980).
14. Esaki, N., Seraneeprakarn, V., Tanaka, H., Soda, K., *J. Bacteriol. 170*, 751 (1988).
15. Gottschalk, G., Andreesen, J .R., Hippe, H. In *The prokaryotes. A Handbook on Habitats, Isolation, and Identification of Bacteria*; Starr, M. P., Stolp, H., Truper, H. G., Balow, A., Schlegel, H. G., Eds.; Springer-Verlag KG, Berlin, 1981.
16. Goldberg, M. E., Baldwin, R. L., *Biochemistry 6*, 2113 (1967).
17. Miles, E. W., Hatanaka, M., Crawford, I. P., *Biochemistry 7*, 2742 (1968).
18. Esaki, N., Tanaka, H., Miles, E. W., Soda, K., *Agric. Biol. Chem. 47*, 2861 (1983).
19. Esaki, N., Tanaka, H., Miles, E. W., Soda, K., *FEBS Lett. 161*, 207 (1983).
20. Chocat, P., Esaki, N., Tanaka, H., Soda, K., *Agric. Biol. Chem. 49*, 1143 (1985).
21. Soda, K., Yonaha, K., Misono, H., Osugi, M., *FEBS Lett. 46*, 359 (1974).
22. Tanizawa, K., Masu, Y., Asano, S., Tanaka, H., Soda, K., *J. Biol. Chem. 264*, 2445 (1989).
23. Esaki, N., Shimoi, H., Tanaka, H., Soda, K., *Biotechnol. Bioeng. 34*, 1231 (1989).
24. Ohshima, T., Wandrey, C., Kula, M.-R., Soda, K., *Biotechnol. Bioeng. 27*, 1616 (1985).
25. Ohshima, T., Misono, H., Soda, K., *J. Biol. Chem. 253*, 5719 (1978).
26. Nagata, S., Tanizawa, K., Esaki, N., Sakamoto, Y., Ohshima, T., Tanaka, H., Soda, K., *Biochemistry 27*, 9056 (1988).
27. Esaki, N., Shimoi, H., Yang, Y.-S., Tanaka, H., Soda, K., *Biotechnol. Appl. Biochem. 11*, 312 (1989).

PART III

IMMOBILIZATION TECHNIQUES AND PROCESS CONSIDERATIONS FOR BIOCATALYTIC PROCESSES

IMMOBILIZED CELLS AND THEIR APPLICATION TO AMINO ACID PRODUCTION

Steven F. Karel

CONTENTS

12.1 INTRODUCTION

Immobilized-cell technology has been reviewed extensively in recent years, even as the rate of publication of primary source material has tapered off. Several monographs and conference proceedings have appeared discussing immobilized cells [1-7]. The reviews by Birnbaum *et al.* [8] and by Klein and Wagner [9] provide an excellent introduction to methods and applications; for a more recent summary of applications, I recommend the review by Tanaka and Nakajima [10]. Many features of natural systems, especially with regard to adherent cells and biofilms, are quite similar to those found in "artificially" immobilized cells [11, 12]. In previous reviews, my colleagues and I have attempted to provide a summary of the engineering principles governing the properties of immobilized cells [13, 14]. This present chapter will discuss some basic principles that are central to designing immobilized-cell processes, with emphasis on recent findings, and will also review the application of immobilized-cell technology to amino acid production.

Whole-cell immobilization has been defined as the confinement of cells to a defined region of space, with the preservation of some desired catalytic activity [13]. This definition comprehends a wide variety of both natural and industrial processes. Three points implicit in this definition need to be clarified in any discussion of immobilized cells:

1. **The size of the region of confinement.**
   By convention, one usually considers only those schemes in which there is no bulk mixing in the region of confinement; as a result, solute transport occurs predominantly by diffusion. By this definition, a cell-recycle reactor is not considered an immobilized-cell process. A typical dimension of an immobilized-cell region is between 50 μm and 5 mm.
2. **The range of cell activities preserved.**
   Catalyst preparations that have been labeled as Rimmobilized cells" range from disrupted cells, in which a single enzyme activity is utilized, to intact living cells growing at rates comparable to those in a well-mixed fermentor. In the former extreme, whole-cell immobilization provides a convenient method to produce a relatively impure immobilized enzyme. In the latter extreme, the role of immobilization may be simply to increase the concentration of cells in a reactor, and hence the volumetric productivity of a reactor. Both extremes have been applied to amino acid production.
3. **The degree to which cell escape is prevented.**
   Most processes require partial, but incomplete, prevention of cell escape from the immobilized cell layer. In situations where growth does not occur, a high degree of confinement may be necessary for good longevity. In situations where growth does occur, it is quite conceivable that a high degree of confinement may be detrimental if the process must operate at steady state over a long period of time [15].

Numerous advantages of immobilization have been championed. Among these, the most generally accepted is that immobilization provides a partial separation between cells and their products, reducing downstream processing costs. In some cases, this rapid separation may prevent product degradation. Almost by definition, immobilization improves catalyst retention within a reactor. This may allow operation at high catalyst (cell) concentrations, perhaps improving the volumetric productivity of a reactor. These are essentially the same claims that can be made for immobilized enzymes. Advantages of immobilized-cell processes that are intended to use a single enzymatic activity of the cells can include the relative simplicity of preparing immobilized cells compared to immobilizing a purified enzyme, and the possibility that cofactors will be retained along with the enzyme. In terms of disadvantages, the fact that other undesirable enzymatic activities (for example, proteases) may also be present in the preparation is a major concern. At present, it is not possible to predict whether an immobilized-cell or an immobilized-enzyme preparation will be more stable, and the relative stabilities must be determined empirically for a particular system. Results from the literature should always be examined critically to determine if mass-transfer limitations are contributing to the apparent stability (see comments in Section 12.3.6 on *pseudostability*).

It has occasionally been claimed that immobilized-cell processes are advantageous because they enhance the specific activity of the cells, relative to their behavior in solution. This idea, that immobilization *per se* enhances growth or product formation in cells, is discussed in detail below. A much more reasonable case can be made for the advantages of immobilized-cell processes in stabilizing mixed cultures. In at least some systems, plasmid-containing strains are maintained against competing strains over longer times in immobilized-cell reactors than in well-mixed continuous reactors. This phenomenon is likely to result from a process of outward growth in which cells trapped deep within the cell layer act as a reservoir of plasmid-containing cells that "reinoculate" the outer layers of the immobilized-cell layer as the cells at the surface grow and are removed from the layer [16]. Mixed cultures containing several distinct species may also tend to be more stable in immobilized-cell reactors, for the reason cited above, and as a result of the variety of different environmental conditions that prevail at different positions within a biofilm. Recent experiments indicate that mixed-culture biofilms are spatially heterogeneous and that simple models assuming a one-dimensional biofilm may need to be modified [17, 18]. On the other hand, this complexity may be essential in maintaining efficient cell populations in waste treatment.

The major disadvantage that is inherent in immobilized-cell processes is that the solution and the cells are not well-mixed. This usually leads to mass-transfer limitations, either in substrate supply or product removal. Normally, these reduce the overall rate of the desired reaction. Understanding mass-transfer limitations is the crucial factor in the design and analysis of an immobilized-cell process. In some cases, especially where substrate inhibition is involved, the mass-transfer limitations may be advantageous. The spatial heterogeneity also interferes with the ability to monitor, and hence to control, an immobilized-cell process. Another minor disadvantage is the requirement for an extra process step to immobilize the cells.

12.2 METHODS FOR IMMOBILIZING CELLS

A simple classification of immobilization processes is based on the relationship of the cells to the supporting phase. Cells can be immobilized:

on a support (*attachment*),
in a porous matrix (*entrapment*),
behind a barrier (*containment*), or
as the support (*aggregation*).

Examples of each of these classes are found in amino acid production.

12.2.1 Attachment

Most attachment processes depend on the inherent tendency of cells to adhere to foreign surfaces. The initial stage of adhesion, in which cell coverage is less than a monolayer, is usually mediated by either hydrophobic or ionic interactions of the cell surface with a support. Most biofilm and immobilized-cell systems involve systems substantially thicker than a monolayer. In this case, cohesive interactions between the cells can be as important as, if not more important than, the interactions with the surface. These latter interactions may involve extracellular polysaccharides or proteins [12, 19].

Occasionally, attachment has been achieved through covalent linkage of cells to surfaces. The most frequently used cross-linking agent is glutaraldehyde. While covalent linkage is often necessary to immobilize enzymes, the larger surface area of cells means that the large number of weak interactions spread over the surface is more than adequate to ensure immobilization.

12.2.2 Entrapment

Entrapment methods can be distinguished into two categories: those in which cells are introduced into a preformed support, and those in which a gel is formed around the cells. In the case of preformed supports, the pore size must obviously be larger than the cells in order for the cells to be loaded into the support [20]. Immobilization in the support is thus dependent on an initial attachment of the cells, followed by the subsequent growth of the cells. The cohesion of the cells to form microcolonies in the support leads to high cell densities. A high concentration of cells reduces cell mobility, which limits the leakage of cells from the support. Preformed supports include porous glass and ceramics, as well as plastic and steel meshes [21, 22]. By choosing a particular size of support particle and the specific gravity of the support, a range of sedimentation rates can be obtained. This is especially useful in the design of fluidized-bed reactors.

The use of polymer gels to entrap cells is probably the most popular of the immobilization techniques. Very generally applicable methods for the immobilization of essentially all types of cells, with retention of the ability to grow, have been developed using gel entrapment [23]. The two gel media used most frequently are alginates and κ-carrageenan. In the case of alginate, the cells are mixed with a dissolved solution of sodium alginate and forced through a small orifice to form drops, which are gelled in a solution containing multivalent cations (typically Ca^{2+}, but Ba^{2+} and Al^{3+} have also been used). Various different compositions of alginate are available, and the physical properties of the immobilized-cell matrix can be tailored through choice of alginate and counterion [24]. Typically, small droplets are prepared, although other forms can be cast. A number of novel devices have been described that allow the rapid preparation of small gel particles of uniform size, for example through the use of vibrating nozzles [25, 26].

Entrapment with alginate or carrageenan gels are currently the methods of choice for whole-cell immobilization. While cell retention is not absolute, it is usually adequate, especially in the absence of cell growth. When cells grow in these gels, microcolonies are formed which subsequently expand and stretch the gel matrix [13]. Cell escape from the gels is typically thought to be a result of microcolonies bursting through to the gel surface, allowing cell leakage.

Many other polymerization techniques are also available. Polyacrylamide gels were among the first methods used for immobilized cells and for applications to amino acid production [27]. Acrylamide polymerization has toxic effects on cells [28], but in the case of simple enzymatic processes this is unlikely to be a serious problem. The choice of a particular polymer may be influenced by the physical properties of the resultant catalyst (rigidity, uniformity of size, porosity), or by the solubility of compounds in the gels. Polyurethanes and silicone polymers have been used with the intent of increasing the permeabilities of hydrophobic substrates or products within the gel matrix, especially for steroid biotransformations [29, 30].

12.2.3 Containment

Another versatile technique involves the use of semipermeable membranes to contain cells. Containment can be distinguished from entrapment by the fact that the membrane acts as a barrier between a cell-laden region and the environment, instead of having the cells dispersed within the structure of the membrane. Both microporous and ultrafiltration membranes have been used for this purpose [31, 32]. The use of commercial synthetic membranes offers the possibility of near-absolute entrapment (with sterile effluents, at least initially). In addition, combinations of membrane composition and permeability can be used to develop dual-membrane reactors allowing simultaneous nutrient supply from gas and liquid streams [32, 33].

Methods for the *in situ* synthesis of microcapsular membranes around cells have also been developed [34–36]. A typical process involves the initial synthesis of an alginate gel bead containing the cells, which then serves as a scaffold for the synthesis of the size-selective membrane, using polylysine.

12.2.4 Aggregation

In some cases, cells can be aggregated into pellets of an appropriate size for use in an immobilized-cell process without the addition of a support. This could be the result of mycelial growth, or of the flocculation of the cells. Cross-linking agents such as glutaraldehyde may also be used [37]. Two classic examples of this process are in the production of immobilized cells for glucose isomerase [38], and in activated-sludge reactors for wastewater treatment [39].

12.3 REACTION AND DIFFUSION IN IMMOBILIZED-CELL SYSTEMS

The most important unifying factor that relates the variety of processes that are given the label "immobilized-cell" is the need to understand diffusional limitations. The inclusion of a large concentration of a catalyst in a small volume almost invariably gives rise to some form of mass-transfer limitation. In the majority of cases, the diffusion of one of the substrates to the cells will be rate-limiting. Less frequently, multiple substrates may be simultaneously limiting the reaction, or the diffusion of an inhibitory product away from the cells may be important. Depending on the immobilization scheme, the site of the mass-transfer limitation could be inside or outside the immobilized-cell aggregate. Recognizing all of the possible ways that mass transfer may be limiting is extremely important, especially in trying to interpret the results of experiments that claim to rule out or rule in mass-transfer limitations.

One possible means of averting diffusional limitations is to force convection to occur within the layer containing the immobilized cells. This can be achieved either by a steady flow across the layer, or by a fluctuating flow near the surface, due to turbulence outside the layer [40, 41]. Membrane reactors are ideal for imposing cross-layer flows [42, 43], but gel slabs have also been used [44]. In the case of membranes, using selective membranes can lead to added benefits in discriminating between substrates and products in the effluent; ultrafiltration and reverse osmosis membranes have been used in the production of ethanol [42]. The effects of cell compression and channeling in these reactors have yet to be elucidated experimentally, and prediction of the flow rate through the composite of cells and support is difficult. A certain degree of channeling may actually enhance performance by reducing the pressure drop, yet it may not affect mass transfer substantially, depending on the length scale of the channeling.

12.3.1 Modeling Reaction and Diffusion

In spite of the lack of experimental data on transport processes and reaction rates, a vast literature on modeling the combined processes of reaction and diffusion in immobilized-enzyme and -cell processes was produced in the 1970s and 1980s [12, 13, 45–47]. The similarities of these processes to those in supported metal catalysts and other chemical engineering processes allows wholesale carryover of the previous literature [48]. Models of increasing complexity have continued to be applied, especially to growth-related immobilized-cell processes [49–51], which differ from conventional processes in that the catalyst concentration is continually changing, and in that the growth of cells can easily result in cell movement or cell layer expansion. Such models generally require a detailed computer simulation of the growth history of the immobilized-cell layer.

The basic structure used in most of these models is that the cell layer is represented as a pseudohomogeneous region containing a dispersed catalyst. The concentration of the catalyst is important in determining the *intrinsic rate* of the reaction at any position in space. The transport of substrates is usually characterized by an *effective diffusivity*, $\mathcal{D}_j$, which takes into account the resistances to diffusion posed by the support, as well as by the cells. The general reaction-diffusion equations for each substrate or product j can be written as:

$$\frac{\partial C_j}{\partial t} = \nabla \cdot \mathcal{D}_j \nabla C_j - \sum_{k=1}^{N} \varepsilon_{jk} R_k \tag{1}$$

where C_j is the concentration of j in the immobilized-cell aggregate, ε_{jk} is the stoichiometric coefficient defined as the number of moles of substrate j consumed in reaction k ($\varepsilon_{jk} > 0$ for substrates, < 0 for products), R_k is the volumetric rate of each independent reaction k, and N is the number of independent reactions occurring.

The full form of eq 1 is rarely used without simplifications. The transient term on the left-hand side is usually dropped by assuming the system reaches a pseudosteady state [13]. Even if the cell mass profile in the layer changes over time due to growth, this process is slow compared to the diffusion of substrates. The second set of assumptions necessary to allow simulations is that the diffusivities and volumetric reaction rate constants are usually taken to be functions only of the cell concentration.

Two further assumptions are usually made if general or analytical solutions are desired. These are: first, that the cell concentration does not vary with position; and second, that there is a single independent reaction occurring in the layer. This reaction may involve many different substrates and products, but at a fixed stoichiometry. These assumptions greatly simplify the problem, although their applicability varies from system to system. If these assumptions are not implemented, detailed computer simulations of the immobilized-cell layer are required to solve the problem. As a result, it is suggested that these assumptions be applied as an initial guess in essentially all problems. In this approximation, eq 1 can be rewritten as:

$$\mathcal{D}_j \nabla^2 C_j = \varepsilon_j R(\bar{C}) \tag{2}$$

where $\bar{C}$ represents the vector of concentrations of all the substrates and products involved in the reaction. Boundary conditions are also required. At the boundary with the solution, it is best to take:

$$C_j^{\text{int}} = K_j C_j^{\text{ext}} \tag{3}$$

where K_j is the partition coefficient for species j. This partition coefficient is often set equal to unity and the solubility effects are lumped into the measurement or estimate of the effective diffusivity. At other boundaries, a no-flux condition is usually applied. Eqs 2 and 3 still require a catalog of exact solutions for various different reaction rate equations and geometries [13].

For any two substrates, eqs 2 and 3 can be combined, and the resulting analysis suggests that the ratio:

$$\psi_{ij} = \frac{\varepsilon_j \mathcal{D}_i C_i^{\text{ext}} K_i}{\varepsilon_i \mathcal{D}_j C_j^{\text{ext}} K_j} \tag{4}$$

determines which of the substrates is diffusion-limiting; if $\psi_{ij} < 1$, then substrate i will tend to be exhausted before substrate j [13, 51a]. Product inhibition can be handled in a similar manner, if the external substrate concentration C^{ext} is replaced by a characteristic concentration at which the product inhibits the reaction of interest.

To predict if mass transfer limits the rate of reaction, a useful first step is to evaluate the Thiele modulus, ϕ_j, which is defined by:

$$\phi_j^2 = \frac{\varepsilon_j R(\overline{KC}^{\text{ext}}) L^2}{\mathcal{D}_j K_j C_j} \tag{5}$$

Here $R(\overline{KC}^{\text{ext}})$ is the volumetric reaction rate, evaluated at the conditions that would be found in the gel in the absence of mass transfer limitations (that is, including any solubility effects), and L is the characteristic length scale of the cell layer. An appropriate definition for L that minimizes the differences between planar, cylindrical, and spherical geometries is to take L as the ratio of the aggregate volume, V_a, to the aggregate surface area in contact with the external solution (for example, $r/3$ for a sphere).

To determine if mass transfer is likely to be limiting, one wishes to know whether ϕ_j is greater or less than 1. For $\phi_j << 1$, mass transfer limitations with respect to substrate j are predicted not to occur; for $\phi_j >> 1$, mass transfer limitations are expected to be severe. In the near vicinity of 1, a detailed solution for an appropriate geometry and for appropriate kinetics should be consulted. The results of such solutions have traditionally been expressed in terms of the effectiveness factor:

$$\eta \equiv \frac{\int_{V_a} R(\overline{C})\, dV}{V_a\, R(KC^{\text{ext}})} = \frac{\Re_{\text{tot}}}{V_a\, R(KC^{\text{ext}})} \tag{6}$$

which is the ratio of the actual overall reaction rate ($\Re_{\text{tot}}$) observed to the rate that would be predicted in the absence of mass transfer limitations (denominator). η is predicted to be a function of ϕ; the limits are such that $\eta \rightarrow 1$ for small ϕ, and that η goes as $1/\phi$ for large ϕ. The limiting behavior of η can be adjusted using the generalized effectiveness factor [13, 48].

When experimental data for overall rates are available, it is often convenient to evaluate the observable Weisz modulus [52]:

$$\Phi = \eta\, \phi^2 = \frac{\Re_{\text{tot}}\, L^2}{KC^{\text{ext}}\, \mathcal{D}} \tag{7}$$

Given the asymptotic behavior of η and ϕ, the value of Φ relative to 1 is also a good test for mass transfer limitations. The advantage of this modulus is that it does not require an estimate of the intrinsic reaction rate, only the effective diffusivity.

Eventually, for sufficiently large values of ϕ, external mass transfer limitations are likely to be important. External mass transfer is usually characterized by a mass transfer coefficient k_j that describes the transport of j to the surface, or its dimensionless equivalent, the Biot number:

$$\beta_j \equiv \frac{k_j\, L}{\mathcal{D}_j\, K_j} \tag{8}$$

Obviously, k_j will be a function of flow rates outside the cell layer. For sufficiently high reaction rates, external mass transfer limitations become the dominant feature of the process, and, for large ϕ, η will vary as β/ϕ^2.

12.3.2 Recommendations for Modeling

To formulate a minimal model for an immobilized-cell process based on the above considerations, the following set of steps is recommended:

1. Estimate the reaction stoichiometries for substrates and products of the reactions of interest. The formation of compounds that may inhibit the process, including acid or base production, must be taken into consideration [15, 52a].
2. Estimate the diffusivities of each of the substrates and products in an aqueous solution. Even if only rough guesses for the diffusivities in solution can be obtained, this is often sufficient, in combination with a knowledge of the stoichiometry, to allow the identification of which substrates, products, or both are limiting, by evaluation of the parameter Φ_{ij} for each substrate, with reference to one of the other substrates. The identification of the limiting substrate can often be tested by adjusting the concentration of that nutrient in the medium.
3. Estimate the intrinsic activity of the immobilized cells per volume of cell layer in the catalyst. This may require an accounting of changes in the cell mass and activity in the immobilized-cell aggregate. Given the prevalence of mass-transfer limitations, it is generally not sufficient to estimate this from overall activity measurements performed on immobilized cells, unless extensive control experiments are undertaken.
4. From these estimates, calculate values of ϕ and β, and estimate an effectiveness factor using results from the literature [13].

If greater accuracy is required, detailed simulations along the lines of the references cited above should be carried out.

12.3.3 Diffusivity Measurements

The usefulness of the models described above is limited by the availability of data for the effective diffusivities of solutes in immobilized-cell matrices, and of a means to correlate such data for extrapolation. The matrix may consist of several different dispersed phases. One important phase in which mass transfer may be slow is the cell-containing region.

Diffusivity measurements in cell layers have been reviewed by Libicki *et al.* [53] and subsequently by Fan *et al.* [54], and new results continue to appear [55–57]. Unfortunately, while there is a clear trend that increasing cell concentrations lead to decreased diffusivities, as expected on the grounds that the cell membranes are a substantial barrier to diffusion, there is no consensus on the cell concentration at which this occurs, and the details of the structure of cell layers may play a significant role. The measured effective diffusivity of a solute typically ranges from being quite close to the solution diffusivity at low cell densities, falling to a value 10–20% of the free values at the highest cell densities studied (200–400 g dry mass/l). In some cases, however, much of the drop occurs before the cell concentration reaches 50 g/l. Surprisingly, no systematic variation can be observed in the degree to which the effective diffusivity is reduced as a function of the polarity or size of the solutes studied. Numerous suggestions have been made as to functions that could be used to correlate the diffusivity as a function of cell mass concentration or volume fraction [53–55, 58–60]. Given the scatter in the data, it is not clear that any particular functional dependence on cell mass concentration should be preferred. The author's personal preference is that for the rough estimation of Thiele moduli, the conservative rule of thumb that the effective diffusivity in the cell layer is approximately one-third of the solution value should be used as a first guess. If further information on cell concentration is available, the Hashin-Shtrikman bounds based on the excluded volume of the cells should be used for a second iteration [53]. As always, there is no good substitute for experimental data obtained with the system of interest.

It is harder to predict the rates of diffusion for ionic species than for neutral solutes, since the diffusion of all ionic species is coupled through electrostatic effects [61]. In addition, there may be interactions between the ions in solution and fixed charges on the support or on the cells. Some such effects have been modeled in the context of immobilized enzymes [62, 63]. It seems likely that the diffusivities for many ionic species in immobilized-cell aggregates, including amino acids, could be considerably lower than those for neutral solutes.

12.3.4 Intrinsic Activities of Immobilized Cells

The second requirement for a reaction-diffusion model is a knowledge of the intrinsic reaction rate catalyzed by cells within the layer. As a first approximation, it seems simplest to assume that on a cell mass basis, the intrinsic reaction rate is the same for free cells exposed to the same concentrations of substrates and products [15]. While much of the literature suggests that this may be an oversimplification, the evidence for specific effects of surfaces or entrapment on cells is ambiguous.

This question of whether immobilization has an effect on the intrinsic kinetics of cellular reactions, other than the effect of mass transfer on the local solute concentrations, is perhaps the most contentious issue in immobilized-cell technology. This issue is part of a larger question in microbial physiology, namely, whether the intrinsic kinetics of cell metabolism can be altered due to interactions with surfaces. This debate predates the concept of immobilized cells, having been originally brought forward in the case of cells adhering to surfaces or in biofilms, with many claims for acceleration of growth due to adhesion [64–66]. With good reason, the debate has focused on rate *accelerations* caused by immobilization; it would be exceptionally difficult to prove that a decrease in the rate of a reaction was <u>not</u> due to mass transfer limitations at some level. Recent reviews of experiments designed to test these claims suggest that the evidence for true growth acceleration is not well established, either for adherent cells [67], or for artificially immobilized cells [14]. On the other hand, it is clear that cells can sense that they are adhering to a surface and synthesize specific products as a result; a particularly nice demonstration of this is the induction of lateral flagellar genes in *Vibrio parahaemolyticus* [68].

At present, the most compelling evidence for a significant change in intrinsic metabolic properties due to immobilization involves the fermentative growth of the yeast *Saccharomyces cerevisiae*. The yield of cell mass production on glucose consumed is observed to change in immobilized cells. This results in the acceleration of the specific rates of glucose consumption and ethanol production, although the rate of cell mass synthesis actually declines. Bailey and coworkers have done extensive experiments on this system [69–72]. Using NMR techniques, they have documented changes in rates in the glycolytic pathway in the immobilized cells, as well as apparent changes in chromosome number. These are long-term changes in cell behavior, since they can be observed in cells eluted from the immobilized-cell layer. Recent results from another laboratory also suggest that immobilization increases the degree of saturation of phospholipid fatty acids in the same system [73]. Until the root cause of these metabolic alterations is understood, it seems premature to extrapolate these findings to other immobilized-cell systems.

12.3.5 Tests of Reaction-Diffusion Models in Immobilized Cells

Our detailed understanding of how to model immobilized-cell systems, especially in the case of growing cells, comes from the use of methods to determine spatial distributions of substrates, products, or cells in an immobilized-cell aggregate. While overall measurements of reaction rate are undoubtedly useful, their interpretation in terms of mass transfer is rarely unambiguous. Unlike overall measurements, the spatial distribution provides a direct test for the presence of mass-transfer limitations.

Direct measurements of substrate concentrations within an immobilized-cell layer have been made using microelectrodes. The tip diameter of electrodes can be as small as a few micrometers, and the tip can be slowly inserted into a cell layer with fine control of position. Most of the measurements that have been made have used O_2-sensitive electrodes [41, 74], although microelectrodes for the measurement of pH [75] and of certain ions, such as NH_4^+ and NO_3^- [76] can also be manufactured. While electrode insertion into a cell layer is a necessarily invasive technique, and cannot be used in reactors, these *in situ* measurement techniques are unsurpassed in their capacity to study transient phenomena, which are particularly useful for diffusivity estimation in immobilized-cell systems.

In growing cell systems, the distribution of cells or of components within the cells will be determined by the cell growth history and will thus allow analysis of the state of the catalyst, and the indirect estimation of the reaction rate history of the catalyst.

The size of the cell-containing region is only indirectly a measure of mass-transfer limitations; both the local rates and effective diffusivities change as the cell concentration increases, so that the cell distribution depends on the history of the immobilized-cell region. Analysis of a series of micrographs of immobilized-cell aggregates, prepared at various times in the history of an aggregate, may allow the reconstruction of the behavior [77]. Electron microscopy can also reveal morphological changes in the cells, including extracellular polymers, lysis, cell size, and shape. Fluorescent stains for RNA, DNA, and protein are available and can be detected sensitively in a fluorescence microscope [78]. RNA content is related to the growth rate and can be used to infer the growth rate history of the cells, although the rate of degradation of RNA in starving cells should be considered in such calculations. Staining techniques that visualize the distribution of enzyme activity in a fixed section of a cell aggregate have also begun to be applied. The distribution of alkaline phosphatase in alginate-gel beads containing phosphate-limited *E. coli* cells has been used to show that cells in the interior of a bead can still respond to environmental stimuli by synthesizing appropriate proteins [79]. The use of fluorescently labeled antibodies to stain sections of cell aggregates is another technique that provides related information.

Light microscope autoradiography is perhaps the most versatile of the histological techniques applied to immobilized cells [15, 32]. Radioisotope precursors for the synthesis of all major biological polymers are readily available. With appropriate choice of a pulse-chase labeling protocol and measurements of inlet and outlet isotope concentrations, parameters of immobilized-cell growth have been obtained. These include:

1. the net rate of cell mass synthesis,
2. the yield coefficients,
3. the cell mass degradation rate,
4. the size of the region in which cell growth takes place,
5. the specific rates of substrate consumption and cell growth in the growth region, and
6. the rate and pattern of cell movement within the immobilized-cell layer.

12.3.6 Catalytic Stability of Immobilized Cells

In applications of immobilized-cell technology, a paramount concern is the longevity of the catalyst under process conditions. The factors governing this stability are quite different in enzymatic and growth processes. For growth-related processes, it is often wise to allow either continual or sporadic regrowth of the cells. In many cases, especially when mixed cultures are used, this leads to very stable systems, exemplified by biofilm systems used in waste treatment [12]. This may be due to the greater capacity for biomass degradation in such systems. With pure cultures, designing a system with exceptionally good cell retention using membranes has been shown to lead to severe operational problems [15]; thus, allowing a certain degree of cell escape from the immobilized-cell layer is also essential for stability. The pattern of growth can be important. If the main route of cell loss is sloughing from the layer surface, then those cells that have grown most recently are the most likely to be lost; this may lead to the enhancement of culture stability, especially with plasmid-bearing cells [16].

For immobilized-cell processes that do not involve cell growth, a major concern in choosing between immobilized-cell and immobilized-enzyme processes is the relative enzymatic stability in the two systems. In the case of aspartase, Chibata and co-workers have found that the immobilized-cell system is more stable than immobilized purified enzyme [27]. Similar results were obtained with a phenylalanine aminotransferase [80]. Cofactor retention in the cells may be an advantage compared to immobilized enzymes. The factors contributing to the increased stability of immobilized enzymes compared to free enzymes are not well understood, but one widely accepted theory is that limiting the mobility of a protein imparts some resistance to thermal denaturation [81, 82]. These effects are likely to be enzyme-specific, and it is not possible to make general conclusions or recommendations.

In considering the observed stability over time of enzyme activities in immobilized-cell catalysts, one should keep in mind the possibility of diffusion limitations leading to "pseudostability". In the case of external mass-transfer limitations, the overall rate of reaction is independent of the intrinsic enzyme activity; in the case of internal mass-transfer limitations, it is proportional to the square root of the intrinsic activity. Under such conditions, the system will *appear* to be much more stable than a free-cell system. Eventually, as the intrinsic activity decays, the system will exit the mass-transfer-limited regime, and the overall rate will reflect the decreasing intrinsic activity. While this pseudostability may be advantageous in allowing operation of a process at nearly constant conditions, it comes at the expense of a lowered initial rate. Given these considerations, it is imperative that apparent stability measurements *not* be extrapolated to times outside the range of measurement.

12.4 APPLICATIONS OF IMMOBILIZED-CELL TECHNOLOGY TO AMINO ACID PRODUCTION

12.4.1 Enzymatic Processes

The first immobilized-cell processes to be successfully implemented on a large scale involved the production of L-aspartate from ammonium fumarate, using the enzyme aspartase [83]. This process involves a single cellular enzymatic activity.

Furthermore, the substrates (NH_4^+ and fumarate) can be used at very high concentrations without inhibiting the enzyme. The enzyme is not cofactor-dependent, and appears to be quite stable. The substrates are typically supplied to the immobilized cells as a 1 M solution, a concentration which is high enough to interfere with cell growth, and which may also result in cell permeabilization [27]. Very high enzyme activities in the catalyst were obtained using hyperproducing mutants, and more recently the enzyme has been cloned and can be overproduced in cells [84]. These factors make aspartase an ideal enzyme for use in either an immobilized-cell or an immobilized-enzyme process.

The original implementation of the process involved the immobilization of bacterial cells by entrapment in polyacrylamide gels [85]. Subsequent research led to the substitution of κ-carrageenan for polyacrylamide, and to the use of the hardening agents hexamethylenediamine and glutaraldehyde [27]. These cross-linkers lead initially to a small decrease in the activity of the catalyst preparation, but appear to vastly improve the longevity of the catalyst, perhaps by reducing leakage of the protein, preventing the physical attrition of the catalyst, or both. Polymers such as locust bean gum, polyethyleneimine, and Chinese gallotannin have also been used by the Tanabe Seiyaku researchers in a similar fashion to stabilize immobilized-cell activities, either with aspartase or with fumarase from *Brevibacterium flavum* [86–88]. Recent work from this group has focused on improving the reactor configuration for the aspartase process. Improving the heat-transfer capacity of the reactor and controlling the temperature were found to reduce thermal inactivation of the enzyme and improve the operation of the system [88a].

A number of other groups have investigated different immobilization schemes using the aspartase system (see Table 12.1). These efforts have generally focused on improving the physical properties of the catalyst to allow increased flow rates and increased productivities. For example, the vermiculite system is reported to employ residence times approximately one-half of those used in the Tanabe Seiyaku system, with a corresponding twofold increase in the volumetric productivity [97, 129]. Only modest improvements in productivity or catalyst stability can be reasonably expected, given the efficiency of the current systems.

An extension of the aspartase process involved the use of a second immobilized-cell system to decarboxylate L-aspartate to L-alanine. Immobilized *Pseudomonas dacunhae* was found to be a suitable source of L-aspartate-β-decarboxylase [90]. As before, κ-carrageenan was used as the immobilizing agent. A spin-off of this process is the production of D-aspartate using the same L-aspartate β-decarboxylase from *P. dacunhae* immobilized in κ-carrageenan, only changing the feed stream to a racemic mixture of DL-aspartate [95]. With increasing complexity of enzymatic transformations comes a requirement to ensure that side activities are eliminated. In the case of the two-step process to L-alanine, competing enzymes such as fumarase and alanine racemase needed to be eliminated from each of the two immobilized-cell preparations. Mild heat treatments under acidic conditions were found to be effective [128,

TABLE 12.1 Applications of Immobilized-Cell Technology in Amino Acid Production

Amino acid	Biological system*	Immobilization Technique	Reference
Alanine	*Pseudomonas dacunhae* L-aspartate-β-decarboxylase	Polyurethane gels	Fusee and Weber, 1984 [89]
		κ-Carrageenan	Yamamoto *et al*., 1980 [90]
		Cross-linked gelatin on nylon mesh	Burdick & Schaeffer, 1988 [91]
	Coimmobilized *Pseudomonas dacunhae* L-aspartate-β-decarboxylase and *Escherichia coli* aspartase	κ-Carrageenan	Takamatsu *et al*., 1985 [92]
	Corynebacterium dismutans	Polyacrylamide	Sarkar & Mayaudon, 1983 [93]
Arginine	*Serratia marcescens*	κ-Carrageenan	Fujimura *et al*., 1984 [94]
D-Aspartate	*Pseudomonas dacunhae* L-aspartate-β-decarboxylase	κ-Carrageenan	Senuma *et al*., 1989 [95]
Aspartate	*Escherichia coli* aspartase	Polyacrylamide	Chibata *et al*., 1974 [85]
		κ-Carrageenan	Chibata, 1979 [27]
			Hamer & Richenburg, 1988 [96]
		κ-Carrageenan + locust bean gum	Nishida *et al*., 1979 [86]
		Vermiculite	Hamilton *et al*., 1985 [97]
		Cross-linked gelatin on nylon mesh	Burdick & Schaeffer, 1988 [91]
		Polyazetidine gel on ion-exchange resin beads	Wood & Calton, 1984 [98]

Amino acid	Biological system*	Immobilization Technique	Reference
Aspartate	*Escherichia coli* aspartase	Hollow fibers	Smith, 1989 [99]
		Polyurethane	Fusee *et al.*, 1981 [100]
			Fusee, 1987 [101]
		Ion-exchange resin	Kawabata *et al.*, 1990 [102]
	Alcaligenes metalcaligenes aspartase	Cells cross-linked with polyethyleneimine-glutaraldehyde	Vojtisek *et al.*, 1986 [102a]
Glutamate	*Corynebacterium glutamicum*	κ-Carrageenan gel cubes	Kim & Ryu, 1982 [103]
		Polyacrylamide gels	Slowinski & Charm, 1973 [104]
	Brevibacterium flavum	Collagen membrane	Constantinides *et al.*, 1981 [105]
	Brevibacterium flavum protoplasts	Agar gels on filters	Karube *et al.*, 1985 [106]
	Brevibacterium ammoniagenes	Alginate gels	Lu *et al.*, 1988 [107]
	Cryptococcus albidus glutaminase (glutamine degradation)	Silica-alginate gels	Fukushima & Motai, 1990 [108]
Isoleucine	*Serratia marcescens*	κ-Carrageenan	Wada *et al.*, 1980 [109]
Lysine	*Bacillus subtilis*	Alginate gels	Israelides *et al*. 1989 [110]
	Corynebacterium sp.	Alginate gels	Nasri *et al.* , 1989 [111]
Methionine	*Aspergillus ochraeus* aminoacylase	Mycelial pellets, cross-linked with glutaraldehyde	Hirano *et al.*, 1977 [37]

Amino acid	Biological system*	Immobilization Technique	Reference
Phenylalanine	Coimmobilized *Corynebacterium* acetamidocinnamate amidohydrolase and *Paracoccus denitrificans* aminotransferase	κ-Carrageenan	Nishida *et al*., 1987 [112]
	Rhodotorula rubra phenylalanine ammonia-lyase	Vermiculite	Hamilton *et al*., 1985 [97]
		Alginate gels treated with glutaraldehyde	Evans *et al*., 1987 [113]
	Corynebacterium equi phenylalanine dehydrogenase	Alginate gels	Evans *et al*., 1987 [114]
	Pseudomonas fluorescens L-phenylalanine aminotransferase	Alginate gels treated with glutaraldehyde	Evans *et al*., 1987 [115]
	Pseudomonas putida L-phenylalanine aminotransferase	Chitosan	Ziehr *et al*., 1987 [80]
	Paracoccus denitrificans L-phenylalanine aminotransferase	κ-Carrageenan gels	Nakamichi *et al*.1989 [116]
	Escherichia coli L-phenylalanine aminotransferase	Polyazetidine gel on ion-exchange resin beads	Calton *et al*., 1986 [117], Calton *et al*., 1987 [118]
	Escherichia coli	Hollow fibers	Paterson *et al*., 1986 [119]
Serine	*Corynebacterium glycinophilum*	Alginate gels	Tanaka *et al*., 1989 [120]
	Pseudomonas AM1	Hollow fibers	Klier *et al*., 1986 [121]

Amino acid	Biological system*	Immobilization Technique	Reference
Tryptophan	*Escherichia coli* tryptophan synthetase	Polyacrylamide gels	Bang *et al.*, 1983 [122]
	Escherichia coli tryptophanase	Chitosan	Vorlop & Klein, 1981 [123]
		Polyacrylamide	Decottignies-Le Maréchal *et al.*, 1979 [124]
	Escherichia coli tryptophan synthetase & *Pseudomonas putida* amino acid racemase	Alginate gels	Ishiwata *et al.*, 1990 [125]
Tyrosine	*Erwinia herbicola* β-tyrosinase	Collagen	Yamada *et al.*, 1978 [126]
	Escherichia intermedia tyrosine phenol-lyase	Polyacrylamide gels	Para *et al.*, 1985 [127]

* Note: where no enzyme name is given, the process employs immobilized living cells

128a]. For alanine production, the use of coimmobilized cells was found not to be competitive with the use of separately immobilized catalysts for alanine production [92]. Apparently the potential for using limitations to mass transfer of aspartate (the intermediate product) to increase its local concentration is more than offset by the advantage gained from adjusting the operating pH between the columns used for aspartate production and for decarboxylation. In other systems, as in the production of L-phenylalanine from acetamidocinnamic acid, the use of coimmobilized cells has been shown to have advantages over the use of separately immobilized cells [112].

In addition to the aspartate/alanine system, the only other amino acid for which enzymatic immobilized-cell processes have received substantial attention is L-phenylalanine [130]. Here, several different enzymes and starting materials have been investigated (see Table 12.1). In this case, the starting products are more expensive, and cannot be used in as high a concentration as is the case with aspartase. For these reasons, the enzymatic immobilized-cell process may have difficulty competing with more classical fermentation processes.

12.4.2 Immobilized-Living-Cell Processes

Several different amino acids have been produced by immobilized cells in growth processes (see Table 12.1). Today, almost 20 years after the first of these processes was described in the literature [104], there is little enthusiasm for their use, at least judging from the rate of publication, in spite of the commercial importance of amino acid fermentations.

Generally, these processes involve the immobilization of cells, followed by incubation in the presence of a medium that allows growth with concomitant amino acid production. The idea of uncoupling growth from amino acid production by using selective inhibitors has been tested in the case of lysine production; initial results from this process were quite discouraging [110]. The major problem cited with these processes is oxygen mass-transfer limitation. Oxygen transport has been cited as the limiting factor in using immobilized cells for the production of arginine [94], isoleucine [109], lysine [111], phenylalanine [119], and serine [121]. Given the importance of the control of oxygen tension in the microbial production of amino acids, especially glutamate [131], and the exceptional difficulty in monitoring and controlling O_2 tension within an immobilized-cell layer, immobilized-cell processes would need to offer massive increases in cell concentrations to be a viable process alternative. In general, anaerobic processes are much more amenable to the use of immobilized living cells than aerobic processes.

Instead of immobilizing the cells, other processes can be employed that give similar advantages. In particular, the use of fed-batch or cell-recycle reactors with operating strategies that are designed to increase the cell concentration may offer some of the same advantages without incurring the problems of mass-transfer limitations. These techniques have been applied to the production of leucine [131a], lysine [132], and phenylalanine [133], among others.

12.5 CONCLUSIONS

Whole-cell immobilization is a mature technology that has been successfully applied to a wide variety of processes. In particular, the major success in the use of artificially immobilized cells has been in the production of aspartic acid. New techniques to characterize immobilized cells have given us confidence in our ability to model the behavior of these systems, and widely applicable methods for immobilization have enabled the rapid development of new immobilized-cell catalysts. Still, in order for immobilized-cell processes to emerge from their current limited niche in biotechnology, methods need to be developed to improve mass transport within immobilized-cell layers.

12.6 REFERENCES

1. Chibata, I., Wingard, L. B., Jr. *Immobilized Microbial Cells*; Academic Press, New York, 1983.
2. Mattiasson, B. *Immobilized Cells and Organelles*; CRC Press, Boca Raton, FL, 1983.
3. Woodward, J. *Immobilized Cells and Enzymes: A Practical Approach*; IRL Press, Oxford, 1985.
4. Tampion, J., Tampion, M. D. *Immobilized Cells: Principles and Applications*; Cambridge Univ. Press, Cambridge, 1987.
5. Mosbach, K. *Immobilized Cells and Enzymes. Part B*; *Methods Enzymol. 135*, (1987).
6. Phillips, C. R., Poon, Y. C. *Immobilization of Cells*; Springer-Verlag, Berlin, 1988.
7. de Bont, J. A. M., Visser, J., Mattiasson, B., Tramper, J. *Physiology of Immobilized Cells*; Elsevier, Amsterdam, 1990.
8. Birnbaum, S., Larsson, P. O., Mosbach, K. In *Solid Phase Biochemistry: Analytical and Synthetic Aspects*; Scouten, W. H., Ed.; Wiley-Interscience, New York, 1983; pp 679-762.
9. Klein, J., Wagner, F., *Appl. Biochem. Bioeng. 4*, 11-51 (1983).
10. Tanaka, A., Nakajima, H., *Adv. Biochem. Eng./Biotechnol. 42*, 97-131 (1990).
11. Savage, D. C., Fletcher, M. *Bacterial Adhesion. Mechanisms and Physiological Significance*; Plenum, New York, 1985.
12. Bryers, J. D., *Biotechnol. Progress 3*, 57-68 (1987).
13. Karel, S. F., Libicki, S. B., Robertson, C. R., *Chem. Eng. Sci. 40*, 1321-1354 (1985).
14. Karel, S. F., Salmon, P. M., Stewart, P. S., Robertson, C. R. In *Physiology of Immobilized Cells*; de Bont, J. A. M., Visser, J., Mattiasson, B., Tramper, J., Eds.; Elsevier, Amsterdam, 1990; pp 115-126.
15. Karel, S. F., Robertson, C. R., *Biotechnol. Bioeng. 34*, 337-356 (1989).

16. Sayadi, S., Nasri, M., Barbotin, J. N., Thomas, D., *Biotechnol. Bioeng. 33*, 801-808 (1989).
17. Bryers, J. D., Banks, M. K. In *Physiology of Immobilized Cells*; de Bont, J. A. M., Visser, J., Mattiasson, B., Tramper, J., Eds.; Elsevier, Amsterdam, 1990; pp 49-61.
18. Siebel, M. A., Characklis, W. G., *Biotechnol. Bioeng. 37*, 778-789 (1991).
19. Characklis, W. G., Cooksey, K. E., *Adv. Appl. Microbiol. 29*, 93-138 (1983).
20. Messing, R. A., Oppermann, R. A., *Biotechnol. Bioeng. 21*, 49-58 (1979).
21. Black, G. M., Webb, C., Matthews, T. M., Atkinson, B., *Biotechnol. Bioeng. 26*, 134-141 (1984).
22. Gbewonyo, K., Wang, D. I. C., *Biotechnol. Bioeng. 25*, 967-983 (1983).
23. Nilsson, K., Flygare, S., Linse, S., Birnbaum, L., Mosbach, K., Schroder, U. Jeppsson, U., Larsson, P.-O., Brodelius, P., *Eur. J. Appl. Microbiol. Biotechnol. 17*, 319-326 (1983).
24. Smidsrod, O., Skjåk-Bræk, G., *Trends Biotechnol. 8*, 71 (1990).
25. Hulst, A. C., Tramper, J., Van 't Riet, K., Westerbeek, J. M. M., *Biotechnol. Bioeng. 27*, 870-876 (1985).
26. Matulovic, U., Rasch, D., Wagner, F., *Biotechnol. Lett. 8*, 485-490 (1986).
27. Chibata, I., *ACS Symp. Ser. 106*, 187-202 (1979).
28. Starostina, N. G., Lusta, K. A., Fikhte, B. A., *Eur. J. Appl. Microbiol. Biotechnol. 18*, 264-270 (1983).
29. Fukui, S., Tanaka, A., *Adv. Biochem. Eng. Biotechnol. 29*, 1-33 (1983).
30. Kawakami, K., Tsuruda, S., Miyagi, K. In *Physiology of Immobilized Cells*; de Bont, J. A. M., Visser, J., Mattiasson, B., Tramper, J., Eds.; Elsevier, Amsterdam, 1990; pp 439-448.
31. Inloes, D. S., Michaels, A. S., Robertson C. R., Matin, A., *Appl. Microbiol. Biotechnol. 23*, 85-91 (1985).
32. Karel, S. F., Robertson, C. R., *Biotechnol. Bioeng. 34*, 320-336 (1989).
33. Robertson, C. R., Kim, I. H., *Biotechnol. Bioeng. 27*, 1012-1020 (1985).
34. Lim, F. In *Biomedical Applications of Microencapsulation*; Lim, F., Ed.; CRC Press, Boca Raton, FL, 1983; pp 137-154.
35. Lamberti, F. V., Evangelista, R. A., Blysniuk, J., Wheatley, M. A., Sefton, M. V., *Appl. Biochem. Biotechnol. 10*, 101-104 (1984).
36. King, G. A., Daugulis, A. J., Faulkner, P., Goosen, M. F. A., *Biotechnol. Progress 3*, 231-240 (1987).
37. Hirano, K.-I., Karube, I., Suzuki, S., *Biotechnol. Bioeng. 19*, 311-321 (1977).
38. Boersma, J. G., Vellenga, K., De Wilt, H. G. J., Joosten, G. E. H., *Biotechol. Bioeng. 21*, 1711-1724 (1979).
39. Atkinson, B., Daoud, I. S., *Adv. Biochem. Eng. 4*, 42-124 (1976).
40. Siegrist, H., Gujer, W., *Water Res. 19*, 1369-1378 (1985).
41. Wittler, R., Baumgartl, H., Lubbers, D. W., Schugerl, K., *Biotechnol. Bioeng. 28*, 1024-1036 (1986).
42. Vasudevan, M., Matsuura, T., Chotani, G. K., Vieth, W. R., *Ann. N. Y. Acad. Sci. 506*, 345-356 (1986).
43. Fowler, J. D., Robertson, C. R., *Appl. Environ. Microbiol. 57*, 102-113 (1991).
44. Mitani, Y., Nishizawa, Y., Nagai, S., *J. Ferment. Technol. 62*, 249-253 (1984).

45. Grady, C. P. L., Jr. In *Fixed-Film Biological Processes for Wastewater Treatment*; Wu, Y. C., Smith, E. D., Eds.; Noyes Data Corp., Park Ridge, NJ, 1983; pp 75-134.
46. Klein, J., Vorlop, K.-D., Wagner, F., *Ann. N. Y. Acad. Sci. 434*, 437-449 (1984).
47. Radovich, J. M., *Enzyme Microb. Technol. 7*, 2-11 (1985).
48. Aris, R. *The Mathematical Theory of Diffusion and Reaction in Permeable Catalysts*; Clarendon Press, Oxford, 1975.
49. Wanner, O., Gujer, W., *Biotechnol. Bioeng. 28*, 314-328 (1986).
50. Namkung, E., Rittmann, B. E., *Biotechnol. Bioeng. 29*, 269-278 (1987).
51. Sayles, G. D., Ollis, D. F., *Biotechnol. Bioeng. 34*, 160-170 (1989).
51a. Leypoldt, J. K., Gough, D. A., *Biotechnol. Bioeng. 24*, 2705-2719 (1982).
52. Weisz, P. B., *Science 179*, 433-440 (1973).
52a. Stewart, P. S., Robertson, C. R., *Appl. Environ. Microbiol. 54*, 2464-2471 (1988).
53. Libicki, S. B., Salmon, P. M., Robertson, C. R., *Biotechnol. Bioeng. 32*, 68-85 (1988).
54. Fan, L. S., Leyvaramos, R., Wisecarver, K. D., Zehner, B. J., *Biotechnol. Bioeng. 35*, 279-286 (1990).
55. Mignot, L., Junter, G.-A., *Appl. Microbiol. Biotechnol. 33*, 167-171 (1990).
56. Mignot, L. Junter, G.-A., *Appl. Microbiol. Biotechnol. 32*, 418-423 (1990).
57. Brito, E., Juan, J. D., Dominguez, F., Casas, L. T., *J. Ferment. Bioeng. 69*, 135-137 (1990).
58. Klein, J., Schara, P., *Appl. Biochem. Biotechnol. 6*, 91-117 (1981).
59. Furusaki, S., Seki, M., *J. Chem. Eng. Japan 18*, 389-393 (1986).
59a. Fukui, M., Yamashita, K., *J. Ferment. Technol. 63*, 167-173 (1985).
60. Westrin, B. A., *Appl. Microbiol. Biotechnol. 34*, 189-190 (1990).
61. Cussler, E. L. *Diffusion. Mass Transfer in Fluid Systems*; Cambridge University Press, Cambridge, 1984.
62. Shuler, M. L., Aris, R., Tsuchiya, H. M., *J. Theor. Biol. 35*, 67-76 (1972).
63. Hamilton, B. K., Stockmeyer, L. J., Colton, C. K., *J. Theor. Biol. 41*, 547-560 (1973).
64. Fletcher, M., Marshall, K. C., *Adv. Microb. Ecol. 6*, 199-236 (1982).
65. Fletcher, M., *Appl. Environ. Microbiol. 52*, 672-676 (1986).
66. Hattori, R., *J. Gen. Appl. Microbiol. 18*, 319-330 (1972).
67. van Loosdrecht, M. C. M., Lyklema, J., Norde, W., Zehnder, A. J. B., *Microbiol. Rev. 54*, 75-87 (1990).
68. Belas, R., Simon, M., Silverman, M., *J. Bacteriol. 167*, 210-218 (1986).
69. Doran, P. M., Bailey, J. E., *Biotechnol. Bioeng. 29*, 892-897 (1987).
70. Doran, P. M., Bailey, J. E., *Biotechnol. Bioeng. 28*, 73-85 (1986).
71. Galazzo, J. L., Bailey, J. E., *Biotechnol. Bioeng. 33*, 1283-1289 (1989).
72. Galazzo, J. L., Bailey, J. E., *Biotechnol. Bioeng. 36*, 417-426 (1990).
73. Hilge-Rotmann, B., Rehm, H.-J., *Appl. Microbiol. Biotechnol. 34*, 502-508 (1991).
74. Hooijmans, C. M., Briasco, C. A., Huang, J., Geraats, B. G. M., Barbotin, J. N., Thomas, D., Luyben, K. C. A. M., *Appl. Microbiol. Biotechnol. 33*, 611-618 (1990).
75. Kim, J. Y., Lee, Y. H., *Biotechnol. Bioeng. 34*, 131-136 (1989).
76. de Beer, D., van den Heuvel, J. C., Sweerts, J. P. R. A. In *Physiology of Immobilized Cells*; de Bont, J. A. M., Visser, J., Mattiasson, B., Tramper, J., Eds.; Elsevier, Amsterdam, 1990; pp 613-624.
77. Burrill, H. N., Bell, L. E., Greenfield, P. F., Do, D. D., *Appl. Environ. Microbiol. 46*, 716-721 (1983).

78. Ollis, D. F., Monbouquette, H., Kuhn, R., Peretti, S. W. In *Physiology of Immobilized Cells*; de Bont, J. A. M., Visser, J., Mattiasson, B., Tramper, J., Eds.; Elsevier, Amsterdam, 1990; pp 637-648.
79. Salmon, P. M.; *Mass Transport Phenomena in Reactors Containing Entrapped Enzymes or Bacterial Cells*; Ph. D. Thesis, Stanford University, Stanford, CA, 1989.
80. Ziehr, H., Kula, M.-R., Schmidt, E., Wandrey, C., Klein, J., *Biotechnol. Bioeng. 29*, (1987).
81. Klibanov, A. M., *Anal. Biochem. 93*, 1-25 (1979).
82. Klibanov, A. M., *Science 219*, 722-727 (1983).
83. Chibata, I., Tosa, T., *Ann. Rev. Biophys. Bioeng. 10*, 197-216 (1981).
84. Komatsubara, S., Taniguchi, T., Kisumi, M., *J. Biotechnol 3*, 281-291 (1986).
85. Chibata, I., Tosa, T. Sato, T., U.S. Pat. 3,791,926 (1974).
86. Nishida, Y., Sato, T., Tosa, T., Chibata, I., *Enzyme Microb. Technol. 1*, 95 (1979).
87. Takata, I., Kayashima, K., Tosa, T., Chibata, I., *J. Ferment. Technol. 60*, 431-437 (1982).
88. Takata, I., Tosa, T., Chibata, I., *Appl. Microbiol. Biotechnol. 19*, 85-90 (1984).
88a. Seko, H., Takeuchi, S., Yajima, K., Senuma, M., Tosa, T., *Ann. N. Y. Acad. Sci. 589*, 540-552 (1990).
89. Fusee, M. C., Weber, J. E., *Appl. Environ. Microbiol. 48*, 694-698 (1984).
90. Yamamoto, K., Tosa, T., Chibata, I., *Biotechnol. Bioeng. 22*, 2045 (1980).
91. Burdick, B. A., Schaeffer, J. R., *Biotechnol. Bioeng. 31*, 390-395 (1988).
92. Takamatsu, S., Tosa, T., Chibata, I., *J. Chem. Eng. Japan 18*, 66-70 (1986).
93. Sarkar, J. M., Mayaudon, J., *Biotechnol. Lett. 5*, 201-206 (1983).
94. Fujimura, M., Kato, J., Tosa, T., Chibata, I., *Appl. Microbiol. Biotechnol. 19*, 79-84 (1984).
95. Senuma, M., Otsuki, O., Sakata, N., Furui, M., Tosa, T., *J. Ferment. Bioeng. 67*, 233-237 (1989).
96. Hamer, J. W., Rickenberg, C. B., *AIChE J. 34*, 626-632 (1988).
97. Hamilton, B., Hsiao, H.-Y., Swann, W., Anderson, M., Delente, J., *Trends Biotechnol. 3*, 64 (1985).
98. Wood, L. L., Calton, G. J., *Bio/Technology* 2, 1081-1084 (1984).
99. Smith, W. J.; *Gas-Liquid Slug-Flow Oxygen Transport and Noninvasive Biomass Estimation in Hollow Fiber Reactors*; Ph.D. Thesis, Stanford Univ., Stanford, CA, 1989.
100. Fusee, M., Swann, W., Calton, G. J., *Appl. Environ. Microbiol. 42*, 672 (1981).
101. Fusee, M. C. In *Immobilized Enzymes and Cells. Part C*; Mosbach, K., Ed.; Academic Press, Orlando, 1987; pp 463-471.
102. Kawabata, N., Nishimura, S., Yoshimura, T., *Biotechnol. Bioeng. 35*, 1000-1005 (1990).
102a. Vojtisek, V., Guttmann, T., Barta, M., Netrval, J., *Biotechnol. Bioeng. 28*, 1072-1079 (1986).
103. Kim, H. S., Ryu, D. D. Y., *Biotechnol. Bioeng. 24*, 2167-2174 (1982).
104. Slowinski, W., Charm, S., *Biotechnol. Bioeng. 15*, 973 (1973).
105. Constantinides, A., Bhatia, D., Vieth, W. R., *Biotechnol. Bioeng. 23*, 899-916 (1981).
106. Karube, I., Kawarai, M., Matsuoka, H., Suzuki, S., *Appl. Microbiol. Biotechnol. 21*, 270-272 (1985).
107. Lu, W.-M., Chen, W.-C., *Proc. Natl. Sci. Counc. ROC (A) 12*, 400-406 (1988).
108. Fukushima, Y., Motai, H., *J. Ferment. Bioeng. 69*, 189-191 (1990).

109. Wada, M., Kato, J., Chibata, I., *Eur. J. Appl. Microbiol. Biotechnol. 8*, 241-247 (1979).
110. Israilides, C. J., Weir, A. N. C., Bull, A. T., *Applied Microbiology and Biotechnology 32*, 134-136 (1989).
111. Nasri, M., Dhouib, A., Zorfuani, F., Kriaa, H., Ellouz, R., *Biotechnol. Lett. 11*, 865-870 (1989).
112. Nishida, Y., Nakamichi, K.,, Nabe, K., Tosa, T., *Enzyme Microb. Technol. 9*, 479-483 (1987).
113. Evans, C. T., Choma, C., Peterson, W., Misawa, M., *Biotechnol. Bioeng. 30*, 1067-1072 (1987).
114. Evans, C. T., Bellamy, W., Gleeson, M., Aoki, H., Hanna, K., Peterson, W., Conrad, D., Misawa, M., *Bio/Technology 5*, 818 (1987).
115. Evans, C. T., Peterson, W., Choma, C., Misawa, M., *Appl. Microbiol. Biotechnol. 26*, 305-312 (1987).
116. Nakamichi, K., Nabe, K., Nishida, Y., Tosa, T., *Appl. Microbiol. Biotechnol. 30*, 243-246 (1989).
117. Calton, G. J., Wood, L. L., Updike, M. H., Lantz, L. I. I., Hamman, J. P., *Bio/Technology 4*, 317-320 (1986).
118. Calton, G. J., Wood, L. L., Campbell, M. L. In *Immobilized Enzymes and Cells. Part C*; Mosbach, K., Ed.; Academic Press, Orlando, 1987; pp 497-503.
119. Paterson, S. L., Fane, A. G., Fell, C. J. D., Rogers, P. L. R., *Biotechnol. Lett. 8*, 561-566 (1986).
120. Tanaka, T., Yamamoto, K., Towprayoon, S., Nakajima, H., Sonomoto, K., Yokozeki, K., Kubota, K., Tanaka, A., *Appl. Microbiol. Biotechnol. 30*, 564-568 (1989).
121. Klier, J., Cabral, J. M. S., Cooney, C. L., *Appl. Biochem. Biotechnol. 13*, 181-187 (1986).
122. Bang, W. G., Behrendt, U., Lang, S., Wagner, F., *Biotechnol. Bioeng. 25*, 1013-1025 (1983).
123. Vorlop, K.-D., Klein, J., *Biotechnol. Lett. 3*, 9-14 (1981).
124. Decottignies-Le Maréchal, P. Calderón-Seguin, R., Vandecasteele, J. P., Azerad, R., *Eur. J. Appl. Microbiol. Biotechnol. 7*, 33-44 (1979).
125. Ishiwata, K. I., Fukuhara, N., Shimada, M., Makiguchi, N., Soda, K., *Biotechnol. Appl. Biochem. 12*, 141-149 (1990).
126. Yamada, H., Yamada, K., Kumagai, H., Hino, T., Okamura, S. In *Enzyme Engineering*; Pye, E. K., Weetall, H. H., Eds.; Plenum Press, New York, 1978; Vol. 3, pp 57-62.
127. Para, G., Lucciardi, P., Baratti, J., *Appl. Microbiol. Biotechnol. 21*, 273-279 (1985).
128. Takamatsu, S., Umemura, I., Yamamoto, K., Sato, T., Tosa, T., Chibata, I., *Eur. J. Appl. Microbiol. Biotechnol. 15*, 147-152 (1982).
128a. Umemura, I., Takamatsu, S., Sato, T., Tosa, T., Chibata, I., *Appl. Microbiol. Biotechnol. 20*, 291-295 (1984).
129. Chibata, I., Tosa, T., Takamatsu, S. In *Immobilized Enzymes and Cells. Part C*; Mosbach, K., Ed.; Academic Press, Orlando, 1987; pp 472-479.
130. de Boer, L., Dijkhuizen, L., *Adv. Biochem. Eng./Biotechnol. 41*, 1-27 (1990).
131. Crueger, W., Crueger, A. *Biotechnology: A Textbook of Industrial Microbiology*; Sinauer Associates, Sunderland, MA, 1984.
131a. Wichmann, R., Wandrey, C., Grosse-Wiesmann, W. J., *Appl. Microbiol. Biotechnol. 32*, 373-379 (1990).

132. Hadj Sassi, A., Queric, M. P., Deschamps, A. M., Lebeault, J. M., *Biotechnol. Lett. 10*, 583-586 (1988).
133. Forberg, C., Haggstrom, L., *J. Biotechnol 8*, 291-300 (1988).

IMMOBILIZED ENZYMES: TECHNIQUES AND APPLICATIONS

J. David Rozzell

CONTENTS

13.1 INTRODUCTION

Immobilized enzymes have captured the interest of biotechnologists since the 1950s, but it was with the work by Katchalski-Katzir and Chibata and co-workers in the 1960s that research activities in this area began to accelerate, culminating in the First Enzyme Engineering Conference in 1971. Significant efforts towards improved immobilized-enzyme preparations continue today as new uses emerge. The first industrial application of enzymes in immobilized form was for amino acid production, as reported by Chibata and co-workers at Tanabe Seiyaku in Japan in 1969 [1]. This group immobilized L-aminoacylase for use in a packed-bed reactor in the resolution of various DL-amino acids into their corresponding optically pure enantomeric forms. Since that time, enzymes in immobilized form have become increasingly important as catalysts for the production of amino acids, as well as numerous other substances.

By way of definition, immobilized-enzyme biocatalysts consist of the enzyme, in varying degrees of purity, attached to or otherwise retained by a support matrix. At one extreme, intact dead cells (which are effectively bags of enzymes) may be bound to a support for use as a catalyst; at the other extreme, partially purified or purified forms of the enzyme(s) of interest may be immobilized. Driving the development of this technology is the fact that the immobilization of an enzyme can improve the economics of its application, improve the quality of the product produced, or both. By changing from batch to continuous operation, one can often significantly reduce the economics of an enzyme-catalyzed reaction. Other advantages, such as improved control of the reaction, leading to better uniformity of the product and greater ease of product recovery, are also often achieved through the immobilization of an enzyme.

In assessing the economics of a process using a biological catalyst (enzyme), the critical issue is not the cost of the biocatalyst itself but rather the contribution of the biocatalyst to the cost of the final product. Biocatalyst costs themselves depend on various components, including enzyme or cell production, support matrix, auxiliary reagents, and the loss of activity associated with immobilization. However, the important factors which determine the cost contribution of the biocatalyst are the yield of product, the volumetric productivity achieved in the process, the product concentration attained, and the useful lifetime of the biocatalyst under operational conditions. This chapter will survey immobilization methods, with special attention being paid to those which have been found useful in amino-acid production. The methods discussed here are not meant to be exhaustive but rather illustrative of what has been developed.

13.2 GENERAL IMMOBILIZATION METHODS

Five general techniques have been described for immobilizing enzymes:

1. adsorption;
2. covalent attachment;
3. cross-linking;
4. entrapment of an enzyme in a polymeric matrix; and
5. encapsulation or confinement of an enzyme in a membrane [2].

Each will be discussed along with its advantages and disadvantages. Although each immobilization technique is conceptually distinct, it is important to realize that there is often a certain amount of overlap or even combining of techniques in actual practice. Furthermore, the results of over thirty years of work by numerous researchers suggest that there is no one universally applicable immobilization technique; rather, a range of methodologies is available and must be evaluated on a case by case basis. There is still some art mixed with the science of enzyme immobilization. Certain enzymes may be immobilized successfully with one method but not with others. Each immobilization method has both advantages and disadvantages which must be considered within the context of the enzyme to be immobilized and the overall process in which the enzyme is to be used.

13.2.1 Adsorption

Adsorption is one of the most economical and operationally simple processes by which enzymes can be immobilized, and this simplicity is no doubt responsible for its attractiveness as an immobilization method. Adsorption was the technique used by Chibata and co-workers in their first commercial process involving an immobilized enzyme. However, even though this immobilization procedure may be straightforward to carry out, the interactions involved in adsorption are complex and not completely understood. A good definition of adsorption, given by Messing, is the adhesion of an enzyme to the surface of a support which has not been specifically modified for covalent attachment [3].

One of the principal advantages of adsorption is its ease to perform; simply contacting an aqueous solution of the enzyme with the support is all that is required. Occasionally, with judicious selection of the support, the desired enzyme will have an affinity for the support relative to other proteinaceous material such that a partial purification can be accomplished during the act of immobilization. Enzyme in varying states of purity (including whole cells and enzyme mixed with cell debris) may be used. Additionally, adsorbed immobilized enzyme preparations lend themselves readily to regeneration and reuse of the support material, if desired.

An enzyme may be immobilized by bonding to the external or internal surface of a support, or both. An important advantage of immobilizing the enzyme on a nonporous, external surface is that diffusional limitations are minimized. However, the disadvantages of adsorption on a nonporous material are:

1. a relatively low surface area for bonding may lead to low catalyst activity;
2. leaching, already a problem, may be worse because the enzyme is more susceptible to physical abrasion and sloughing off due to shear forces in the bulk solution; and
3. the enzyme is more exposed to microbial or proteolytic attack.

Although immobilization of an enzyme to the internal surface of a porous carrier has one marked disadvantage—that of diffusional limitations to the rate of the reaction catalyzed—certain advantages may be gained from this approach. An enzyme bound within a porous material is protected from sloughing off through abrasion and attack by adventitious microbial contamination. It has also been reported that, generally speaking, a more stable enzyme system may result from immobilization within a porous material [1].

Typically, support particles used in adsorption of enzyme will have diameters in the range of 0.5 to 100 micrometers. Smaller particles than this result in pressure drops that are unacceptably high; larger particles have such a reduced surface area for contact with substrate solutions that volumetric productivities are too low.

Supports used in the past for adsorption of enzymes include bentonite clay, chitosan, vermiculite, DEAE-cellulose and composites, ion exchange resins, titania, alumina, silica, glass, and granular activated carbon. With each support, the immobilization process depends on factors such as preconditioning of the surface with metal ions prior to contact with the enzyme, optimization of pH and ionic strength, and time. A typical procedure is given by Chibata *et al.* for the immobilization of porcine aminoacylase on DEAE-Sephadex [4], and illustrates the simplicity of the method.

At room temperature (20–25 °C), 100 g of DEAE-Sephadex A-25 (3.5 ± 0.5 meq/g, bead type) suspended in 1 l of distilled water is added to 5 l of 0.1 N NaOH, and the suspension is stirred for 3 h. After filtration, the precipitate is washed with water until the washing solution is neutralized, stirred with 5 l of 0.1 M phosphate buffer (pH 7.0) for 3 h, and kept standing for 16 h. The suspension is filtered, and the resulting precipitate is stirred with 1.67 l of aminoacylase solution (total enzyme activity: 167,000 μmol/h) for 3 h at room temperature. After filtration, 5 l of distilled water is added to the precipitate; the preparation is stirred for 1 h, then filtered. For further washing, 5 l of 0.2 M sodium acetate is added to the precipitate; the preparation is stirred for 1 h, then filtered. The precipitate is washed with distilled water until no protein is detected in the washing solution. These washing procedures are carried out at room temperature. The resulting precipitate is ready for use in the enzyme-catalyzed reaction.

In order to store the immobilized preparation for a long period or to measure its enzyme activity, it is suspended in 2.5 l of distilled water and lyophilized. Using this procedure, 106 g of immobilized DEAE-Sephadex-aminoacylase is obtained. The activity of the preparation is reported to be about 700 μmol/h·g of preparation, under standard assay conditions.

The above method formed the basis for the first industrial process for the production of an amino acid by Tanabe Seiyaku Company in Japan. *N*-Acetyl-DL-methionine was passed through the column, where the L-isomer was deacetylated stereoselectively. The L-methionine was separated from the unreacted *N*-acetyl-D-methionine, which itself was recovered, racemized and recycled through the process. This process was successfully commercialized in 1969 and is considered to be the first commercial-scale immobilized-enzyme process.

In this example, the useful lifetime of the immobilized enzyme was approximately 30 days, and the support could be regenerated each month by adsorbing fresh enzyme on the column. It was reported that the same batch of support matrix could be used for a period of five years, with repetitive regeneration cycles.

Other enzymes useful for amino acid production which have been immobilized by the adsorption method include aspartase on DEAE-cellulose and silica gel [5], leucine aminopeptidase on hydroxylapatite [6], glutamic-aspartic transaminases on DEAE-Sepharose [7], and α-amino-ε-caprolactam racemase on DEAE-Sephadex [8]. There are numerous reports of the immobilization of other enzymes such as amylase, glucose isomerase, glucose oxidase, and proteases on materials such as activated carbon, bentonite, and alumina. Given the data base, it is probably fair to surmise that an appropriate material may be found that will allow the immobilization by adsorption of virtually any enzyme; however, the strength of the adsorptive forces binding the enzyme to the support will vary considerably from one enzyme to another. Adoption of such a catalyst in a process will likely depend on whether the mechanical properties and the operational lifetime of the catalyst are sufficient for commercial application.

13.2.2 Covalent Attachment

Covalent attachment of enzymes to surfaces is often employed when leaching of enzyme activity from the support is a concern. This method generally offers the advantage of an immobilized enzyme system that is more permanently anchored, and may also show greater stability and the ability to withstand a broader spectrum of pH conditions, ionic strengths, and temperatures. Enzymes immobilized by covalent attachment are also more resistant to attack by proteolysis. The main disadvantage of this type of attachment is its somewhat greater complexity and higher cost to prepare. Cost notwithstanding, when covalent attachment results in a significantly more stable enzyme system or when the absence of enzyme in the product solution is of critical importance, it still may be the method of choice.

There are three different techniques by which covalent attachment can be effected. The first is through exposure of the enzyme to a support which has been preactivated to accommodate covalent binding. Operationally, once the activated support has been prepared, the immobilization proceeds like adsorption, but the result is an enzyme covalently bonded to the support matrix. The second technique involves exposure of the enzyme to the support in the presence of an activating or cross-linking reagent. Inevitably, some chemical modification or cross-linking of enzyme molecules occurs during an immobilization of this type which can lead to a loss in catalytic activity. A third possibility—much less commonly used—is to preactivate the enzyme and expose it to a support functionalized for covalent binding. The risk of inactivation of the enzyme by chemical processes during the preactivation procedure is significantly higher, and the procedure is less reproducible. Thus, virtually all practical systems for immobilization fall into one of the first two types.

Frequently, covalent coupling is preferred to other processes in cases where the enzyme is multimeric or contains prosthetic groups. There may be a reduced tendency to disrupt the complex nature of these enzymes since specific bonds can be formed with the functional group to bind the enzyme through multiple points of attachment to the support.

A list of the amino acid functional groups which are chemically reactive enough to participate in covalent binding reactions may be found in Table 13.1.

Of those functional groups listed, -NH_2, -CO_2H, and -SH are involved in most immobilization procedures due to their nucleophilicity. The phenolic ring of tyrosine is also extremely reactive in diazo-coupling reactions, and its hydroxyl group can be an excellent nucleophile at basic pH. The guanidino group of arginine can react with aldehydes. Histidine displays a lower nucleophilicity, but can sometimes react with supports activated with tosylates, tresylates, or other good leaving groups.

Functional groups of a wide range of types have been used for the covalent attachment of enzymes. The variety of chemistries available for covalent attachment allows the conditions of immobilization to be tailored to each enzyme system. The microenvironment of the enzyme may also be tailored by appropriate modification of the support surface; hydrophobic residues or ionically charged groups may be used to alter the support to affect in a desirable way the enzyme-catalyzed reaction of interest [9].

The range of support materials that has been used for covalent attachment, includes porous glass, porous ceramics, sand, charcoal, modified cellulose, polymeric resins, and metallic oxides. A few examples are described here to illustrate typical covalent attachment methodology.

TABLE 13.1 Useful Functional Groups for Covalent Attachment of Enzymes to a Support Matrix

Functional Group	Corresponding Amino Acid
$-NH_2$	Lysine, N-terminus
$-CO_2H$	Glutamic Acid, Aspartic Acid, C-terminus
$-SH$	Cysteine
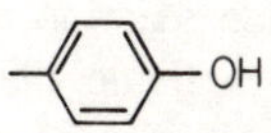	Tyrosine
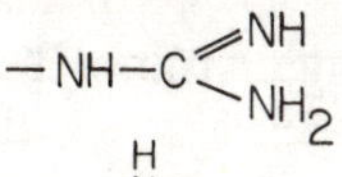	Arginine
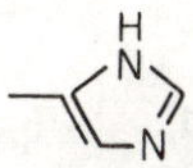	Histidine
$-CH_2OH$	Serine

13.2.2.1 Specific Examples of Covalent Coupling: Binding to Activated Carbohydrate Supports

One of the most commonly used procedures for the covalent coupling of enzymes to carbohydrate support matrices is based on a preactivation of a support with cyanogen bromide [10, 11]. The mechanism of this reaction has been studied extensively by Wilchek and Kohn [12, 13].

The activated carbohydrate will couple generally to an amino group of lysine on the protein or the free N-terminus of the protein to yield a covalently bound product. Supports which have been preactivated with cyanogen bromide can be prepared in advance and stored for periods of up to one year at freezer temperatures. Preactivated supports are also available commercially. Coupling of an enzyme to a CNBr-activated support requires no more than exposure of the enzyme to the support in aqueous solution for a few hours, followed by washing. This method, while extremely popular in lab-scale reactions, has not been widely used in large-scale systems due to the requirements to handle toxic reagents and the poorer mechanical stability of most carbohydrate gels and polymers compared to other support materials. The bond between the enzyme and the support is also potentially susceptible to hydrolytic cleavage.

13.2.2.2 Specific Examples of Covalent Coupling: Carbodiimide Coupling

Carbodiimide reagents activate carboxyl groups through the formation of an *O*-acylisourea intermediate, which reacts rapidly with nucleophilic functional groups of proteins to form amide, thioester, or ester linkages. Hydrolysis is a competing side reaction. This method has been described for the immobilization of transaminases useful for amino acid production [14]. The reaction typically forms a stable amide linkage between enzyme-bound carboxyl groups and support-bound primary amines, and the rate of the reaction is relatively rapid. One disadvantage of carbodiimide reagents is their relatively high cost. Chemical modification of the enzyme leading to a loss of catalytic activity is another potentially negative side reaction.

13.2.2.3 Specific Examples of Covalent Coupling: Amine-Bearing Supports

Amine-bearing supports are among the most useful and most utilized supports for the covalent attachment of enzymes, and a variety multifunctional reagents has been utilized for the covalent attachment of enzymes to supports bearing pendent amines. Difunctional reagents have most frequently been employed for binding. Glutaraldehyde, one of the least expensive reagents available in bulk, reacts in complex fashion to form Schiff bases with amine groups on the support and produces pendent aldehydes and α,β-unsaturated carbonyl functionalities through which proteins may attach. Other useful difunctional agents include diimidate esters and diisocyanates.

Inorganic supports bearing amine functionality have been widely used for the immobilization of proteins. Perhaps the most frequently used technique for the covalent attachment to inorganic surfaces is the preactivation and functionalization with aminosilane reagents [15, 16, 17]. This technique involves attaching an aminosilane to the inorganic surface, leaving the amine available for covalent attachment of the enzyme. The most frequently used silane, popularized through the developments at Corning Glass Works, is γ-aminopropyltriethoxysilane [16]. The terminal amino group may be functionalized by bifunctional coupling agents such as glutaraldehyde, resulting in a support which has been preactivated for enzyme attachment. Coupling of enzyme may also be achieved through the use of carbodiimide reagents, in which case it is through enzymic carboxyl groups that attachment is achieved.

Another process for immobilizing on amino-functionalized inorganic supports involves isocyanate bonding [18]. If the enzyme is attached under alkaline conditions, a substituted urea bond is formed between an amine on the protein surface and the isocyanate. If moderately acidic conditions are employed, then the isocyanate reacts with a hydroxyl group on the enzyme and a urethane bond is formed. Isothiocyanates have also been used successfully [16].

Polyethyleneimine is a common polyamine, useful in providing amino group functionality for attachment. It has the advantage that it is inexpensive and that a wide range of supports including those other than inorganic particles may be used. Examples are alumina [19], carbon [20], diatomaceous earth [21], and poly(vinyl chloride)-silica composites [22, 23].

A typical procedure involving polyethylenimine activation, which has been used for the immobilization of recombinant transaminase, aspartase, and aspartate-β-decarboxylase in the industrial-scale production of L-amino acids, has been described by Rozzell [24]. The activity retained after immobilization approached 90%, with the operational half-lives of the immobilized biocatalysts ranging from 2 to 6 months.

Recently, Flaschel and co-workers have developed a method for the covalent attachment of enzymes involving a mineral or carbon particle coated with chitosan, providing a hydrophilic surface of attachment [25]. Chitosan, which is deacetylated chitin, contains available amino groups for chemical activation and is easily obtained at relatively low cost. Activation with a bifunctional reagent such as glutaraldehyde provides a stable immobilized-enzyme preparation. The use of rigid, incompressible particles on which the chitosan is deposited allows this catalyst to be used in both fixed-bed and fluidized-bed reactors. Yields of activity after immobilization of up to 90% have been reported.

13.2.2.4 Specific Examples of Covalent Coupling: Oxirane or Epoxy-Activated Polymers

Epoxy-activated polymers fall into the catagory of preactivated supports for covalent attachment, and they have gained attention as commercially useful support matrices for immobilized enzymes [26]. One advantage is the irreversible reaction by which enzymes may be attached to a support through epoxides; as the epoxide opens, in a displacement reaction involving a nucleophilic group on the enzyme, a nonhydrolyzable linkage is formed. Another advantage is the ability to activate a wide range of different surfaces with epoxides. Yet an additional advantage is that binding through epoxides does not appreciably change the charge state of the enzyme once it is bound. For example, the pK_a of the secondary amine formed after binding of an enzyme through a lysine side chain is not too different from that of the lysine side chain prior to coupling.

An epoxide-activated support has been developed by Rohm Pharma (Darmstadt, FRG), and is sold commercially under the trade name of *Eupergit*. The support is based on a methacrylic polymer bead bearing epoxide functionality. Eupergit has been successfully applied to the immobilization of many enzymes including penicillin acylase for use in the production of 6-aminopenicillanic acid [27].

13.2.3 Cross-Linking

Cross-linking of enzyme molecules with other proteins affords an insoluble preparation that can be readily handled or manipulated in a continuous reactor. These immobilized preparations are often produced in a particulate form, for use in packed bed reactors.

As enzymes are often stabilized by a concentrated protein environment, copolymerization of enzymes with other inexpensive proteins such as gelatin (using bifunctional reagents like glutaraldehyde) has been used effectively by companies such as Novo and Gist-Brocades to produce long-lived pellets of immobilized enzymes [28]. As an example of a commercially useful method, Novo takes cells which have been recovered from the fermentation broth in the form of a paste and extrudes them in the form of spaghetti. The strands are cut into uniform pellets and hardened in glutaraldehyde. This method is particularly useful because it gives particles of a controlled size for use in packed-bed reactor systems.

Although glutaraldehyde is the most popular cross-linking reagent, dimethyl adipimidate and diisocyanates have also been used. Diamines may also be used in conjunction with carbodiimides, which activate carboxyl groups on the proteins to react with amines.

The enzyme activity of immobilized preparations prepared by cross-linking with multifunctional reagents will depend upon the degree of cross-linking. If high concentrations of cross-linking agents are employed, more bonds will be formed, which will generally result in lower activities. The additional bonds formed at high concentrations may block the entrance of the substrate to the active sites and also restrict the conformational mobility of the enzyme severely. The effect of the quantity of the cross-linking agent, diazobenzidine, on the catalytic activity of papain was demonstrated by Silman and co-workers [29]. A similar effect of glutaraldehyde cross-linking on the activity of carboxypeptidase A was reported by Quiocho and Richards [30, 31]. Generally speaking, an increased level of cross-linking will lead to a more stable enzyme preparation, but often at the cost of part of its catalytic activity.

13.2.4 Entrapment in a Polymeric Matrix

Entrapment is perhaps best viewed as the physical confinement of an enzyme in a polymeric network or matrix. In this technique, an enzyme is typically added to a solution of monomer, and the resulting polymerization entraps the enzyme molecules in the matrix.

This immobilization technique has been extensively explored since the early work of Bernfeld and Wan using polyacrylamide [32]. The most frequently employed matrix for entrapment has in fact been polyacrylamide, which has had significant early success in Japan. However, various other materials have also been used. Bauman and co-workers reported the immobilization of cholinesterase in polymerized starch [33]. Vieth and Venkatasubramanian investigated the entrapment of enzymes in collagen matrices [34]. Whitesides has published on the use of PAN (polyacrylonitrile) gels [35, 36].

One of the major concerns with respect to entrapped enzymes is that of leaching. The enzyme may migrate out of the pore if the pore is too large. In many cases, this leaching may be overcome by simply cross-linking the enzyme after entrapment with a bifunctional reagent such as glutaraldehyde.

The opposite effect is also a concern with entrapment: that of pore diffusion limitations. If the substrate is a rather large molecule, such as a protein, it may be restricted from entry into the pore and thus be inaccessible to the enzyme.

Dead cells containing enzymes have been very successfully immobilized by entrapment. The size of the cells prevents loss of catalyst due to diffusion out of the pores of the matrix. Fusee described the production of L-aspartic acid using polyurethane-immobilized cells containing aspartase [37]. Calton *et al.* reported the production of L-aspartic acid, L-phenylalanine, and L-alanine using cells entrapped in a polyazetidine matrix [38]. Chibata and co-workers have commercialized processes for L-aspartic acid and L-alanine using cells entrapped in carrageenan gels and hardened with hexamethylenediamine and glutaraldehyde [39]. Calcium alginate gel has similarly been used.

A novel method to overcome the problem of leaching out of activity was developed as a part of the polyazetidine method by Calton and co-workers. Cells containing enzyme are mixed with a polyazetidine polymer and cured by drying or mild heating. The polymer chains contain a reactive N-containing ring which opens under nucleophilic attack by cellular material or proteins. The result is believed to be a combination of entrapment and covalent attachment, rendering a more stable immobilized enzyme preparation in a hydrophilic environment, which is less prone to leaching [40]. Further enhancements to this technique have been developed by Novo [41].

13.2.5 Encapsulation or Confinement in a Membrane

Encapsulation is distinguished from entrapment methods by the fact that a solution of the enzyme is separated from the bulk solution by a membrane. In this approach, pioneered by Chang [42, 43], enzymes are encapsulated within membranes that are impermeable to the enzymes but permeable to the enzyme substrate. Such semi-permeable microcapsules have been used as artificial cells in which the enzymes, cofactors, organelles, and other bioactive materials are retained [30]. Encapsulation offers the opportunity to immobilize larger quantities of enzyme per unit volume of immobilized preparation than any other procedure [3]. Perhaps the biggest disadvantage of this immobilization technique is that only relatively small substrate molecules can be utilized with the intact membranes. Depending on the type of membrane used, encapsulation can also be a relatively expensive way to immobilize enzymes due to the high cost of membranes.

Chang has described the encapsulation of enzymes in a variety of membranes [44]. Two of the membranes that he used most successfully were cellulose nitrate and nylon. In another example, the Snamprogetti Company in Italy has entrapped aminoacylase and hydantoinase, for the production of amino acids in hollow fibers of cellulose acetate [45]. The half-lives were greater than one month under operating conditions. Other enzymes have been similarly immobilized for producing a range of different products, showing wide applicability of this general method.

Another membrane system which has shown promising results in biocatalytic applications has been developed by Sepracor [46, 47]. This system has been particularly effective in facilitating contact between an aqueous phase and a non-water-miscible phase. The enzyme is entrapped against the membranes and held in place by a slight positive pressure. As long as the stability of the enzyme is high enough, this system can operate continuously with little down time. In addition, old, deactivated enzyme can be flushed out and new enzyme loaded while the membrane cartridge is in place.

13.3 CONCLUSIONS

Perhaps the most important conclusion one can draw from the past thirty years of work on enzyme immobilization is that, due to the range of methods that has been developed, it can be asserted with a high degree of confidence that a successful immobilization method can be developed for virtually any enzyme. That does not mean, however, that one method will be successful for almost all enzymes. In fact, the method of choice will likely vary from one case to another. All five methods surveyed and described here have been found useful in the past in certain instances. Even more importantly, a number of these methods have been proven successful at the commercial scale, thus giving the researcher several options to choose from which have withstood the "test of practicality" at the manufacturing scale.

Products manufactured today using immobilized enzymes include high-fructose corn syrup, 6-aminopenicillanic acid, certain fatty acids and derivatives, L-malic acid, and isomaltulose. Specifically, a number of amino acids are also made using immobilized enzyme catalysts, including L-aspartic acid from fumaric acid and ammonia, L-alanine from L-aspartic acid, D-4-hydroxyphenylglycine from the corresponding hydantoin, L-ornithine and L-citrulline from L-arginine, L-tryptophan from indole and L-serine. Depending on the economics, technology has been developed for other amino acids as well, including L-lysine, L-phenylalanine, L-tyrosine, L-serine, L-4-phenyl-2-aminobutanoic acid, L-norvaline, and L-DOPA. Genetic engineering has played a key role in a number of cases in making enzymes available in quantity at costs that previously could not be achieved.

Future applications will likely have an increasing dependence on genetic engineering as a tool to enhance enzyme-catalyzed processes by allowing improvements in the properties of the enzymes: modified substrate specificity, improved stability, and more suitable temperature or pH optima. Figure 13.1 illustrates in flow chart form how this may be done.

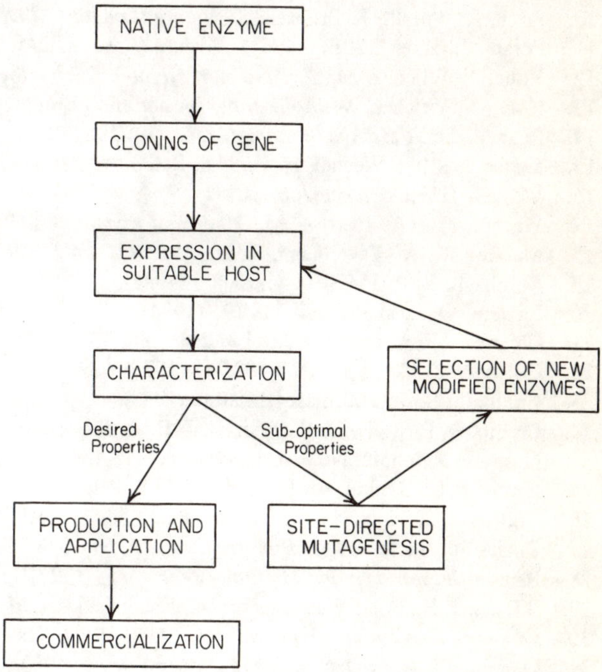

Figure 13.1. Using Genetic Engineering to Improve the Catalytic Properties of Enzymes.

The field of immobilized biocatalysts is rich with technology awaiting exploitation. Furthermore, biocatalysis is not standing still. The use of enzymes in non-aqueous solutions is rapidly expanding the scope of applications for biological catalysts. Continuing advances in genetic engineering, as mentioned above, and also in reactor design, should help to accelerate the realization of the potential benefits of this technology.

13.4 REFERENCES

1. Chibata, I. *Immobilized Enzymes*; John Wiley & Sons, New York, 1978.
2. Mosbach, K., *Methods Enzymol. 44*, v-vi (1974).
3. Messing, R. A., *Methods Enzymol. 44*, 148 (1974).
4. Chibata, T. *et al.*, *Methods Enzymol. 44*, 746 (1974).
5. Tosa, T., Soto, T., Mori, T., Matuo, Y., Chibata, I., *Biotechol. Bioeng. 15*, 69 (1973).
6. Schwabe, C., *Biochemistry 8*, 795 (1969).
7. Rozzell, J. D., unpublished results.
8. Fukumura, T., Jpn. Pat. 74-15795 (1974).
9. Goldstein, L., *Biochemistry 11*, 4072-4084 (1972).
10. Axen, R., Porath, J., Ernback, S., *Nature (London) 214*, 1302 (1967).
11. Porath, J., Axen, R., Ernback, S., *Nature (London) 215*, 1491 (1967).
12. Kohn, J., Wilchek, M., *Enzyme and Microbial Technology 4*, 161-163 (1982).
13. Kohn, J., Wilchek, M., *Appl. Biochem and Biotechnology 9*, 285-305 (1984).
14. Rozzell, J. D., *Methods Enzymol. 136*, 479 (1987).
15. Messing, R. A., Weetall, H. H., U.S. Pat. 3,519,538 (1970).
16. Weetall, H. H., *Science 166*, 615 (1969).
17. Weetall, H. H., Filbert, A. M., *Methods Enzymol. 34*, 59 (1974).
18. Messing, R. A., Yaverbaum, S., U.S. Pat. 4,071,409 (1978).
19. Rohrbach, R. P., U.S. Pat. 4,525,456 (1985).
20. Lantero, O. J., U.S. Pat. 4,438,196 (1984).
21. Chiang, J. P., Lantero, O. J., U.S. Pat. 4,713,333 (1987).
22. Goldberg, B. S., U.S. Pat. 4,102,746 (1978).
23. Goldberg, B. S., U.S. Pat. 4,169,014 (1978).
24. Crump, S. P., Meier, J. S., Rozzell, J. D. In *Biocatalysis*; Abramowicz, D. A., Ed.; Van Nostrand Reinhold, New York, 1990; pp 115-133.
25. Leuba, J.-L., Renker, A., Flaschel, E., U.S. Pat. 4,918,016 (1990).
26. Bigwood, M. P., Naples, J. O., U.S. Pat. 4,582,860 (1986).
27. Bihari, V., Buchholz, K., *Biotech. Letters 6*, 571-576 (1984).
28. Paulsen, P., *Enzyme and Microbial Technology 3*, 271 (1981).
29. Silman, I. H., Albu-Weissenberg, M., Katchalski, E., *Biopolymers 4*, 441 (1966).
30. Quiocho, F. A., Richards, F. M., *Proc. Natl. Acad. Sci. USA 52*, 833 (1964).
31. Quiocho, F. A., Richards, F. M., *Biochemistry 5*, 4062 (1986).
32. Bernfield, P., Wan, J., *Science 142*, 678 (1963).
33. Bauman, E. U., Goodson, L. U., Guilbault, G. G., Kramer, D. N., *Anal. Chem. 37*, 1378 (1965).
34. Vieth, W. R., Venkatsubramanian,K., *Methods Enzymol. 44*, 243 (1976).
35. Pollack, A., Blumerfeld, H., Baughlin, R. L., Whitesides, G. M., *J. Am. Chem. Soc. 102*, 6324-6336 (1980).
36. Crans, D. C., Kazlavskas, R. J., Hirschbein, B. L., Wong, C.-H., Abril, O., Whitsides, G. M., *Methods Enzymol. 136*, 263 (1987).
37. Fusee, M. C., *Methods Enzymol. 136*, 463 (1987).
38. Calton, G. J., Wood, L. L., Campbell, M. L., *Methods Enzymol. 136*, 497 (1987).
39. Chibata, I., Tosa, T., Takamatsu, S., *Methods Enzymol. 136*, 472 (1987).

40. Wood, L. L., Calton, G. J., *Biotechnology 1*, 1091 (1985).
41. Wumpelmann, M., Mollgaard, M., U.S. Pat. 4,892,825 (1990).
42. Chang, T. M. S. *Artificial Cells*; Thomas, Springfield, Illinois, 1972.
43. Chang, T. M. S., *Methods Enzymol. 136*, 67 (1987).
44. Chang, T. M. S. *Biomedical Applications of Immobilized Enzymes and Proteins*; Plenum, New York, 1977.
45. Marconi, W., Morisi, F., *Applied Biochemistry and Bioengineering 2*, Wingard, L. B., Jr. *et al.*, Eds.; Academic Press, New York, 1979.
46. McConville, F. X., Lopez, J. L., Wald, S. A. In *Biocatalysis*; Abramowicz, D. A., Ed.; Van Nostrand Reinhold, New York, 1990; 167-177.
47. Matson, S. L., Lopez, J. L., Wald, S. A., Bratzler, R. L. In *Industrial Use of Enzymes: Technical and Economic Barriers*; Wolnak, B., Scher, M., Eds.; B. Wolnak and Associates, Chicago, IL, 1990; pp 70-81.

ENZYME KINETICS AND REACTOR DESIGN

Erwin Flaschel

CONTENTS

The primary intention of this chapter is to bring together the basic principles of enzyme reaction engineering. If this chapter reduced by one-half the number of books and specialized articles usually needed in order to get an overview on the subject, its aim would have been attained. Enzymes applied in amino acid production are of very diverse natures, so a general view may be more helpful than discussions of specific examples. Since a finite-length chapter cannot go into all of the fine details, the reader is provided with some entry points into the literature for more specific topics.

14.1 MODELING ENZYME REACTION KINETICS

Modeling enzyme kinetics means analyzing experimental data to formulate a unique mathematical expression which relates the changes in reactant concentrations with time to the concentrations of the respective reactants. The underlying mechanistic view of enzyme catalysis may be given in an abstract way schematically as a three-step process involving a substrate, S, a product, P, and the active site of an enzyme, E:

$$\mathbf{S + E \overset{1}{\rightleftarrows} ES \overset{2}{\rightleftarrows} EP \overset{3}{\rightleftarrows} E + P} \tag{S1}$$

The first step involves the association of the substrate with the enzyme. The complex formed, ES, may dissociate, or the substrate may be transformed in the second step into the product whilst being associated with the enzyme. The third step involves the dissociation of the product from the complex, EP, by liberation of the catalyst. The reaction scheme given in S1 may be applicable to enzymic racemization, for which there is only one substrate and one product, and for which all three steps represent reversible processes.

The entire fundamental modeling according to S1 is already rather complex. Therefore, let us consider the simplest scheme for an enzymic reaction, consisting of an association/dissociation process of substrate and enzyme, and a combined step of irreversible reaction and dissociation of the product:

$$\mathbf{S + E \underset{k_1^-}{\overset{k_1^+}{\rightleftarrows}} ES \overset{k_2}{\longrightarrow} E + P} \tag{S2}$$

If we assumed first-order reactions in terms of the species involved, our knowledge of the enzyme-substrate-complex concentration would lead to a model relating the rate of change of the product concentration to the time:

$$\frac{dP}{dt} = k_2\, ES \tag{1.1}$$

However, for the concentration of the complex, ES, only a rate equation may be derived:

$$\frac{dES}{dt} = k_1^+ S \cdot E - (k_1^- + k_2)\, ES \tag{1.2}$$

Both equations have to be combined with the mass balances for the enzyme and the reactants:

$$E_0 = E + ES \tag{1.3}$$

$$S_0 = S + ES + P \tag{1.4}$$

with the assumption that there is no product before the reaction starts ($P_0 = 0$). These expressions may be gathered into a system of two differential equations:

$$\frac{dES}{dt} = k_1^+ (S_0 - ES - P)(E_0 - ES) - (k_1^- + k_2)\, ES \tag{1.5}$$

$$\frac{dP}{dt} = k_2\, ES \tag{1.1}$$

This system of differential equations must be integrated numerically, with appropriate initial conditions (for example: $S = S_0$, $E = E_0$ at $t = 0$), since an analytical solution does not exist. Thus, the kinetic model cannot be expressed by a classical rate law. Although the dynamics of enzymic reactions presents a very interesting subject [K7], fortunately, it need be applied only in very rare or complex cases. However, it should always be kept in mind that our thinking in terms of rate laws of enzymic reactions does not reflect the entire scale of phenomena involved. We should even be astonished at our success in applying simple rate laws.

14.1.1 Derivation of the Basic Rate Laws

Deriving rate laws for enzymic catalysis implies the identification of rate-limiting steps, lumping together of model parameters, and neglecting terms in the mass balance owing to differences in the concentration levels of the species involved.

For practical purposes, the concentration of the enzyme complexes can be neglected in the mass balance of the reactants (eq 1.4), because normally the total enzyme concentration amounts to some mmol m^{-3}, whereas the reactant concentration may attain some kmol m^{-3}.

A.-J. Brown (1902) and V. Henri (1901-1904), as well as L. Michaelis and Miss Maud L. Menten (1913), were the first to propose the reaction scheme given as S2 for the catalytic action of invertase, recognizing that the formation of an intermediate complex of enzyme and substrate was essential. They further recognized that the decomposition of the intermediate complex was rate-limiting. In consequence, they assumed that the adsorption/desorption interaction of substrate and enzyme was so rapid that this step could be described by a simple spontaneous equilibrium.

Thus, eq 1.2 may be rewritten as:

$$\frac{dES}{dt} = k_1^+ S \cdot E - k_1^- ES \doteq 0 \tag{1.6}$$

while still keeping:

$$E_0 = E + ES \tag{1.3}$$

$$\frac{dP}{dt} = k_2 ES \tag{1.1}$$

It is convenient in enzyme kinetics to use dissociation equilibrium constants. Therefore, eq 1.6 may be divided by the rate constant k_1^+:

$$S \cdot E - K_s ES = 0 \tag{1.7}$$

with:

$$K_s = \frac{k_1^-}{k_1^+} = \frac{S \cdot E}{ES} \tag{1.8}$$

The eqs 1.7 and 1.3 may form a system of linear equations, which can be solved for the intermediate complex concentration:

$$ES = \frac{S}{K_s + S} E_0 \tag{1.9}$$

In combination with eq 1.1, the rate law is obtained as:

$$\frac{dP}{dt} = -\frac{dS}{dt} = k_2 E_0 \frac{S}{K_s + S} \tag{1.10}$$

This equation is called the *Michaelis-Menten equation* or the *quasi-equilibrium rate law* (the latter because it is based on the spontaneous equilibrium assumption).

G. E. Briggs and J. B. S. Haldane (1925) refined the model by including the rate of disappearance of the intermediate complex due to the reaction as given in eq 1.2. If the rate of change of the intermediate complex concentration can be neglected ($dES/dt \cong 0$), then an apparent dissociation constant can be defined as:

$$K_m = \frac{k_1^- + k_2}{k_1^+} = K_s + \frac{k_2}{k_1^+} = \frac{S \cdot E}{ES} \tag{1.11}$$

and the same structure of rate law is obtained:

$$\frac{dP}{dt} = -\frac{dS}{dt} = k_2 E_0 \frac{S}{K_m + S} \tag{1.12}$$

This model is also frequently called the *Michaelis-Menten equation* or the *quasi-steady-state approximation*. The parameter K_m is known as the *Michaelis constant*. The term $k_2 E_0$ is often lumped together and called the *maximum reaction velocity*, V_{max}. Both approaches, by Michaelis and Menten, and by Briggs and Haldane, yield the same mathematical model, but that of Briggs and Haldane is based on safer ground, and the definition of the value of K_m reminds us always that we should care about the validity of the underlying assumption. It demands that the global rate of change of the intermediate complex concentration must be negligible compared to the rate of complex formation. This may be true in most cases when the substrate concentration is much higher than the enzyme concentration.

14.1.2 The Michaelis-Menten Equation

Because of the interest in transforming a given substrate into a product, let us reconsider the Michaelis-Menten equation and introduce the substrate conversion:

$$X = \frac{S_0 - S}{S_0} \quad \text{or} \quad S = S_0(1 - X) \tag{1.13}$$

Then eq 1.12 may be written as:

$$r_m = -\frac{1}{E_0}\frac{dS}{dt} = k_2 \frac{\frac{S}{K_m}}{1 + \frac{S}{K_m}} = k_2 \frac{\frac{S_0}{K_m}(1 - X)}{1 + \frac{S_0}{K_m}(1 - X)} \tag{1.14}$$

Some special situations may be of general interest. Half of the maximum reaction velocity is obtained when the substrate concentration is equal to the value of K_m. When the ratio S/K_m is large compared to unity, then the reaction is of zero order and the rate is maximal:

$$r_m = k_2 \quad \text{for} \quad \frac{S}{K_m} \gg 1 \tag{1.15}$$

When the ratio S/K_m is small compared to unity, the reaction is of first order with low overall rate:

$$r_m = \frac{k_2}{K_m} S_0 (1 - X) \quad \text{for} \quad \frac{S}{K_m} \ll 1 \tag{1.16}$$

Since kinetic measurements are commonly performed in batch mode, the original data are received as degree of substrate conversion as a function of enzyme concentration and time. In consequence, the integral form of the rate law is of equal importance. We derive the integral form of the Michaelis-Menten equation by writing eq 1.14 in the form:

$$-\frac{dS}{dt} = -\frac{d}{dt} S_0 (1 - X) = S_0 \frac{dX}{dt} = k_2 E_0 \frac{\frac{S_0}{K_m}(1 - X)}{1 + \frac{S_0}{K_m}(1 - X)} \tag{1.17}$$

and rearranging:

$$S_0 \int_0^X \frac{1 + \frac{S_0}{K_m}(1 - X)}{\frac{S_0}{K_m}(1 - X)} dX = k_2 E_0 \int_0^t dt \tag{1.18}$$

This leads to:

$$K_m \int_0^X \frac{dX}{1 - X} + S_0 \int_0^X dX = k_2 E_0 \int_0^t dt \tag{1.19}$$

and finally:

$$E_0 t = \frac{1}{k_2} S_0 X - \frac{K_m}{k_2} \ln (1 - X) \tag{1.20}$$

or:

$$E_0 t = Et = \frac{1}{k_2} S_0 X + \frac{K_m}{k_2} \ln \left(\frac{1}{1 - X} \right) \tag{1.21}$$

Obviously, the enzyme concentration and the time of reaction may be interchanged; they are of equal influence. This is an outcome of the quasi-steady-state approximation leading to proportionality of the reaction rate with the enzyme concentration.

The first term in eq 1.21 gives the normalized time of reaction, Et, for attaining a given conversion, X, if the reaction were of zero order. The second term is due to first-order reaction. Depending on the ratio S_0/K_m and the final conversion, the order of a reaction may change during a batch reaction from zero to one.

14.1.3 Common Kinetic Models

Classical kinetic modeling for enzymic reactions is based on the assumptions put forward by Michaelis and Menten, and Briggs and Haldane. The balance around the intermediate complex leading to product formation is normally modeled according to Briggs and Haldane by lumping the rate-determining rate constant into the Michaelis constant, K_m. Other intermediate complexes are modeled as if they were in spontaneous equilibrium. The so-called Michaelis-Menten equation may hold for a whole variety of enzymic transformations, but normally only for initial conditions and at low substrate levels. The most common features encountered are product and substrate inhibition, as well as inhibition by substrate analogs, particularly in the case of racemic substrates. These influences are commonly modeled by assuming formation of dead-end complexes, characterized by their entering into the mass balance of enzyme species but not leading to product formation. Let us consider the following general scheme, separated into two parts for simplicity:

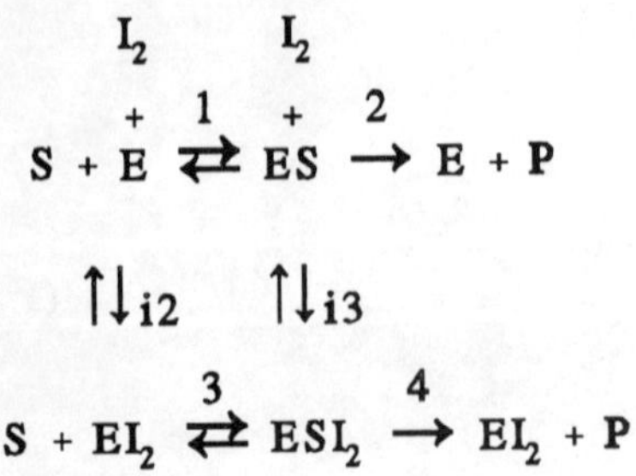

(S3a)

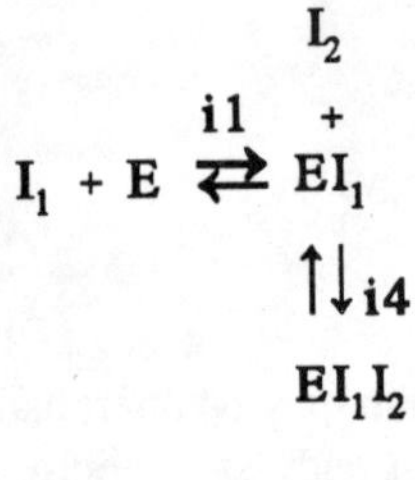

(S3b)

The inhibitor I_1 is supposed to occupy the active site of the enzyme: it is a *competitive inhibitor* because it is in competition with the substrate for the active site. The inhibitor I_2 binds with the enzyme, but it does not block the active site. If it associated only with the free enzyme, E, causing a type of allosteric dead-end inhibition which cannot be distinguished from a competitive one, I_2 would be called an *anticompetitive inhibitor*. If it bound only to the complex, ES, it would be called an *uncompetitive inhibitor*. However, if it bound with both enzyme species with equal probability it would be called a *noncompetitive inhibitor*. In the case of unequal affinity, a *mixed-type inhibition* would occur. If the species ESI_2 were not a dead-end complex, the inhibition by I_2 would, in addition, be called *partial*.

For the derivation of a rate law for scheme S3, the rates of transformation of all five reactant enzyme complexes must be taken into account, along with the mass balances for these enzyme species. According to the nomenclature in S3, the following expressions are obtained:

$$\begin{array}{llllll} SE - K_m ES & & & & & = 0 \\ I_1 E & - K_{i1} EI_1 & & & & = 0 \\ I_2 E & & - K_{i2} EI_2 & & & = 0 \\ & I_2 ES & & - K_{i3} ESI_2 & & = 0 \\ & & I_2 EI_1 & & - K_{i4} EI_1 I_2 & = 0 \\ E + ES & + EI_1 & + EI_2 & + ESI_2 & + EI_1 I_2 & = E_0 \end{array} \tag{1.22}$$

Since we assume spontaneous equilibria for each inhibition step, process 3 in scheme S3 is obviously redundant. The rate of reaction would be obtained as:

$$\frac{dP}{dt} = k_2 ES + k_4 ESI_2 \tag{1.23}$$

Therefore, an equation for ES and ESI_2 must be derived. Since eq 1.22 represents a system of linear equations, it can be represented as:

$$\mathbf{M}\,\mathbf{e} = \mathbf{y} \tag{1.24}$$

with the vector of enzyme species:

$$\mathbf{e}^T = (E, ES, EI_1, EI_2, ESI_2, EI_1 I_2) \tag{1.25}$$

and the vector of results:

$$\mathbf{y}^T = (0, 0, ..., E_0) \tag{1.26}$$

For convenience, the first five equations in 1.22 may each be divided by their respective (quasi-)equilibrium constant to compose the coefficient matrix:

$$\mathbf{M} = \begin{bmatrix} \frac{S}{K_m} & -1 & 0 & 0 & 0 & 0 \\ \frac{I_1}{K_{i1}} & 0 & -1 & 0 & 0 & 0 \\ \frac{I_2}{K_{i2}} & 0 & 0 & -1 & 0 & 0 \\ 0 & \frac{I_2}{K_{i3}} & 0 & 0 & -1 & 0 \\ 0 & 0 & \frac{I_2}{K_{i4}} & 0 & 0 & -1 \\ 1 & 1 & 1 & 1 & 1 & 1 \end{bmatrix} \tag{1.27}$$

Equation 1.24 can be solved by any method applicable for linear equation systems. Since the vector **y** has only one element unequal to zero, Cramer's rule is appropriate for solving the system for the different enzyme species e_i by replacing the ith row of **M** by the vector **y** and solving the determinant of matrix $\mathbf{M}_i$. The determinant of **M** need not be known because new enzyme concentrations may be defined:

$$\mathbf{e}' = \det(\mathbf{M}_i) = \det(\mathbf{M}) \cdot \mathbf{e} \tag{1.28}$$

Summing up the new terms gives:

$$\sum e_i' = \det(\mathbf{M}) \cdot \sum e_i \tag{1.29}$$

Dividing one equation by the other shows the equivalence:

$$\frac{e_i'}{\sum e_i'} = \frac{e_i}{\sum e_i} = \frac{e_i}{E_0} \tag{1.30}$$

The real concentrations may then be obtained by the following transformation:

$$e_i = \frac{e_i'}{\sum e_i'} E_0 \tag{1.31}$$

Applying Cramer's rule to our actual system defined by eqs 1.24–1.27 gives the vector of enzyme concentrations:

$$\mathbf{e}'^{\mathrm{T}} = \left(1\,,\ \frac{S}{K_m}\,,\ \frac{I_1}{K_{i1}}\,,\ \frac{I_2}{K_{i2}}\,,\ \frac{I_2}{K_{i3}}\frac{S}{K_m}\,,\ \frac{I_2}{K_{i4}}\frac{I_1}{K_{i1}}\right) \tag{1.32}$$

corresponding to the entities defined by eq 1.25 normalized according to eq 1.28. The rate of reaction according to eq 1.23 then becomes:

$$\frac{dP}{dt} = \frac{\left(k_2 + k_4\frac{I_2}{K_{i3}}\right)\frac{S}{K_m}}{1 + \frac{S}{K_m} + \frac{I_1}{K_{i1}} + \frac{I_2}{K_{i2}} + \frac{I_2}{K_{i3}}\frac{S}{K_m} + \frac{I_2}{K_{i4}}\frac{I_1}{K_{i1}}}\,E_0 \tag{1.33}$$

or in the more familiar form:

$$\frac{dP}{dt} = \frac{\left(k_2 + k_4\,\frac{I_2}{K_{i3}}\right)S_0(1 - X)}{K_m\left[1 + \frac{I_1}{K_{i1}}\left(1 + \frac{I_2}{K_{i4}}\right) + \frac{I_2}{K_{i2}}\right] + \left(1 + \frac{I_2}{K_{i3}}\right)S_0(1 - X)}\,E_0 \tag{1.34}$$

The different inhibiting influences are now visible. All influences on the free enzyme species, E, are summed together to increase the value of K_m. The additional effect of the inhibitor I_2 on the competitive inhibition of inhibitor I_1 appears as an enhancement factor for the latter effect. All inhibitory influences on the intermediate complex, ES, tend to increase the value of the substrate concentration in the denominator.

Simple models like that given by eq 1.34 may be transformed into a Michaelis-Menten model structure:

$$\frac{dP}{dt} = k_2' E_0 \frac{S_0(1 - X)}{K_m' + S_0(1 - X)} \tag{1.35}$$

with, in this case:

$$k_2' = \frac{K_{i3}}{K_{i3} + I_2}\left(k_2 + k_4\,\frac{I_2}{K_{i3}}\right) \tag{1.36}$$

$$K_m' = K_m\,\frac{K_{i3}}{K_{i3} + I_2}\left[1 + \frac{I_1}{K_{i1}}\left(1 + \frac{I_2}{K_{i4}}\right) + \frac{I_2}{K_{i2}}\right] \tag{1.37}$$

The advantage of this rearrangement is the ability to use the same formalism to obtain the integral model according to eq 1.20:

$$Et = \frac{S_0}{k_2'} X + \frac{K_m'}{k_2'} \ln\left(\frac{1}{1 - X}\right) \tag{1.38}$$

The model equations 1.33 or 1.34 contain a variety of possible influences of inhibitors and may therefore be tailored according to a particular need. The model may be changed, so that the intermediate complex, EI_1I_2, represents a dead-end complex ($k_4 = 0$); or the inhibitors can be redefined as reactants; or inhibitory influences can be canceled ($K_{ij} \rightarrow \infty$).

Two cases of general interest will be discussed, both characterized by the fact that the enzyme species EI_1I_2 represents a dead-end complex. The first case takes the reaction product as the general inhibitor ($I_1 = I_2 = P = S_0X$). The model for this case may be written according to eq 1.34:

$$S_0 \frac{dX}{dt} = k_2 E_0 \frac{S_0(1 - X)}{K_m\left[1 + \frac{S_0 X}{K_{i1}}(1 + \frac{S_0 X}{K_{i4}}) + \frac{S_0 X}{K_{i2}}\right] + (1 + \frac{S_0 X}{K_{i3}}) S_0 (1 - X)} \tag{1.39}$$

The integral model of this case is obtained by stepwise integration:

$$\begin{aligned} k_2 E_0 \int_0^t dt = {} & K_m \int_0^X \frac{dX}{1 - X} + K_m (\frac{1}{K_{i1}} + \frac{1}{K_{12}}) S_0 \int_0^X \frac{X\,dX}{1 - X} \\ & + \frac{K_m S_0^2}{K_{i1} K_{i4}} \int_0^X \frac{X^2 dX}{1 - X} + S_0 \int_0^X dX + \frac{S_0^2}{K_{i3}} \int_0^X X\,dX \end{aligned} \tag{1.40}$$

This leads to:

$$\begin{aligned} k_2 E_0 t = {} & \left\{1 - K_m\left[\frac{1}{K_{i1}}(1 + \frac{S_0}{K_{i4}}) + \frac{1}{K_{i2}}\right]\right\} S_0 X \\ & + K_m \left\{1 + S_0\left[\frac{1}{K_{i1}}(1 + \frac{S_0}{K_{i4}} + \frac{1}{K_{i2}})\right]\right\} \ln\left(\frac{1}{1 - X}\right) \\ & + \left(\frac{1}{K_{i3}} - \frac{K_m}{K_{i1} K_{i4}}\right) \frac{1}{2} S_0^2 X^2 \end{aligned} \tag{1.41}$$

If we take the factor associated with the product concentration, S_0X:

$$q = 1 - K_m\left[\frac{1}{K_{i1}}(1 + \frac{S_0}{K_{i4}}) + \frac{1}{K_{i2}}\right] \tag{1.42}$$

we may define new lumped parameters:

$$k_2' = \frac{k_2}{q} \tag{1.43}$$

$$K_m' = \frac{K_m}{q}\left(1 + S_0\left[\frac{1}{K_{i1}}(1 + \frac{S_0}{K_{i4}}) + \frac{1}{K_{i2}}\right]\right) \tag{1.44}$$

$$K' = \frac{1}{2q}\left(\frac{1}{K_{i3}} - \frac{K_m}{K_{i1}K_{i4}}\right) \tag{1.45}$$

which allow us to define a general integral model according to the Michaelis-Menten equation (eq 1.21):

$$Et = \frac{1}{k_2'}S_0 X + \frac{K_m'}{k_2'}\ln\left(\frac{1}{1 - X}\right) + \frac{K'}{k_2'}S_0^2 X^2 \tag{1.46}$$

The benefit of this rearrangement is evident when the rate law, eq 1.39, is written in terms of these parameters:

$$S_0\frac{dX}{dt} = k_2' E_0 \frac{S_0(1 - X)}{K_m' + S_0(1 - X) + 2K'S_0^2 X(1 - X)} \tag{1.47}$$

A Michaelis-Menten type of rate law results, which must be extended by only one element. Most of the basic enzyme kinetic models may be transformed to fulfill both eqs 1.46 and 1.47. For the classical types of product inhibition ($K_{i4} = \infty$), the respective lumped parameters are given in Table 14.1.

TABLE 14.1 General Lumped Parameters for Product Inhibition Models

Inhibition	k_2'	K_m'	K'
none			
$K_{i1} = K_{i2} = K_{i3} = \infty$	k_2	K_m	0
competitive			
$K_{is} = K_{i3} = \infty$ $K_{ic} = K_{i1}$	$k_2 \dfrac{K_{in}}{K_{in} - K_m}$	$K_m \dfrac{K_{ic} + S_0}{K_{ic} - K_m}$	0
noncompetitive			
$K_{i1} = \infty$ $K_{in} = K_{i2} = K_{i3}$	$k_2 \dfrac{K_{in}}{K_{in} - K_m}$	$K_m \dfrac{K_{in} + S_0}{K_{in} - K_m}$	$\dfrac{1}{2K_{in}}$
uncompetitive			
$K_{i1} = K_{i2} = \infty$ $K_{iu} = K_{i3}$	k_2	K_m	$\dfrac{1}{2K_{iu}}$
mixed			
$K_{i2} = \infty$ $K_{ic} = K_{i1}$, $K_{iu} = K_{i3}$	$k_2 \dfrac{K_{ic}}{K_{ic} - K_m}$	$K_m \dfrac{K_{ic} + S_0}{K_{ic} - K_m}$	$\dfrac{1}{2K_{iu}}$

The second case to be discussed involves a reaction product as the inhibitor, I_1, and the substrate itself as the second inhibitor, I_2:

$$I_1 = S_0 X \ ; \qquad I_2 = S_0(1 - X) \tag{1.48}$$

Dead-end inhibition is assumed ($k_4 = 0$). Under these conditions, the rate law may be established according to eq 1.34:

$$S_0 \frac{dX}{dt} = k_2 E_0 \frac{S_0(1 - X)}{K_m \left[1 + \dfrac{S_0 X}{K_{i1}}\left(1 + \dfrac{S_0}{K_{i1}}(1 - X)\right) + \dfrac{S_0}{K_{i2}}(1 - X)\right] + \left[1 + \dfrac{S_0}{K_{i3}}(1 - X)\right] S_0(1 - X)} \tag{1.49}$$

We define a general factor as before:

$$q = 1 + K_m\left(\frac{1}{K_{i2}} - \frac{1}{K_{i1}}\right) + \frac{S_0}{K_{i3}} \tag{1.50}$$

The lumped parameters according to eq 1.47 are:

$$k_2' = \frac{k_2}{q} \tag{1.43}$$

$$K_m' = \frac{K_m}{q}\left(1 + \frac{S_0}{K_{i1}}\right) \tag{1.51}$$

$$K' = \frac{1}{2q}\left(\frac{K_m}{K_{i1} K_{i4}} - \frac{1}{K_{i3}}\right) \tag{1.52}$$

Two classical inhibition schemes may be derived from these formulae. The first consists of a pure substrate inhibition ($K_{i1} = K_{i2} = K_{i4} = \infty$, and $K_{is} = K_{i3}$):

$$q = 1 + \frac{S_0}{K_{is}} \qquad k_2' = \frac{k_2}{q} \tag{1.53, 1.43}$$

$$K_m' = \frac{K_m}{q} \qquad K' = -\frac{1}{2q\, K_{is}} \tag{1.54, 1.55}$$

The other case consists of a pure competitive product inhibition accompanied by a substrate inhibition ($K_{i2} = K_{i4} = \infty$, and $K_{ic} = K_{i1}$, $K_{is} = K_{i3}$):

$$q = 1 + \frac{S_0}{K_{is}} - \frac{K_m}{K_{ic}} \qquad k_2' = \frac{k_2}{q} \tag{1.56, 1.43}$$

$$K_m' = \frac{K_m}{q}\left(1 + \frac{S_0}{K_{ic}}\right) \qquad K' = -\frac{1}{2q\, K_{is}} \tag{1.57, 1.55}$$

Up to now, only irreversible enzyme reactions have been discussed. However, reversible reactions occur quite often, and not only in the case of isomerases or transferases. We shall discuss a rather simple scheme:

$$\mathrm{S + E} \overset{1}{\rightleftarrows} \mathrm{ES} \overset{2}{\rightleftarrows} \mathrm{EP} \overset{3}{\rightleftarrows} \mathrm{E + P} \tag{S4}$$

A substrate is reversibly transformed into a single product. It is assumed that the reaction step 2 may be rate-limiting. According to the scheme of Briggs-Haldane, the following system of linear equations results:

$$\begin{bmatrix} S & -K_m^S & \dfrac{k_2^-}{k_1^+} \\ P & \dfrac{k_2^+}{k_3^-} & -K_m^P \\ 1 & 1 & 1 \end{bmatrix} \begin{bmatrix} E \\ ES \\ EP \end{bmatrix} = \begin{bmatrix} 0 \\ 0 \\ E_0 \end{bmatrix} \tag{1.58}$$

with:

$$K_m^S = \frac{k_1^- + k_2^+}{k_1^+} \quad \text{and} \quad K_m^P = \frac{k_3^+ + k_2^-}{k_3^-} \tag{1.59, 1.60}$$

The rate of product formation is obtained as:

$$\frac{dP}{dt} = -\frac{dS}{dt} = k_2^+ ES - k_2^- EP \tag{1.61}$$

And with the solution of eq 1.58:

$$r_m = \frac{k_2^+\left(K_m^P - \dfrac{k_2^-}{k_3^-}\right)S - k_2^-\left(K_m^S - \dfrac{k_2^+}{k_1^+}\right)P}{K_m^S K_m^P - \dfrac{k_2^- k_2^+}{k_1^+ k_3^-} + \left(K_m^P + \dfrac{k_2^+}{k_3^-}\right)S + \left(K_m^S + \dfrac{k_2^-}{k_1^+}\right)P} \tag{1.62}$$

With vanishing reaction rate at equilibrium, the equilibrium constant of the reaction is obtained:

$$K_{eq} = \frac{P_{eq}}{S_{eq}} = \frac{k_2^+\left(K_m^P - \dfrac{k_2^-}{k_3^-}\right)}{k_2^-\left(K_m^S - \dfrac{k_2^+}{k_1^+}\right)} \tag{1.63}$$

If we take into account the definitions of the values of K_m (1.59, 1.60), we obtain the following general equation, known as the *Haldane equation*:

$$K_{eq} = \frac{P_{eq}}{S_{eq}} = \frac{k_1^+ k_2^+ k_3^+}{k_1^- k_2^- k_3^-} \qquad (1.64)$$

For convenience, let us assume that the rate constants of step 2 are much less than those of steps 1 and 3. Under these conditions, the Michaelis-Menten formalism may be applied to obtain the classical expression:

$$r_m = \frac{k_2^+ K_s^P S_0(1 - X) - k_2^- K_s^S S_0 X}{K_s^S (K_s^P + S_0 X) + K_s^P S_0 (1 - X)} \qquad (1.65)$$

with:

$$K_s^S = \frac{k_1^-}{k_1^+} \quad \text{and} \quad K_s^P = \frac{k_3^+}{k_3^-} \qquad (1.66, 1.67)$$

If we now take eq 1.63 in its reduced form:

$$K_{eq} = \frac{P_{eq}}{S_{eq}} = \frac{X_{eq}}{1 - X_{eq}} = \frac{k_2^+ K_s^P}{k_2^- K_s^S} \qquad (1.68)$$

and solve for the equilibrium conversion:

$$X_{eq} = \frac{K_{eq}}{1 + K_{eq}} \qquad (1.69)$$

we can write the rate expression (1.65) as:

$$S_0 \frac{dX}{dt} = \frac{k_2^+ E_0}{X_{eq}} \frac{S_0 (X_{eq} - X)}{K_s^S \left(1 + \dfrac{S_0 X}{K_s^P}\right) + S_0 (1 - X_{eq}) + S_0 (X_{eq} - X)} \qquad (1.70)$$

On integration, the following lumped parameters are obtained:

$$k_2' = \frac{k_2^+}{X_{eq}} \frac{K_s^P}{K_s^P - K_s^S} = \frac{k_2^+ K_s^P + k_2^- K_s^S}{K_s^P - K_s^S} \qquad (1.71)$$

$$K_s' = \frac{K_s^P}{K_s^P - K_s^S}\left[\frac{K_s^S}{K_s^P}(K_s^P + S_0 X_{eq}) + S_0(1 - X_{eq})\right] \tag{1.72}$$
$$= K_s^S\frac{K_s^P + S_0}{K_s^P - K_s^S} + S_0(1 - X_{eq})$$

These parameters can serve to represent the integral kinetic equation in the form equivalent to eq 1.21:

$$Et = \frac{S_0}{k_2'}X + \frac{K_s'}{k_2'}\ln\left(\frac{X_{eq}}{X_{eq} - X}\right) \tag{1.73}$$

and the rate law is accordingly:

$$r_m = \frac{S_0}{E_0}\frac{dX}{dt} = k_2'\frac{S_0(X_{eq} - X)}{K_s' + S_0(X_{eq} - X)} \tag{1.74}$$

14.1.4 Multiple-Substrate Reactions

Until now we have considered only simple reactions characterized by the transformation of one substrate into one product or into two products with only one of these being important. Enzyme reaction mechanisms are classified according to the number of substrates and products involved, as well as according to the sequence of interaction. The terms *Uni*, *Bi*, and *Ter* are used to indicate the number of species involved. The terms *Random*, *Ordered*, and *Ping Pong* indicate whether there is no sequential preference for one reactant to bind with the enzyme, there is a strict sequential order, or there is a sequential order with the release of a first product before the next substrate may bind, respectively. The Ping Pong mechanism, in contrast to both Random and Ordered, does not involve the formation of a ternary complex in the case of a two-substrate reaction. To illustrate the nomenclature, let us take a reaction of type Bi Bi. Then the general Random Bi Bi mechanism is the most complicated one, and can be illustrated schematically by:

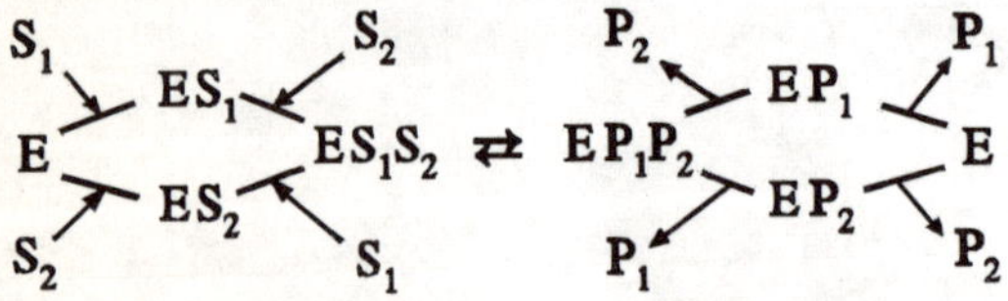

(S5)

The bars indicate potentially reversible reaction steps. Yeast alcohol dehydrogenase, several kinases, and phosphorylases follow this mechanism. Reasonable kinetic expressions may be derived if rapid equilibrium for the enzyme complexes is assumed, with the reaction path between both ternary complexes being rate-limiting.

Rate equations for such complex reaction mechanisms involving multiple substrates and products are preferentially obtained by the schematic representation according to King and Altman [1, 2] or by graphical approaches [3, 4]. The King and Altman method is based mathematically on the determinant method, applying Cramer's rule in order to solve the equation systems. Thus, the approach is similar to that formulated under eqs 1.22–1.34. Computer programs have been written for solving the systems of equations, leading to symbolic solutions [5, 6]. The Ordered Bi Bi mechanism may be presented schematically as:

$$\mathbf{E} \xrightarrow{\;S_1 \downarrow\;} \mathbf{ES_1} \xrightarrow{\;S_2 \downarrow\;} \mathbf{ES_1S_2} \rightleftarrows \mathbf{EP_1P_2} \xrightarrow{\;P_1 \uparrow\;} \mathbf{EP_2} \xrightarrow{\;P_2 \uparrow\;} \mathbf{E} \qquad \text{(S6)}$$

Quite a few dehydrogenases and several other enzymes, in particular those depending on coenzymes, have been shown to follow this mechanism. In the case of dehydrogenases the coenzyme has to bind first.

The Ping Pong Bi Bi mechanism can be briefly presented as:

$$\mathbf{E} \xrightarrow{\;S_1 \downarrow\;} \mathbf{ES_1} \xrightarrow{\;P_1 \uparrow\;} \mathbf{E^*} \xrightarrow{\;S_2 \downarrow\;} \mathbf{E^*S_2} \xrightarrow{\;P_2 \uparrow\;} \mathbf{E} \qquad \text{(S7)}$$

Transaminases, glucose oxidase and several other enzymes are known to follow this mechanism, characterized by the absence of a ternary complex. The first substrate is transformed by covalently modifying the enzyme (for example, by group transfer). This modified enzyme species then transforms a second substrate, in the course of which time the original species is regenerated. Most hydrolases, like serine proteases, can be described by this mechanism because they form a covalent intermediate (for example, acylated enzyme) which then is regenerated by water or other nucleophiles. A noteworthy example of this mechanism is the synthesis of peptides by transacylation with proteinases [7].

The only model discussed in this section will be one for coenzyme-dependent reactions. The simplest scheme is obtained for an Ordered Bi Bi mechanism with a nonpreferential decomposition of the ternary complex:

$$\mathbf{E + C} \underset{}{\overset{K_s^C}{\rightleftarrows}} \mathbf{EC}$$

$$\mathbf{EC + S} \overset{K_s}{\rightleftarrows} \mathbf{ECS} \overset{k_2}{\longrightarrow} \mathbf{E + C^* + P} \qquad \text{(S8)}$$

where C denotes the active coenzyme and C* the transformed coenzyme. If we assume equilibrium for the formation of the intermediate complexes and a rate-limiting step only in the combined reaction and dissociation of the ternary complex, we can solve the two equations according to the Michaelis-Menten approach and obtain the rate expression:

$$S_0 \frac{dX}{dt} = C_0 \frac{dX_C}{dt} = \frac{k_2 E_0 \frac{C_0}{K_s^C} \frac{S_0}{K_s} (1 - X_C)(1 - X)}{1 + \frac{C_0}{K_s^C}(1 - X_C)\left(1 + \frac{S_0}{K_s}(1 - X)\right)} \tag{1.75}$$

with X_C and X denoting the conversion of the coenzyme and the substrate, respectively. The development of these conversions with time may be obtained if we eliminate either X_C or X in eq 1.75. Since the amount of coenzyme transformed must correspond to that of the substrate, the relation sought is:

$$X_C = \frac{S_0}{C_0} X \tag{1.76}$$

Introducing this into eq 1.75 yields:

$$S_0 \frac{dX}{dt} = \frac{k_2 E_0 \frac{C_0}{K_s^C} \frac{S_0}{K_s}\left(1 - \frac{S_0}{C_0} X\right)(1 - X)}{1 + \frac{C_0}{K_s^C}\left(1 - \frac{S_0}{C_0} X\right) + \frac{C_0}{K_s^C} \frac{S_0}{K_s}\left(1 - \frac{S_0}{C_0} X\right)(1 - X)} \tag{1.77}$$

This equation shows clearly that the initial molar ratio of substrate and coenzyme must be at least unity to attain complete substrate conversion. It can be integrated by parts to obtain the time dependence of the substrate conversion:

$$Et = \frac{1}{k_2} S_0 X + \frac{K_s}{k_2} \ln\left(\frac{1}{1 - X}\right) + \frac{K_s}{k_2} \frac{K_s^C}{C_0} b \tag{1.78}$$

where the term b depends on the ratio S_0/C_0:

$$b = \frac{X}{(1 - X)} \qquad \text{for} \qquad S_0 = C_0 \tag{1.79a}$$

$$b = \frac{C_0}{S_0 - C_0} \ln\left(\frac{1 - X}{1 - \frac{S_0}{C_0} X}\right) \qquad \text{for} \qquad S_0 \neq C_0 \tag{1.79b}$$

When there is less coenzyme than substrate ($S_0 > C_0$), the term b tends to infinity at $X = C_0/S_0$, which is the limiting conversion according to eq 1.75.

14.1.5 Limitations on Yield and Optical Purity

As was discussed in the last section, product yield may simply be limited by insufficient levels of cosubstrates or coenzymes. In a more general manner, the maximum product yield may be limited due to secondary reactions taking place in parallel or in series, and the inherent reversibility of reactions. A typical example for which the yield of a primary product is limited by secondary reactions in parallel as well as in series is found in transesterification. The substrate ester is subject to hydrolysis, as is the product ester. Both hydrolytic reactions limit the maximum product ester yield. Uncommon transfer reactions of hydrolases suffer from limitation due to competing hydrolysis, in general.

However, thermodynamics may limit the yield in the case of reversible reactions. Some simple rules for influencing the yield of reversible enzymatic transformations may be deduced from the law of mass action. Let us take a simple isomerization for which a single substrate yields a single product:

$$\mathrm{S} \rightleftarrows \mathrm{P} \tag{S9}$$

The equilibrium constant of this reaction is defined as:

$$K_{eq} = \frac{P_{eq}}{S_{eq}} = \frac{X_{eq}}{1 - X_{eq}} \tag{1.80}$$

The equilibrium conversion, which is equal to the equilibrium product yield, cannot be influenced by the composition of the reaction system. If we consider a decomposition reaction of the type:

$$\mathrm{S} \rightleftarrows \mathrm{P_1} + \mathrm{P_2} \tag{S10}$$

the equilibrium constant may be defined as:

$$K_{eq} = \frac{P_{1eq}\, P_{2eq}}{S_{eq}} = S_0 \frac{X_{eq}^2}{1 - X_{eq}} \tag{1.81}$$

This equation may be rewritten in the form:

$$\frac{K_{eq}}{S_0} = \frac{X_{eq}^2}{1 - X_{eq}} \tag{1.81a}$$

Since the right-hand part of this equation is a steadily increasing function of the substrate conversion, it can serve as a measure of product yield. Its derivative with respect to the concentration level of substrate, S_0, is obviously negative. Therefore, for reversible reactions characterized by an increasing number of species, the maximum product yield decreases with increasing initial substrate concentration.

For reactions with a decreasing number of reactant species the opposite is true, as can be deduced from the following simple example:

$$2\,\mathrm{S} \rightleftarrows \mathrm{P} \tag{S11}$$

$$K_{eq} = \frac{P_{eq}}{S_{eq}^2} = \frac{1}{S_0}\,\frac{X_{eq}}{(1 - X_{eq})^2} \tag{1.82}$$

$$K_{eq}\,S_0 = \frac{X_{eq}}{(1 - X_{eq})^2} \tag{1.82a}$$

The general case for two-substrate reactions would appear more like:

$$\mathrm{S}_1 + \mathrm{S}_2 \rightleftarrows \mathrm{P}_1 + \mathrm{P}_2 \tag{S12}$$

If we assume that P_2 is the desired product and its yield is equivalent to the conversion of substrate S_2, and the yield of P_1 corresponds to the conversion of S_1, then the basic relation would be:

$$K_{eq} = \frac{P_{1eq}\,P_{2eq}}{S_{1eq}\,S_{2eq}} = \frac{X_{1eq}}{1 - X_{1eq}}\,\frac{X_{2eq}}{1 - X_{2eq}} \tag{1.83}$$

From this, the product yield may be obtained as:

$$Y_{2eq} = X_{2eq} = \frac{1}{1 + \frac{1}{K_{eq}}\,\frac{X_{1eq}}{1 - X_{1eq}}} \tag{1.84}$$

We can maximize the yield only by decreasing the equilibrium conversion of the cosubstrate S_1. Since X_{1eq} and X_{2eq} are closely related:

$$S_{1,0}\,X_{1eq} = S_{2,0}\,X_{2eq} \tag{1.85}$$

we can achieve this by increasing the concentration level of substrate S_1. This is what is normally called "pushing" an equilibrium. A reversible reaction can as well be "pulled" if the desired product is eliminated by some means like precipitation, distillation, or extraction, while the reaction takes place.

The influence of temperature on the reaction equilibrium is normally quite small. The equilibrium constant depends on temperature according to van't Hoff's law:

$$\frac{d \ln K_{eq}}{dT} = \frac{\Delta H_R}{RT^2} \tag{1.86}$$

For exothermic reactions ($\Delta H_R < 0$) the equilibrium constant, and with it the maximum yield, decreases with temperature. For endothermic reactions ($\Delta H_R > 0$) we may increase the maximum yield by operating at higher temperature.

The specificity of enzymatic reactions plays a crucial role in the optical resolution of racemic substrates. The theory developed by Sih and co-workers [8] will therefore be briefly discussed. An enzyme may act upon a racemic substrate without exclusive preference for one of the enantiomers. The reaction shall be of a Michaelis-Menten type:

$$\mathrm{E} + \mathrm{S_L} \underset{}{\overset{K_m^{\mathrm{L}}}{\rightleftarrows}} \mathrm{ES_L} \xrightarrow{k_2^{\mathrm{L}}} \mathrm{E} + \mathrm{P_L}$$

$$\mathrm{E} + \mathrm{S_D} \underset{}{\overset{K_m^{\mathrm{D}}}{\rightleftarrows}} \mathrm{ES_D} \xrightarrow{k_2^{\mathrm{D}}} \mathrm{E} + \mathrm{P_D} \tag{S13}$$

This reaction scheme may then be modeled as:

$$r_v^{\mathrm{L}} = S_0 \frac{dX_{\mathrm{L}}}{dt} = \frac{k_2^{\mathrm{L}} E_0 \dfrac{S_0}{K_m^{\mathrm{L}}}(1 - X_{\mathrm{L}})}{1 + \dfrac{S_0}{K_m^{\mathrm{L}}}(1 - X_{\mathrm{L}}) + \dfrac{S_0}{K_m^{\mathrm{D}}}(1 - X_{\mathrm{D}})} \tag{1.87}$$

$$r_v^{\mathrm{D}} = S_0 \frac{dX_{\mathrm{D}}}{dt} = \frac{k_2^{\mathrm{D}} E_0 \dfrac{S_0}{K_m^{\mathrm{D}}}(1 - X_{\mathrm{D}})}{1 + \dfrac{S_0}{K_m^{\mathrm{L}}}(1 - X_{\mathrm{L}}) + \dfrac{S_0}{K_m^{\mathrm{D}}}(1 - X_{\mathrm{D}})} \tag{1.88}$$

with $S_0 = S_{\mathrm{L}0} = S_{\mathrm{D}0}$ (racemic).

The ratio of the two rates is given by:

$$\frac{dX_{\mathrm{L}}}{dX_{\mathrm{D}}} = \frac{k_2^{\mathrm{L}}/K_m^{\mathrm{L}}(1 - X_{\mathrm{L}})}{k_2^{\mathrm{D}}/K_m^{\mathrm{D}}(1 - X_{\mathrm{D}})} \tag{1.89}$$

This equation can be integrated:

$$\int_0^{X_{\mathrm{L}}} \frac{dX_{\mathrm{L}}}{1 - X_{\mathrm{L}}} = \frac{k_2^{\mathrm{L}}/K_m^{\mathrm{L}}}{k_2^{\mathrm{D}}/K_m^{\mathrm{D}}} \int_0^{X_{\mathrm{D}}} \frac{dX_{\mathrm{D}}}{1 - X_{\mathrm{D}}} \tag{1.90}$$

with the result:

$$\frac{\ln(1 - X_L)}{\ln(1 - X_D)} = \frac{k_2^L / K_m^L}{k_2^D / K_m^D} = \Phi \tag{1.91}$$

Competing enantiomers differ in their so-called *specificity constants*, k_2/K_m, the ratio of which is a constant, Φ. The aim of this modeling is to obtain an expression relating two commonly used quantities: the product yield (here equivalent to substrate conversion) and the *enantiomeric excess* of the L- product:

$$\varphi_{PL} = \frac{X_L - X_D}{X_L + X_D} \tag{1.92}$$

We define a global degree of substrate conversion:

$$X = \frac{2 S_0 - S_L - S_D}{2 S_0} = \frac{X_L + X_D}{2} \tag{1.93}$$

and redefine the two substrate conversions by combining eqs 1.92 and 1.93:

$$X_L = (1 + \varphi_{PL}) X \tag{1.94}$$

$$X_D = (1 - \varphi_{PL}) X \tag{1.95}$$

Introduction of both equations into eq 1.91 yields:

$$\frac{\ln[1 - X(1 + \varphi_{PL})]}{\ln[1 - X(1 - \varphi_{PL})]} = \Phi \tag{1.96}$$

We may obtain an equivalent expression, but now in terms of the conversion of the L- substrate only, by replacing X_D in eq 1.91 by eq 1.92:

$$\frac{\ln(1 - X_L)}{\ln\left(1 - X_L \dfrac{1 - \varphi_{PL}}{1 + \varphi_{PL}}\right)} = \Phi \tag{1.97}$$

We may carry out the same analysis to gain information about the enantiomeric excess of the remaining substrate, S_D. The corresponding expressions are:

$$\varphi_{SD} = \frac{(1 - X_D) - (1 - X_L)}{(1 - X_L) + (1 - X_D)} = \frac{X_L - X_D}{2 - (X_L + X_D)} \tag{1.98}$$

$$X_L = X + \varphi_{SD}(1 - X) \tag{1.99}$$

$$X_D = X - \varphi_{SD}(1 - X) \tag{1.100}$$

$$\frac{\ln[1 - X(1 - \varphi_{SD})]}{\ln[1 - X(1 + \varphi_{SD})]} = \Phi \tag{1.101}$$

or, in terms of X_D:

$$\frac{\ln\left[1 - \dfrac{X_D(1 - \varphi_{SD}) + 2\varphi_{SD}}{1 + \varphi_{SD}}\right]}{\ln(1 - X_D)} = \Phi \tag{1.102}$$

The equations 1.96–1.97 and 1.101–1.102 must be solved numerically in order to obtain the enantiomeric excess as a function of conversion. Simple Newton iteration may be applied, but some precaution is required. The arguments of the logarithmic expressions have to be checked for non-negative values. If such arguments are encountered, the enantiomeric excess has to be altered by some simple algorithm in order to find an appropriate starting value. For more complex mechanisms than that given by scheme S13, the corresponding rate expressions (eqs 1.87, 1.88) have to be integrated numerically in order to predict the enantiomeric excess in discontinuously operated reactors.

14.2 EVALUATION OF KINETIC DATA

In the following three sections, the basic principles of parameter estimation are provided. The first section summarizes the basic mathematical background, and the others discuss different modeling strategies for enzyme kinetics.

14.2.1 Linear and Nonlinear Regression

Linear regression is used for parameter estimation of *linear models*. Linear models are characterized by the fact that their partial derivatives with respect to the parameters do not contain the parameters themselves. Let us take the example of a multivariable model:

$$y_c = a_0 + a_1 v_1 + a_2 v_2 + a_3 v_3 \tag{2.1}$$

The first partial derivatives of this model are:

$$\mathbf{j} = \frac{\partial y_c}{\partial \mathbf{a}^{\mathrm{T}}} = [1, v_1, v_2, v_3] \tag{2.2}$$

With the vector of parameters (^{T} denotes transposition):

$$\mathbf{a}^{\mathrm{T}} = [a_0, a_1, a_2, a_3] \tag{2.3}$$

the model may be written:

$$y_c = \mathbf{j}\ \mathbf{a} \tag{2.4}$$

To estimate parameters, we take a series of measurements, m, relating the independent variables, v, with the dependent variable, y. If the model contains n parameters, the number of experiments, m, should at least exceed n. The model for each data set may then be written as:

$$\mathbf{y}_c = \mathbf{J}\ \mathbf{a} \tag{2.5}$$

where **J**, the *Jacobian matrix*, is the extension of **j**, eq 2.2, for multiple measurements.

The calculated values of the dependent variable, y_c, will deviate from those which have been measured, y, giving rise to a *vector of residuals*:

$$\mathbf{f} = \mathbf{y}_c - \mathbf{y} = \mathbf{J}\ \mathbf{a} - \mathbf{y} \tag{2.6}$$

A convenient starting point for solving the problem is to minimize the sum of squares of the residuals:

$$\mathbf{f}^{\mathrm{T}}\mathbf{f} = \sum_{i=1}^{m} (y_{ci} - y_i)^2 \overset{!}{=} \min \tag{2.7}$$

This implies that the partial derivatives of this function vanish:

$$\frac{\partial \mathbf{f}^{\mathrm{T}}\mathbf{f}}{\partial \mathbf{a}} \overset{!}{=} 0 \tag{2.8}$$

The partial derivatives of the residuals are:

$$\frac{\partial \mathbf{f}}{\partial \mathbf{a}} = \frac{\partial (\mathbf{y}_c - \mathbf{y})}{\partial \mathbf{a}} = \frac{\partial \mathbf{y}_c}{\partial \mathbf{a}} = \mathbf{J}^{\mathrm{T}} \tag{2.9}$$

according to eq 2.5. With this relation, eq 2.8 is rewritten as:

$$\frac{\partial \mathbf{f}^{\mathrm{T}}\mathbf{f}}{\partial \mathbf{a}} = 2\,\mathbf{J}^{\mathrm{T}}\mathbf{f} \overset{!}{=} 0 \tag{2.10}$$

This equation can be transformed into:

$$\mathbf{J}^{\mathrm{T}}\mathbf{f} = \mathbf{J}^{\mathrm{T}}(\mathbf{y}_c - \mathbf{y}) = \mathbf{J}^{\mathrm{T}}(\mathbf{J}\ \mathbf{a} - \mathbf{y}) = 0 \tag{2.11}$$

and finally:

$$[\mathbf{J}^{\mathrm{T}}\mathbf{J}]\ \mathbf{a} = \mathbf{J}^{\mathrm{T}}\mathbf{y} \tag{2.12}$$

Equation 2.12 represents a system of linear equations which can be solved for the parameter vector **a** by commonly used algorithms, like those of Gauss-Jordan, Crout or Choleski. If a statistical analysis is wanted, we may also obtain the solution of eq 2.12 by computing the inverse of $[\mathbf{J}^T\mathbf{J}]$:

$$\mathbf{a} = [\mathbf{J}^T\mathbf{J}]^{-1}\,\mathbf{J}^T\mathbf{y} \qquad (2.13)$$

since $[\mathbf{J}^T\mathbf{J}]^{-1}$ is the *variance-covariance matrix*:

$$\sigma_M^2 = [\mathbf{J}^T\mathbf{J}]^{-1} \qquad (2.14)$$

It contains the variance of the parameters as diagonal elements. The *variance of observations* is defined as:

$$\sigma_{obs}^2 = \frac{\mathbf{f}^T\mathbf{f}}{m - n} \qquad (2.15)$$

With these values the parameters may be given in terms of their *confidence limits*:

$$\mathbf{a} \triangleq \mathbf{a} \pm \Delta\mathbf{a} = \mathbf{a} \pm t_S\sqrt{\sigma_{obs}^2 D(\sigma_M^2)} \qquad (2.16)$$

where $D()$ stands for diagonal, and with t_S, the percentile point of the Student's t distribution, which depends on the degree of freedom $(m - n)$ and the required probability. The square-root term is known as the *standard deviation*. The *covariance coefficients* may be obtained from the nondiagonal elements of the variance-covariance matrix by:

$$C_{ij} = \frac{\sigma_{ij}^2}{\sqrt{\sigma_{ii}^2\,\sigma_{jj}^2}} \qquad (2.17)$$

A covariance coefficient of $C_{ij} = 0$ means that the parameters a_i and a_j are independent. When C_{ij} is equal to 1 or –1, the parameters a_i and a_j are highly related, so that an independent estimation is not possible. Both parameters may be changed in the same direction ($C_{ij} \cong 1$) or in opposite directions ($C_{ij} \cong -1$) without much affecting the sum of squares of the residuals.

The parameter estimation for nonlinear models is complicated by the fact that the first partial derivatives are not sufficient for describing the model. Therefore, iterative methods must be applied because the higher-order partial derivatives cannot be neglected.

Algorithms for nonlinear regression may be roughly divided into three categories: *hill-climbing methods*, *generalized least squares methods*, also called *Taylor series methods*, and *quasi-Newton methods*. The hill-climbing methods extend from very simple ones, like the Fibonacci search and the algorithm of Hooke-Jeeves, to complex ones like the orthogonal search of Rosenbrock [10] or the simplex search of

Nelder-Mead [11]. These algorithms may be of interest for rather small problems, for a first pre-estimation of parameters, or in the case where other methods tend to become unstable.

The Taylor series methods are based on the fact that a first parameter vector must be provided at the location at which the sum of squares of the residuals, the objective function to be minimized, is calculated and developed into a Taylor series for the prediction of a better parameter vector. If $\mathbf{a}_0$ was the parameter vector and it led to the objective function $[\mathbf{f}^T\mathbf{f}]_0$, then $\mathbf{a}_0$ may be extended to a next trial by:

$$\mathbf{a}_1 = \mathbf{a}_0 + \Delta\mathbf{a} \tag{2.18}$$

The iteration step $\Delta\mathbf{a}$ has to be obtained from a Taylor series expansion of the *objective function*:

$$[\mathbf{f}^T\mathbf{f}]_1 = [\mathbf{f}^T\mathbf{f}]_0 + \Delta\mathbf{a}^T\mathbf{g} + \frac{1}{2}\Delta\mathbf{a}^T\mathbf{G}\Delta\mathbf{a} \tag{2.19}$$

This means that the objective function is approximated by a quadratic function. The gradient is obtained according to eq 2.10:

$$\mathbf{g} = \frac{\partial[\mathbf{f}^T\mathbf{f}]_0}{\partial\mathbf{a}} = 2\mathbf{J}^T\mathbf{f} \tag{2.20}$$

and the *Hessian matrix* as:

$$\mathbf{G} = \frac{\partial^2[\mathbf{f}^T\mathbf{f}]_0}{\partial\mathbf{a}^2} = 2\mathbf{J}^T\mathbf{J} + \frac{\partial\mathbf{J}^T}{\partial\mathbf{a}}\mathbf{f} \tag{2.21}$$

The last expression contains the matrix of second-order derivatives of the residuals. Since it would be an unpleasant job to compute their values, it is normally omitted:

$$\frac{\partial\mathbf{J}^T}{\partial\mathbf{a}} \cong 0 \quad \text{and} \quad \mathbf{G} \cong 2\mathbf{J}^T\mathbf{J} \tag{2.22}$$

If we could arrive in one step at the minimum of the objective function, the new sum of squares $[\mathbf{f}^T\mathbf{f}]_1$ should take its minimum value. In consequence, the first derivative of eq 2.19 should be equal to zero at this point:

$$\frac{\partial\mathbf{f}^T\mathbf{f}}{\partial\mathbf{a}} = 2\mathbf{J}^T\mathbf{f} + 2\mathbf{J}^T\mathbf{J}\,\Delta\mathbf{a} \overset{!}{=} 0 \tag{2.23}$$

The iteration step can now be computed from the system of linear equations:

$$\mathbf{J}^T\mathbf{J}\,\Delta\mathbf{a} = -\mathbf{J}^T\mathbf{f} \quad \text{or} \quad \mathbf{G}\,\Delta\mathbf{a} = -\mathbf{g} \tag{2.24}$$

or via matrix inversion:

$$\Delta\mathbf{a} = -[\mathbf{J}^T\mathbf{J}]^{-1}\,\mathbf{J}^T\mathbf{f} = -\mathbf{G}^{-1}\mathbf{g} \tag{2.25}$$

New search steps have to be calculated until the parameter vector does not change any more. This algorithm is known as the *Newton* or the *Gauss-Newton algorithm.* It is quite powerful and straightforward to implement but rather unstable with respect to the initial parameter estimate. This may be circumvented by introduction of a *step-width controller* (a factor for the iteration step $\Delta\mathbf{a}$) which can be initialized with a value in the range of 0 to 1. When a successful step is encountered, the controller value is increased by a factor of $\sqrt{2}$, and for an unsuccessful cycle it is decremented by a factor of 2. If the Jacobian matrix cannot be composed by means of analytical expressions, it may conveniently be calculated by linear interpolation between two function evaluations. More precise numerical methods may be applied, but they require more function evaluations.

More advanced Taylor series methods are based on the introduction of corrective terms for the omitted matrix of second-order derivatives. The algorithms of Levenberg [12] and Marquardt [13] fall into this category. An algorithm especially written for systems for which the partial derivatives are not given analytically is that of Powell [14].

Another approach of minimizing functions and estimating parameters is based on information about the gradient. From a given parameter set $\mathbf{a}$, lower values of the objective function are to be found in the direction $\mathbf{S}$ of the negative gradient, $-\mathbf{g}$. The new search vector, therefore, can be found as:

$$\mathbf{a}_{i+1} = \mathbf{a}_i + \Delta\mathbf{a}_i = \mathbf{a}_i + \alpha_i\mathbf{S}_i \tag{2.26}$$

with:

$$\mathbf{S}_i = -\mathbf{g}_i \tag{2.27}$$

The gradient $\mathbf{g}$ corresponds to that defined by eq 2.20. If the objective function had a quadratic form, then $-\mathbf{g}$ would point exactly in the direction of the global minimum. Since this is rarely the case, there will be only a local minimum in the direction of $-\mathbf{g}$. This local minimum may be found by interpolation of some function values in the direction of the negative gradient. For instance, an Hermite interpolation may yield the necessary information about how to fix the factor α for rescaling $\mathbf{S}$ in order to produce a step width $\Delta\mathbf{a}$ which points to the local minimum in the direction of the negative gradient. However, this *method of steepest descent* converges very slowly, compared to the Newton methods (eq 2.25).

Advanced algorithms, for example, that of Fletcher and Reeves [15], apply the *conjugate gradient method.* The conjugate gradient method is based on the assumption that the objective function is a positive definite quadratic and, hence, of the form:

$$F(\mathbf{a}) = b + \mathbf{a}^{T}\mathbf{w} + \frac{1}{2}\mathbf{a}^{T}\mathbf{G}\mathbf{a} \tag{2.28}$$

with b and **w** being constant, and **G** a positive definite symmetric matrix. If the line searches are performed exactly, that is, the local minimum in the direction of search **S** is found exactly, the gradient at this point, **g**, obeys the relation:

$$\mathbf{g}_i^{\mathsf{T}} \mathbf{S}_{i-1} = 0 \tag{2.29}$$

since the point $\mathbf{a}_i$ is a minimum in the direction $\mathbf{S}_{i-1}$. In consequence, the gradient $\mathbf{g}_i$ has no component in that direction. Conjugate search directions are obtained by:

$$\mathbf{S}_i = -\mathbf{g}_i + \frac{\mathbf{g}_i^{\mathsf{T}} \mathbf{g}_i}{\mathbf{g}_{i-1}^{\mathsf{T}} \mathbf{g}_{i-1}} \mathbf{S}_{i-1} \tag{2.30}$$

We start with a parameter set $\mathbf{a}_0$ and the first search direction $\mathbf{S}_0 = -\mathbf{g}_0$. The factor α of eq 2.26 has to be found in order to fix the step width. At the new point $\mathbf{a}_1$ the search direction $\mathbf{S}_1$ has to be computed by eq 2.30, and the procedure is repeated until the step width falls below a given value.

Since normal conjugate gradient methods omit information on the curvature of the objective function, they suffer from poor convergence properties. Therefore, it would be advantageous to combine the conjugate gradient method with the Newton method, which is exceptionally powerful in the vicinity of the optimum. Such methods are called *quasi-Newton methods*.

Newton's method is based on the iteration:

$$\mathbf{a}_{i+1} = \mathbf{a}_i - \mathbf{G}_i^{-1} \mathbf{g}_i \tag{2.31}$$

according to eq 2.25. But, since the matrix of second derivatives is incomplete, as stated in eq 2.22, the method is very powerful only at the vicinity of the optimum, where the neglected term tends to vanish.

The strategy of the quasi-Newton methods resides in the generation of a sequence of positive definite symmetric matrices, **H**, for defining new search directions:

$$\mathbf{S}_i = -\mathbf{H}_i \mathbf{g}_i \tag{2.32}$$

with which new search vectors are obtained by eq 2.26. The matrix **H** can be regarded as an approximation for the complete set of the inverse Hessian matrix $\mathbf{G}^{-1}$ of eq 2.31. Since information about **H** is scarce at early stages, the search direction will point into the direction of steepest descent, -**g**. At later stages, information on **H** increases so that the search direction will coincide more and more with the Newton step. We define new variables:

$$\Delta \mathbf{a}_i = \mathbf{a}_{i-1} \quad \text{and} \quad \Delta \mathbf{g}_i = \mathbf{g}_{i+1} - \mathbf{g}_i \tag{2.33}$$

where index i refers to the actual search vector and i+1 to the local minimum in the direction of $-\mathbf{g}_i$ from the origin $\mathbf{a}_i$. If the objective function can be described by the quadratic given by eq 2.28, then:

$$\Delta \mathbf{g}_i = \mathbf{G}_{i+1} \Delta \mathbf{a}_i \tag{2.34}$$

should hold, or, by correspondence:

$$\mathbf{H}_{i+1}\,\Delta\mathbf{g}_i = \Delta\mathbf{a}_i \tag{2.35}$$

It is required that $\mathbf{H}_{i+1}$ should be positive definite symmetric and it should be obtained easily from the function evaluations during the procedure. In addition, it should generate conjugate search directions when applied to a quadratic objective function. These requirements are fulfilled if the inverse update is calculated by a quasi-Newton formula such as that of Davidon-Fletcher-Powell (DFP):

$$\mathbf{H}_{i+1} = \mathbf{H}_i + \frac{\Delta\mathbf{a}_i\,\Delta\mathbf{a}_i^{\mathrm{T}}}{\Delta\mathbf{a}_i^{\mathrm{T}}\,\Delta\mathbf{g}_i} - \frac{\mathbf{H}_i\Delta\mathbf{g}_i\Delta\mathbf{g}_i^{\mathrm{T}}\,\mathbf{H}_i}{\Delta\mathbf{g}_i^{\mathrm{T}}\,\mathbf{H}_i\Delta\mathbf{g}_i} \tag{2.36}$$

or that of Broyden, Fletcher, Goldfarb and Shanno (BFGS):

$$\mathbf{H}_{i+1} = \left(\mathbf{I} - \frac{\Delta\mathbf{a}_i\,\Delta\mathbf{g}^{\mathrm{T}}}{\mathbf{g}^{\mathrm{T}}\Delta\mathbf{a}_i}\right)\mathbf{H}_i\left(\mathbf{I} - \frac{\Delta\mathbf{g}_i\Delta\mathbf{a}^{\mathrm{T}}}{\mathbf{g}^{\mathrm{T}}\Delta\mathbf{a}_i}\right) \tag{2.37}$$

where **I** denotes the unit matrix (identity). The BFGS formula is believed to present the most powerful update of all tested quasi-Newton formulae. The algorithms conveniently begin with the matrix **H** equal to the identity matrix **I**. Differences may occur in how to adjust the step-width parameter α in eq 2.26. For a general survey on quasi-Newton methods, the reader may refer to textbooks or specialized articles [16, 17]. A survey on available mathematical and statistical algorithms may be found in the textbook of Rice [P6].

14.2.2 Estimation of Initial Rates

The evaluation of initial-rate data is a very popular tool for obtaining kinetic constants for enzyme characterization, as well as for engineering purposes. Although the initial-rate method may be implemented in a straightforward manner, it suffers from several inherent drawbacks, of which the most important is that a model obtained by this way may not describe the full features of the system's time dependence.

The problems often arise in the first step: the determination of initial rates. The experimental procedure normally consists of measuring the concentration of the substrate or the product over discrete time intervals in a discontinuously operated stirred-tank reactor under controlled reaction conditions. These discrete concentration/time data have to be transformed into an initial rate at the beginning of the experiment, that is, at vanishing conversion. If the data for concentration/time or conversion/time show a strict linear dependence, it can be assumed that this is so even at vanishing conversion, and the initial rate may be obtained by a simple linear regression of a straight line. If the product concentration or substrate conversion, X, is measured, and timing can be assumed reliable with respect to the beginning of the reaction, the very simple model:

$$X = at \tag{2.38}$$

should hold. The parameter a is obtained by regression:

$$a = \frac{\mathbf{t}^{\mathsf{T}}\mathbf{X}}{\mathbf{t}^{\mathsf{T}}\mathbf{t}} = \frac{\sum_{i=1}^{m} t_i X_i}{\sum_{i=1}^{m} t_i^2} = \frac{dX}{dt} \tag{2.39}$$

We would obtain the initial rate by taking into account the enzyme concentration and the initial substrate concentration:

$$r_{m0} = \frac{S_0}{E_0}\frac{dX}{dt} = \frac{S_0}{E_0}a \tag{2.40}$$

If there are doubts with respect to the beginning of reaction due to mixing or other time-consuming operations, the model should account for a time delay:

$$X = a_0 + a\,t \tag{2.41}$$

We obtain the parameter a by linear regression as discussed in Section 14.2.1 or by eq 2.39 after normalizing the data with their mean values:

$$\Delta X_i = X_i - \bar{X} \qquad \Delta t_i = t_i - \bar{t} \tag{2.42}$$

which leads to the model equivalent to eq 2.38:

$$\Delta X = a\,\Delta t \tag{2.43}$$

Often, however, straight lines are not observed, and one sees instead curved ones with decreasing slope. In this case, initial rates must be obtained by mathematical means which allow for extrapolation to time zero. However, departure from linearity can arise from physical causes, such as lag phases due to toxic or activating impurities, subunit association or dissociation, and limitations in the response of analytical devices. In addition, pre-steady-state phenomena are not likely to be observed at technically interesting substrate concentrations.

Nonlinearity in the time dependence of product formation is most often due to the change in concentration of the reactants. Severe substrate inhibition may yield a concave time dependence, whereas product inhibition may give a convex profile. Product inhibition is observed frequently, and its effects often cannot be neglected even at low substrate conversion. There are certainly many mathematical models capable of reproducing nonlinear time dependence of product formation. The behavior required of such a model is that it must be sufficiently stiff and that it must have a realistic finite derivative at time zero.

A second-order polynomial:

$$X = a_0 + a_1 t + a_2 t^2 \tag{2.44}$$

or its form passing through the origin:

$$X = a_1 t + a_2 t^2 \tag{2.45}$$

and its transformation to a straight line:

$$\left(\frac{X}{t}\right) = a_1 + a_2 t \tag{2.46}$$

are easily treated by linear regression, but they are not sufficiently stiff for cases of severe product inhibition and are, therefore, of limited use.

A generally well-behaved function for cases with product inhibition has been derived from the Michaelis-Menten formula [18]:

$$X = \frac{a_1 t}{a_2 + t} \tag{2.47}$$

This model may be rewritten in the linear form:

$$\frac{t}{X} = \frac{a_2}{a_1} + \frac{1}{a_1} t \quad \text{or} \quad \frac{1}{X} = \frac{1}{a_1} + \frac{a_2}{a_1}\frac{1}{t} \tag{2.47a}$$

and evaluated by linear regression. A simple Gauss-Newton routine would rapidly refine these coefficients by nonlinear parameter estimation. The initial rate is obtained as:

$$r_{m0} = \frac{S_0}{E_0}\frac{a_1}{a_2} \tag{2.48}$$

Another practical approach is to use integral kinetic models directly. As shown in Table 14.1 and eq 1.38, the model:

$$Et = \frac{S_0}{k_2'} X + \frac{K_m'}{k_2'} \ln\left(\frac{1}{1 - X}\right) \tag{1.38}$$

should describe the time dependence of a reaction with competitive product inhibition. This model may be evaluated by different means, as will be outlined in Section 14.2.4, and gives initial rates in the form of the rate expression:

$$r_{m0} = \frac{k_2' S_0}{K_m' + S_0} \tag{2.49}$$

Unfortunately, this model is quite sensitive to timing errors, and easily demonstrates strange behavior at the time origin, especially if the first data lie relatively far from the origin.

It should be kept in mind that guessing initial rates may not be as straightforward as it may seem, and that initial-rate profiles estimated by purely mathematical means may differ significantly from those obtained by modeling entire reaction systems by integral kinetic methods. The estimation of initial rates implies, unfortunately, an extrapolation to an experimentally poorly defined origin of time and conversion.

14.2.3 Initial-Rate Kinetics

Most enzyme kinetics modeling is done by measuring initial rates. Despite the potential problems encountered, as discussed in the last section, it is a convenient means for mechanistic studies. Even with few experiments, certain model assumptions can be tested. Since enzyme kinetics are based on nonlinear models, in general, the evaluation of data should be done with nonlinear parameter estimation methods. However, although computers are very common today, the attraction of straight-line correlations persists. Therefore, initial-rate kinetic modeling is typically carried out with linear functions obtained by transformation of the nonlinear models.

The classical linear forms of the Michaelis-Menten model, eq 1.14, are associated with the names of their inventors or proponents, Lineweaver-Burk [19], Eadie-Hofstee [20, 21], and Hanes [22]:

$$\frac{1}{r_{m0}} = \frac{1}{k_2} + \frac{K_m}{k_2}\left\{\frac{1}{S_0}\right\} \tag{2.50}$$

$$r_{m0} = k_2 - K_m\left\{\frac{r_{m0}}{S_0}\right\} \tag{2.51}$$

$$\frac{S_0}{r_{m0}} = \frac{K_m}{k_2} + \frac{1}{k_2}\{S_0\} \tag{2.52}$$

If the left-hand terms are plotted against the transformed variables given in curly braces, straight lines should be obtained. Thus, simple graphical means or linear regression can be used for parameter estimation. Unfortunately, the relative weighting of the data points and the error distribution of the original variables are not preserved. This implies that different parameters will be obtained by each method, in particular when the data are scattered (experimental data are always subject to stochastic variation). It cannot be generally stated that the one or the other transformation is better—these linear forms should be applied only where they are appropriate, as a convenient means for distinguishing among rival models.

If all these transformed models are used for the purpose stated above and used with some precautions, each is justified. Precautions should be taken, depending on whether the models are linear or nonlinear, that data values are measured in such a way as to obtain approximately equal spacing for the dependent variable. For linear models, this is assured by the choice of equidistant values of the independent (transformed) variable. Consequently, the experimental procedure has to be fixed in accordance with the (transformed) model to be used.

Another transformation of the Michaelis-Menten equation has become quite popular, the *direct linear plot* proposed by Eisenthal and Cornish-Bowden [23]. It is based on the calculation of mean values of k_2 and K_m from intersecting lines each defined by a single data point (S_0, r_{m0}).

The method is based on the equations:

$$r_{m0} + K_m \frac{r_{m0}}{S_0} + k_2 \quad \text{and} \quad -S_0 + k_2 \frac{S_0}{r_{m0}} = K_m \tag{2.53, 2.54}$$

Equality in two data pairs i and j out of m leads to [1/2 m (m – 1)] values of:

$$k_2^{ij} = \frac{S_0^i - S_0^j}{\dfrac{S_0^i}{r_{m0}^i} - \dfrac{S_0^j}{r_{m0}^j}} \quad \text{and} \quad K_m^{ij} = \frac{r_{m0}^i - r_{m0}^j}{\dfrac{r_{m0}^j}{S_0^j} - \dfrac{r_{m0}^i}{S_0^i}} \tag{2.55, 2.56}$$

which are used to estimate mean values. This method has the disadvantage that the resulting graphical presentation is too complex for a rapid visual model verification.

The same lack of graphical simplicity is found for other transformations involving the introduction of two new independent variables, for example:

$$S_0 r_{m0} = k_2 \{S_0\} - K_m \{r_{m0}\} \tag{2.57}$$

which can be solved by simple linear regression, but which represents a trajectory in three-dimensional space.

The parameters obtained by methods based on transformation to linear forms should be refined by nonlinear regression. It must be observed that if nonlinear models are being used for direct parameter estimation, fewer measurements are needed to obtain valid estimates if data are collected in the range of variables in which the model is the most sensitive to changes in these variables. This is the case when the first partial derivatives of the model take maximum absolute values. For the Michaelis-Menten model, this means simply that the rate constant can best be obtained if measurements are taken at high substrate concentration ($S_0 >> K_m$), and the value of K_m at low substrate concentration ($S_0 = K_m$). The experimental procedure would be to obtain an initial guess of the value of K_m in order to define appropriate substrate concentrations for refining the parameters. Since working at very low substrate concentrations often may not be feasible, the minimum requirement is that there be data points at substrate concentrations lower than the value of K_m.

The Michaelis-Menten model is normally only the starting point for obtaining information about how the rate of reaction depends on the substrate concentration. Influences, mostly of an inhibitory nature, of the reaction products and substances inherent to the reaction system must be studied as well. The classical procedure would be to add different amounts of one of these substances to the substrate, and to determine each time the apparent rate constant and value of K_m by initial rate experiments. With this information, one may undertake a model discrimination by using linear relationships which can be found in the major textbooks on initial-rate methods.

The underlying idea behind this procedure is that most common enzyme kinetic rate laws can be transformed to linear models. This may be demonstrated by a reaction characterized by simultaneous product and substrate inhibition, the model of which:

$$r_{m0} = k_2 \frac{S_0}{K_m\left(1 + \dfrac{P_0}{K_{ic}}\right) + \left(1 + \dfrac{S_0}{K_{is}}\right) S_0} \tag{2.58}$$

may be transformed to yield, for example:

$$\frac{1}{r_{m0}} = \frac{1}{k_2} + \frac{K_m}{k_2}\left\{\frac{1}{S_0}\right\} + \frac{1}{k_2}\frac{K_m}{K_{ic}}\left\{\frac{P_0}{S_0}\right\} + \frac{1}{k_2 K_{is}}\{S_0\} \tag{2.59}$$

This is a linear model that can be evaluated according to eq 2.12 with:

$$\mathbf{y}^{\mathrm{T}} = \left(\left(\frac{1}{r_{m0}}\right)_1, \left(\frac{1}{r_{m0}}\right)_2, \ldots \left(\frac{1}{r_{m0}}\right)_m\right) \tag{2.60}$$

$$\mathbf{J} = \left(1, \frac{1}{S_0}, \frac{P_0}{S_0}, S_0\right)_{i=1,m} \tag{2.61}$$

$$\mathbf{a}^{\mathrm{T}} = \left(\frac{1}{k_2}, \frac{K_m}{k_2}, \frac{1}{k_2}\frac{K_m}{K_{ic}}, \frac{1}{k_2 K_{is}}\right) \tag{2.62}$$

by multiple linear regression. The original parameters are obtained by appropriate manipulation of the parameters of the linear model. This approach allows a rapid screening of models with few experimental data by applying linear forms of models which are easily derived.

However, a model is appropriate only if it reliably predicts how the reaction will behave in the reactor in which the process will finally be carried out. Therefore, one must verify whether the model obtained is able to make that prediction. Since the initial-rate method neglects the fact that conversion of substrate is the final aim of the process, one should also verify that the integral form of the rate equation is able to describe the time dependence of the substrate conversion in a discontinuously operated stirred-tank reactor.

14.2.4 General Rate Kinetics

Since most kinetic experiments are carried out in discontinuously operated reactors, normally data can be collected over the full range of substrate conversions. However, if modeling is to be performed with rate laws, the substrate conversion/time data must be translated into rate/conversion data. If a continuous analytical method is applicable, there are enough data such that numerical differentiation can be performed with appreciable precision [24].

If only discrete data can be obtained, it may be difficult to get enough data over the whole conversion range in order to be able to extract reliable rates. Numerical methods are available for smoothing scattered data and to differentiate the experimental functions (see, for example, [P6]).

The strategy of working with rate laws is evidently most applicable when reaction rates are obtained directly. This is the case for measurements in continuously operated reactors behaving like ideal stirred-tank reactors (see Section 14.4). Such reactors are the continuously operated stirred-tank reactor (CSTR) and the fixed-bed loop reactor, both of which can be employed in the case of immobilized enzymes. Unfortunately, substrate consumption may be considerable for a series of measurements, hence this kind of experiment is rarely chosen over batch experiments. Kinetic measurements in a CSTR are carried out by measuring the substrate conversion, X, at steady state for different values of the throughput of substrate solution or space time, τ, respectively. The calculated rate must fit the assumed kinetic model, demonstrated here for the simplest case:

$$r_m = S_0 \frac{X}{E\tau} = k_2 \frac{S_0(1 - X)}{K_m + S_0(1 - X)} \tag{2.63}$$

of the Michaelis-Menten model. The term $E\tau$ can be calculated with the equivalent mass of enzyme in the reactor, m_E, and the flow rate of substrate solution through the reactor $\dot{V}$ or $E\tau = m_E / \dot{V}$. If measurements at different substrate levels are not sufficient for modeling, this approach may represent a tedious affair, since steady state in a CSTR is attained only after at least five residence times.

Parameter estimation in the case of general rate data can be accomplished by the same methods used for initial-rate modeling. For instance, the parameters of model 2.58 could be obtained by multiple linear regression of its linear form:

$$\frac{1}{r_m} = \frac{1}{k_2} + \frac{K_m}{k_2 S_0}\left\{\frac{1}{1 - X}\right\} + \frac{K_m}{k_2 K_{ic}}\left\{\frac{X}{1 - X}\right\} + \frac{S_0}{k_2 K_{is}}\{1 - X\} \tag{2.64}$$

or by nonlinear parameter estimation with the original model:

$$r_m = k_2 \frac{S_0(1 - X)}{K_m\left(1 + \frac{S_0}{K_{ic}}X\right) + \left(1 + \frac{S_0}{K_{is}}(1 - X)\right)S_0(1 - X)} \tag{2.65}$$

14.2.5 Integral Kinetics

Integrated rate expressions contain the variables, time and substrate conversion, which are originally obtained by experiments in batch and plug-flow continuous reactors. Since batch operation is the preferred mode for kinetic experiments, the use of integral kinetic models is very attractive because the original data may be used directly, and measurements may be taken over the whole range of substrate conversion. If nonlinear parameter estimation is employed, the information contained in the original data is not reduced as in initial-rate modeling, it is not necessary to apply mathematical procedures for numerical differentiation (in contrast to the general rate law modeling), and the error distribution of the data is preserved. To estimate the parameters, we may apply the same procedures, linear and nonlinear regression. Walker and Schmidt [25] and Foster and Niemann [26] have shown that Henri's law or its extension by eq 1.38 may be transformed into a linear form like:

$$S_0 \frac{X}{Et} = k_2' - K_m'\left\{\frac{\ln\left(\frac{1}{1 - X}\right)}{Et}\right\} \tag{2.66}$$

This equation has interesting features: its left-hand term represents the productivity (see Section 14.4) of the enzyme, and the parameters can be directly obtained from a straight-line plot. As in the case of rate laws, there are other transformations of eq 1.38 leading also to linear forms [27, 28].

Since it has not been explicitly discussed in Section 14.2.2, the method for estimating initial rates by applying eq 2.66 to the data is added here [26]. The initial reaction rate is obtained at vanishing conversion, where:

$$\left\{S_0 \frac{X}{Et}\right\}_{X \to 0} = S_0 \left\{\frac{\ln\left(\frac{1}{1 - X}\right)}{Et}\right\}_{X \to 0} \tag{2.67}$$

because of:

$$\lim_{X \to 0} \frac{X}{\ln\left(\frac{1}{1 - X}\right)} = 1 \tag{2.68}$$

Therefore, if a straight line with slope S_0 is drawn through the origin of the space of transformed variables of eq 2.66, this line and that given by eq 2.66 should intersect at a point from which the initial rate may be obtained graphically according to:

$$r_{m0} = \frac{S_0}{E}\frac{dX}{dt} = \left\{S_0 \frac{X}{Et}\right\}_{X \to 0} = S_0 \left\{\frac{\ln\left(\frac{1}{1 - X}\right)}{Et}\right\}_{X \to 0} \tag{2.69}$$

Mathematically, this point is formed as:

$$r_{m0} = \frac{k_2' S_0}{K_m' + S_0} \tag{2.70}$$

after the parameters have been obtained.

Another strategy of modeling with integral kinetic equations was proposed by Schwert [29]. He uses the intervals of data points. His method is equivalent to integrating eq 1.35 in the limits from X_{i-1}, t_{i-1} to X_i, t_i. With $\Delta X_i = X_i - X_{i-1}$ and $\Delta t_i = t_i - t_{i-1}$, we get the following instead of eq 1.38:

$$E_0 \Delta t_i = \frac{S_0}{k_2'} \Delta X_i + \frac{K_m'}{k_2'} \ln\left(1 + \frac{\Delta X_i}{1 - X_i}\right) \tag{2.71}$$

This can be transformed into:

$$\frac{S_0}{E_0}\frac{\Delta X_i}{\Delta t_i} = k_2' - K_m' \left\{\frac{\ln\left(1 + \frac{\Delta X_i}{1 - X_i}\right)}{E \Delta t_i}\right\} \tag{2.72}$$

for linear regression analysis of the parameters. The main advantage of this method is that initial timing problems are canceled out.

The model discriminating capabilities of integral kinetic methods are comparable to those of rate-law methods. For example, one can easily test the hypothesis of product inhibition by carrying out experiments at different substrate levels. Table 14.1 shows that for competitive product inhibition, the apparent value of K_m (K_m') should depend on the initial substrate concentration, whereas for Michaelis-Menten kinetics both parameters are expected to be constant. For severe product inhibition ($K_{ic} < K_m$), both apparent parameters will become negative, a distinct sign for model discrimination. Very high absolute values of k_2' and K_m' would mean that $K_m \cong K_{ic}$. It should be added that for the rare case when $K_m = K_{ic}$, the formulae of Table 14.1 are no longer valid. For instance, in the case of competitive product inhibition, first-order reaction kinetics would be obtained for $K_m = K_{ic}$:

$$S_0 \frac{dX}{dt} = \left[\frac{k_2 E_0}{K_m + S_0} \right] S_0 (1 - X) \tag{2.73}$$

which, translated into integral form, gives:

$$Et = \left[\frac{K_m + S_0}{k_2} \right] \left\{ \ln \left(\frac{1}{1 - X} \right) \right\} \tag{2.74}$$

This model is of the same structure as that given by eq 2.38, and may be treated accordingly.

For models involving more than two apparent parameters, for example, the one given by eq 1.46, linear forms according to eq 2.66 may be derived:

$$\frac{S_0 X}{Et} = k_2' - K_m' \left\{ \frac{\ln \left(\frac{1}{1 - X} \right)}{Et} \right\} - K' \left\{ \frac{S_0^2 X^2}{Et} \right\} \tag{2.75}$$

the parameters of which may be obtained by multiple linear regression.

However, the simplest linear forms are those given by eqs 1.38 and 1.46 themselves, for example:

$$Et = \frac{1}{k_2'} S_0 X + \frac{K_m'}{k_2'} \ln \left(\frac{1}{1 - X} \right) + \frac{K'}{k_2'} S_0^2 X^2 \tag{1.46}$$

which represents a typical problem for multiple linear regression analysis. Unfortunately, the left-hand term is the independent variable. Therefore, this method should be applied with care because of the much different error distribution on Et compared to X. In particular, data at high substrate conversion should be canceled out since the error on Et can become tremendous, whereas that on X more or less vanishes.

Those strategies discussed for rate-law methods can also be used for integral kinetics modeling. We can evaluate each experiment in order to obtain the apparent kinetic parameters. If we vary the initial concentration of substrate (and inhibitors), we can see how the apparent parameters are correlated with the concentration levels. An appropriate dependence (see, for example, those given in Table 14.1) would then allow the determination of the underlying reaction-rate, Michaelis, and inhibition constants.

The second approach implies treatment of a series of experiments at once by multiple linear regression. For this purpose the apparent parameters must be replaced. This is illustrated for the case of an irreversible enzyme reaction accompanied by competitive product inhibition and substrate inhibition according to eqs 1.46 and 1.55–1.57. Only the initial substrate concentration is varied. The appropriate model in analogy to eq 2.75 is:

$$\frac{S_0 X}{Et} = k_2 - K_m \left\{ \frac{\ln\left(\frac{1}{1 - X}\right)}{Et} \right\} + \frac{K_m}{K_{ic}} \left\{ \frac{S_0}{Et} \left[\ln\left(\frac{1}{1 - X}\right) - X \right] \right\} + \frac{1}{K_{is}} \left\{ \frac{S_0^2}{Et} X \left(1 - \frac{X}{2}\right) \right\} \tag{2.76}$$

Nonlinear regression methods are the best procedure for parameter estimation using integral kinetic models. The objective function to be minimized is the sum of squares of the residuals in terms of substrate conversion. The independent variables are the normalized reaction time or operating time, *Et*, and the initial concentration of substrate and products. Since integral kinetic models, despite rare exceptions like eq 2.74, cannot be solved analytically for the substrate conversion, one must solve them numerically using methods such as Newton iteration. For parameter-estimation algorithms requiring partial derivatives, the derivatives must be obtained, for example, by linear interpolation of two function evaluations at slightly different parameter values.

14.3 MASS-TRANSFER PHENOMENA WITH IMMOBILIZED ENZYMES

Reaction systems consisting of at least two distinct phases may be influenced by mass-transfer phenomena. This has to be taken into account in reactor design, along with the intrinsic kinetics of the enzyme and the characteristics of the reactor. Mass transfer can be important, for example, when one is using immobilized enzymes, in the presence of a gaseous substrate or product, in the presence of solid substrates, or in liquid/liquid two-phase systems. Mass transfer from the dispersed phase into the bulk phase or *vice versa* is called *external mass transfer*, whereas diffusional equilibration phenomena in a dispersed phase are called *internal mass transfer* or, in the case of immobilized enzymes, *pore diffusion*. Internal mass transfer may normally be neglected in the case of dispersed gas and nonporous solids owing to very high and vanishing diffusivity, respectively.

For immobilized enzymes, mass transfer causes the reactant concentration at the catalytic sites to be different from those in the bulk phase. Since reactant concentrations at the active sites cannot be observed directly, we must estimate them on the basis of the observable bulk concentrations by modeling the underlying mass-transfer process. The following sections will deal exclusively with mass transfer associated with immobilized biocatalysts. Other phenomena associated with immobilization will not be discussed, for example, alteration of intrinsic kinetics, partitioning of reactants with the immobilization matrix, effects of local ionization equilibria in the matrix, and vicinity effects with multienzyme systems.

14.3.1 External Mass Transfer

Under appropriate mixing conditions, viscid liquids may be assumed to be homogeneous on macroscopic as well as microscopic scales. However, when an agitated viscid liquid encounters a stagnant surface, a velocity profile develops, with its gradient perpendicular to the surface and with vanishing velocity at the surface. This discontinuity in the velocity distribution causes a discontinuity in the concentration distribution as well, if the surface is a source or sink of matter. Therefore, concentration differences between the surface of an object and a viscid bulk phase may be equilibrated only by diffusional processes across the liquid boundary layer. The most commonly applied approach for modeling mass transfer across a liquid boundary layer at rigid surfaces is based on the stagnant film theory.

We assume that there exists a stagnant boundary layer at the surface (interface), the thickness, δ, of which depends on the local hydrodynamic conditions, and that this layer may be penetrated only by diffusional processes.

Let us take the example of an enzyme immobilized on a rigid support. Since enzymes are usually immobilized on porous carriers, concentration gradients may exist in the boundary layer as well as in the catalyst particle. We shall at first neglect internal concentration gradients by modeling the catalyst as if the enzyme were located exclusively at the catalyst surface. The catalyst may contain active enzyme species with a concentration of E_{t1} (given in mass of enzyme per unit volume of catalyst). The enzyme may follow Michaelis-Menten kinetics with a rate constant k_2 (expressed as usual in moles of substrate transformed per unit time and unit mass of enzyme).

The transformation of substrate by the enzyme creates a substrate concentration gradient between the catalyst particle and the surrounding bulk phase. The establishment of the gradient is counteracted by the diffusional flux of substrate from the bulk phase to the catalyst. At steady state, for constant bulk phase substrate concentration and vanishing local time dependence, the substrate consumption at the catalytic sites must be exactly counterbalanced by the diffusional flux of substrate across the stagnant film. For convenience with respect to nomenclature, we shall take a spherical catalyst particle, although the concepts discussed below may easily be extended to any regular catalyst shape. If the particular geometry of the stagnant film may be neglected, external mass transfer should follow Fick's first law in its simplest form:

$$\frac{1}{A_p}\frac{dn}{dt} = -\mathcal{D}\frac{dS}{dz} \tag{3.1}$$

with A_p the surface area of the catalyst particles and z the distance from the catalyst surface. At steady state the gradient is given by:

$$\frac{dS}{dz} = \frac{1}{\delta}(S_b - S_R) \tag{3.2}$$

with the substrate concentration in the bulk phase S_b and at the catalyst surface S_R. The molar flux across the boundary layer is then given by:

$$\frac{dn}{dt} = -\frac{\mathcal{D}}{\delta}A_p(S_b - S_R) \tag{3.3}$$

The first term on the right-hand side is called the *mass-transfer coefficient*:

$$k_L = \frac{\mathcal{D}}{\delta} \tag{3.4}$$

At steady state the rate of mass transfer across the film must equal the rate of substrate consumption by the enzymatic reaction. For Michaelis-Menten kinetics the mass balance leads to:

$$k_L A_p(S_b - S_R) = V_p k_2 E_I \frac{S_R}{K_m + S_R} \tag{3.5}$$

with V_p the total volume of catalyst particles, and with the reaction taking place at the surface concentration S_R.

We may transform this equation by introducing the dimensionless variables:

$$s_{bm} = \frac{S_b}{K_m} \quad \text{and} \quad s_{Rb} = \frac{S_R}{S_b} \tag{3.6}$$

and the *Damköhler number of the second kind*, defined for a reaction of order n and rate constant k as:

$$\mathrm{Da}_{\mathrm{II}n} = \frac{kS_b^{n-1}}{k_L a_p} \tag{3.7}$$

with the ratio of surface and volume of the catalyst particles $a_p = A_p/V_p$. If the ratio of catalyst surface, A_p, and reactor volume, V_R, is preferred, the enzyme concentration

E_I must be defined for this volume, also. This dimensionless number may be transformed to a more intelligible form:

$$\mathrm{Da}_{\mathrm{II}n} = \frac{V_p k S_b^n}{A_p k_\mathrm{L} S_b} \tag{3.7a}$$

which reveals that it describes the ratio of the maximal rate of reaction and the maximal rate of mass transfer ($S_R = 0$).

With these definitions, we may rewrite eq 3.5 by using either a zero-order or a first-order approximation (see eqs 1.15 and 1.16) for the definition of the Damköhler number:

$$(1 - s_{Rb}) = \mathrm{Da}_{\mathrm{II}0} \frac{s_{Rb} s_{bm}}{1 + s_{Rb} s_{bm}} \tag{3.8}$$

$$s_{bm}(1 - s_{Rb}) = \mathrm{Da}_{\mathrm{II}1} \frac{s_{Rb}}{1 + s_{Rb}} \tag{3.9}$$

with:

$$\mathrm{Da}_{\mathrm{II}0} = \frac{k_2 E_I}{k_\mathrm{L} a_p S_b} \quad \text{and} \quad \mathrm{Da}_{\mathrm{II}1} = \frac{\frac{k_2}{K_m} E_I}{k_\mathrm{L} a_p} \tag{3.10, 3.11}$$

The relative surface concentration is thus obtained by either:

$$s_{Rb} = \frac{b}{2}\left(1 \pm \sqrt{1 + \frac{4}{s_{bm} b^2}}\right) \tag{3.12}$$

with:

$$b = 1 - \mathrm{Da}_{\mathrm{II}0} - \frac{1}{s_{bm}} \tag{3.13}$$

where the sign in front of the square root is the same as that of b, or:

$$s_{Rb} = \frac{1}{2} \frac{\mathrm{Da}_{\mathrm{II}1}}{s_{bm}} \left(\sqrt{1 + 4\left(\frac{s_{bm}}{\mathrm{Da}_{\mathrm{II}1}}\right)^2} - 1\right) \tag{3.14}$$

We may introduce the concept of the *effectiveness factor*, which is defined as:

$$\eta = \frac{r_v^R}{r_v^b} = \frac{\text{reaction rate of actual system}}{\text{reaction rate neglecting mass transfer}} \tag{3.15}$$

which, for our problem, becomes:

$$\frac{S_R}{K_m + S_R} = \eta \frac{S_b}{K_m + S_b} \tag{3.16}$$

or in dimensionless variables:

$$\eta = \frac{(1 + s_{bm})\, s_{Rb}}{1 + s_{Rb}\, s_{bm}} \tag{3.17}$$

If we introduce the solutions of either eq 3.12 or 3.14 for s_{Rb} into this equation, we obtain the effectiveness factor in the case of external mass transfer as a function of the saturation parameter s_{bm} and a Damköhler number. Several regimes may be distinguished according to the order of magnitude of these two parameters.

If the saturation parameter takes finite values, but the reaction rate is much higher than the mass-transfer rate ($Da_{II} >> 1$), the substrate concentration at the catalyst surface will vanish. In consequence, the overall rate of substrate depletion will depend only on external mass transfer:

$$r_v^R = k_L A_p S_b \tag{3.18}$$

This leads to an effectiveness factor of:

$$\eta = \frac{1}{\mathbf{Da}_{II1}}(1 + s_{bm}) = \frac{1}{\mathbf{Da}_{II0}}\left(1 + \frac{1}{s_{bm}}\right) \tag{3.19}$$

In the opposite case ($Da_{II} << 1$), when mass transfer is faster than the enzymatic reaction, the reaction rate is limited by the enzymatic reaction. In consequence, the effectiveness factor tends towards unity because of the equality of conditions on the catalyst surface and the bulk phase.

We may deduce the influence of the saturation parameter s_{bm} if we take both limiting values, zero and infinity. If the value tended to infinity ($S_b >> K_m$), the reaction at bulk-phase conditions would be of zero order and maximum rate. Since a very steep gradient would be necessary in order to change the rate of reaction at the catalytic sites, the effectiveness factor would tend to unity regardless of the order of magnitude of the Damköhler number.

If the saturation parameter tended to vanish ($S_b << K_m$), the reaction would be first-order even under bulk-phase conditions and it would proceed at low rate. Since both reactions, at the surface and in the bulk phase, would proceed with first-order kinetics, the effectiveness factor is simply:

$$\eta = s_{Rb} = \frac{1}{1 + \mathbf{Da}_{II1}} \tag{3.20}$$

For practical purposes, it is of importance to be able to test for an external mass-transfer effect. Since the thickness of the stagnant film, δ, and with it the mass-transfer coefficient, depend on the hydrodynamic conditions in the bulk phase, then the mass-transfer rate should vary with the velocity of the liquid phase relative to the solid catalyst. The best test system is a continuously operated fixed-bed loop reactor operating like an ideal stirred-tank reactor. In such a reactor the fluid velocity in the catalytic bed may be varied independently of the substrate feed flow rate. If, for the same feed flow rate (that is, constant space time) but varied flow rates in the loop, the substrate conversion does not change, external mass-transfer effects may normally be excluded. One may perform a simpler test in a discontinuously operated stirred-tank reactor by carrying out discontinuous reactions at various stirrer speeds. If the time dependence of the substrate conversion changes systematically with increasing stirrer speed, external mass transfer is likely to be important.

For different well-defined reactors (for example, fixed-bed, fluidized-bed and stirred-tank reactors), empirical correlations are available for the estimation of the mass-transfer coefficient as a function of the hydrodynamic conditions and the physicochemical characteristics of the reaction system.

14.3.2 Internal Mass Transfer

As in the case of external mass transfer, intraparticle mass transfer may limit the reaction rate of immobilized biocatalysts due to the competition of substrate transformation with substrate delivery by diffusion. Immobilized biocatalysts usually are porous matrices containing the active species. Substrate may arrive at the active centers only through interconnected pores on tortuous paths. Therefore, the rate of diffusion is normally much lower in such matrices compared with that in a homogeneous aqueous phase. If we neglect restrictions due to very small pore diameters, we may model the effective diffusion coefficient in a porous environment as:

$$\mathfrak{D}_e = \mathfrak{D}\frac{\varepsilon_p}{\tau_p} \tag{3.21}$$

with the diffusion coefficient in homogeneous solution $\mathfrak{D}$ the catalyst porosity ε_p, and the tortuosity factor τ_p. Porosities may fall in the range $0.2 < \varepsilon_p < 0.7$, and tortuosities, $2 < \tau_p < 7$ with a typical range of 3–4. In consequence, diffusivities in porous media are typically approximately seven times less than those observed in solution. Both the porosity and the tortuosity factor must be determined experimentally.

Although the models for diffusion inside porous catalysts may be derived for different catalyst geometries, we will focus on the most common geometry: the spherical bead. We shall assume:

- perfect sphericity of the catalyst,
- uniform diffusivity of the reactants throughout the porous catalyst, with negligible concentration dependence of the diffusion coefficient as well as partitioning of the reactants, and

- active centers uniformly distributed and exhibiting Michaelis-Menten kinetics, which is influenced by the local concentration of substrate only.

In addition, we shall neglect external mass transfer at first.

Let substrate diffuse into a spherical bead of radius R, where it is transformed at a rate r_v, given by the kinetics. Under steady-state conditions, meaning constant bulk substrate concentration and fully developed internal concentration profile, the difference of molar flux across two shell surfaces a distance dr apart from one another must have been consumed by the reaction:

$$\dot{n}_r - \dot{n}_{r+dr} = r_v^r \, dV \tag{3.22}$$

or, explicitly:

$$\left(-\mathcal{D}_e \frac{dS}{dr} 4\pi r^2\right)_r - \left(-\mathcal{D}_e \frac{dS}{dr} 4\pi r^2\right)_{r+dr} = r_v^r \pi r^2 dr \tag{3.23}$$

The change of flux with radial position is thus given by:

$$\frac{\mathcal{D}_e\left[\left(r^2 \frac{dS}{dr}\right)_{r+dr} - \left(r^2 \frac{dS}{dr}\right)_r\right]}{dr} = r^2 r_v^r \tag{3.24}$$

and at infinitesimally small distance in radial position:

$$\frac{1}{r^2}\mathcal{D}_e \frac{d}{dr}\left(r^2 \frac{dS}{dr}\right) = \mathcal{D}_e\left(\frac{d^2S}{dr^2} + \frac{2}{r}\frac{dS}{dr}\right) = r_v^r \tag{3.25}$$

Two boundary conditions are needed to solve this second-order ordinary differential equation. Since ideal spherical geometry is being considered, with uniform distribution of active centers, the radial gradient of substrate concentration must be zero at the center of the bead:

$$\left(\frac{dS}{dr}\right)_{r=0} = 0 \tag{3.26}$$

In addition, the substrate concentration at the surface of the bead must be constant in order to use this simplified material balance:

$$S_R \hat{=} S_b = \text{const.} \tag{3.27}$$

If we introduce Michaelis-Menten kinetics into eq 3.25:

$$r_v^r = k_2 E_I \frac{S_r}{K_m + S_r} \tag{3.28}$$

with the enzyme concentration E_I referred to the volume of the beads, eq 3.25 takes on a nonlinear form which unfortunately cannot be solved analytically. If we assume eq 3.28 describes the intrinsic kinetics of the active centers of the catalyst, we can relate the observed overall reaction rate of the catalyst to the rate which would be observed if the active centers were under conditions found in the bulk phase. This leads to the definition of the *effectiveness factor* for the catalyst:

$$\eta = \frac{r_v^{obs}}{r_v^R} \tag{3.29}$$

The observed reaction rate may be defined by the total flux of substrate into the beads:

$$r_v^{obs} = \frac{A_p}{V_p} \mathcal{D}_e \left(\frac{dS}{dr} \right)_R \tag{3.30}$$

When we combine this with the reaction rate which would be obtained at the bulk phase conditions:

$$r_v^R = k_2 E_I \frac{S_R}{K_m + S_R} \tag{3.31}$$

the effectiveness factor becomes:

$$\eta = \frac{A_p}{V_p} \mathcal{D}_e \left(\frac{dS}{dr} \right)_R \frac{K_m + S_R}{k_2 E_I S_R} \tag{3.32}$$

Unfortunately, the concentration gradient of substrate is not an observable quantity. For convenience, eq 3.25 is usually solved in dimensionless form. By introduction of dimensionless variables:

$$s^* = \frac{S_r}{S_R} \, , \quad r^* = \frac{r}{R} \tag{3.33}$$

eq 3.25 is transformed to:

$$\frac{S_R}{R^2} \frac{d^2 s^*}{dr^{*2}} + \frac{2}{R r^*} \frac{S_R}{R} \frac{ds^*}{dr^*} = \frac{r_v^r}{\mathcal{D}_e} \tag{3.34}$$

or:

$$\frac{d^2 s^*}{dr^{*2}} + \frac{2}{r^*} \frac{ds^*}{dr^*} = \left[R^2 \frac{r_v^R}{\mathcal{D}_e S_R} \right] \frac{r_v^r}{r_v^R} \tag{3.35}$$

For linear kinetics, the term in square brackets is directly lumped together, and its square root is called the *Thiele modulus* [30]. The ratio of reaction rates, r_v^r / r_v^R, can be interpreted as a local effectiveness factor. The appropriate definition of the Thiele modulus, ϕ_T, for linear kinetic models of rate constant k, order of reaction n, and substrate concentration S is:

$$\phi_{Tn}^2 = R^2\left(\frac{k S_R^{n-1}}{\mathcal{D}_e}\right) \tag{3.36}$$

Equation 3.35 for first-order kinetics becomes:

$$\frac{d^2 s^*}{dr^{*2}} + \frac{2}{r^*}\frac{ds^*}{dr^*} = \phi_{T1}^2\, s^* \tag{3.37}$$

and for a zero-order reaction:

$$\frac{d^2 s^*}{dr^{*2}} + \frac{2}{r^*}\frac{ds^*}{dr^*} = \phi_{T0}^2 \tag{3.38}$$

Since biocatalysts usually do not exhibit linear kinetics, there are two ways to use the Thiele modulus—as a zero- or first-order term. Therefore, the rate of reaction, r_v, has to be transformed either as:

$$r_v^r = [k_2 E_I]\, s_{Rm}\, \frac{s^*}{1 + s_{Rm} s^*} \tag{3.39}$$

or as:

$$r_v^r = \left[\frac{k_2}{K_m} E_I\right] S_R\, \frac{s^*}{1 + s_{Rm} s^*} \tag{3.40}$$

with:

$$s_{Rm} = \frac{S_R}{K_m} \tag{3.41}$$

The terms in square brackets correspond to a zero-order rate constant (eq 3.39) and a first-order rate constant (eq 3.40), respectively. Since the first-order Thiele modulus is generally preferred, eq 3.37 for Michaelis-Menten kinetics is:

$$\frac{d^2 s^*}{dr^{*2}} + \frac{2}{r^*}\frac{ds^*}{dr^*} = \phi_{T1}^2\, \frac{s^*}{1 + s_{Rm} s^*} \tag{3.42}$$

with:

$$\phi_{T1}^2 = R^2\left(\frac{k_2 E_I}{K_m \mathcal{D}_e}\right) \tag{3.43}$$

including the transformed boundary conditions:

$$(s^*)_{r^*=1} = 1 \qquad \text{and} \qquad \left(\frac{ds^*}{dr^*}\right)_{r^*=0} = 0 \tag{3.44}$$

Thus, the solution of eq 3.42 depends not only on the Thiele modulus, but also on the saturation parameter, s_{Rm}.

There is some confusion in the literature with respect to the modulus because at least three other definitions exist. Aris [31] extended the definition of the modulus by Thiele for different catalyst shapes. Aris's modulus is therefore most often called the *shape-generalized Thiele modulus*. For better distinction we shall call it simply the *Aris modulus*. With the help of the Aris modulus, the effectiveness factor can be given as a unique function of this modulus when its value is high ($\phi_A > 3$), regardless of the shape of the catalyst particle. His definition, in analogy to the one given in eq 3.36, is:

$$\phi_{An}^2 = \left(\frac{V_p}{A_p}\right)^2 \frac{k S_R^{n-1}}{\mathcal{D}_e} \tag{3.45}$$

The Aris modulus may be written in a more intelligible way as:

$$\phi_{An}^2 = \frac{V_p \; r_v^R}{A_p \mathcal{D}_e \dfrac{S_R}{L_c}} \tag{3.45a}$$

with the characteristic length defined by $L_c = V_p/A_p$. This shows that the mass-transfer modulus is the square root of the ratio of reaction rate under surface conditions and a normalized diffusional mass flux.

For spherical shape and Michaelis-Menten kinetics, the *first-order Aris modulus* becomes:

$$\phi_{A1}^2 = \frac{R^2}{9} \frac{k_2 E_I}{K_m \mathcal{D}_e} \tag{3.46}$$

It is related to the Thiele modulus according to:

$$\phi_{A1} = \frac{1}{3} \phi_{T1} \tag{3.47}$$

The third modulus is the *order-generalized Thiele modulus*. With its help, the effectiveness factor as a function of this modulus is a unique function in the range of high values of the modulus, regardless of the order of reaction. This modulus may be defined either with respect to the Thiele modulus:

$$\phi_{OT}^2 = \frac{n+1}{2} \phi_{Tn}^2 \tag{3.48}$$

or to the Aris modulus:

$$\phi_{OA}^2 = \frac{n+1}{2} \phi_{An}^2 \tag{3.49}$$

The fourth modulus is that proposed by Bischoff [32] for adsorption reaction rate forms like Michaelis-Menten kinetics. Its original definition can be written as:

$$\phi_{BT} = \frac{\phi_{T1}}{\sqrt{2}} \frac{\dfrac{s_{Rm}}{1 + s_{Rm}}}{\sqrt{s_{Rm} - \ln(1 + s_{Rm})}} \tag{3.50}$$

with ϕ_{T1} given by eq 3.43. As with the standard modulus ϕ_T, there exists an analogy with the Aris modulus:

$$\phi_{BA} = \frac{\phi_{A1}}{\phi_{T1}} \phi_{BT} = \frac{1}{3} \phi_{BT} \tag{3.51}$$

and ϕ_{A1} defined as in eq 3.46. The same modulus is sometimes expressed by means of the Aris modulus of zero order:

$$\phi_{BA} = \frac{\phi_{A0}}{\sqrt{2}} \frac{\dfrac{s_{Rm}}{1 + s_{Rm}}}{\sqrt{1 - \dfrac{1}{s_{Rm}} \ln(1 + s_{Rm})}} \tag{3.52}$$

with, according to eq 3.45:

$$\phi_{A0}^2 = \frac{R^2}{9} \frac{k_2 E_I}{\mathcal{D}_e S_R} \tag{3.53}$$

Why the *Bischoff modulus* is of interest will be shown later. This collection of moduli shows that there may be quite some confusion in the literature about associating a given solution of the pore diffusion problem with its appropriate modulus. Since it seems that the Aris-type moduli are preferred in biotechnology, we will use those during the following discussions.

As already mentioned, a general analytical solution for the pore-diffusion problem with a reaction of the Michaelis-Menten type does not exist because of its nonlinearity. If we combine eq 3.35 and eq 3.40, we may write the problem as:

$$\frac{d^2 s^*}{dr^{*2}} + \frac{2}{r^*} \frac{ds^*}{dr^*} = 9 \phi_{A1}^2 \frac{s^*}{1 + s_{Rm} s^*} \tag{3.54}$$

with boundary conditions:

$$(s^*)_{r^*=1} = 1 \quad \text{and} \quad \left(\frac{ds^*}{dr^*}\right)_{r^*=0} = 0 \tag{3.55}$$

We can solve this problem by transforming eq 3.54 to a pair of first-order differential equations and integrating these by Runge-Kutta or other methods, using a shooting procedure. One numerical procedure, useful for conditions where substrate depletion at the center of the bead may not have progressed too far, will be discussed. Since eq 3.54, transformed to a pair of first-order ordinary differential equations, may be difficult to integrate from the origin of the radial position, due to a division by the radius, we will transform variables, as proposed by Lapidus [33]:

$$s^* = z^* \exp\left(-\frac{1}{2}\int_0^{r^*}\frac{2}{r^*}\,dr^*\right) \tag{3.56}$$

This leads to:

$$s^* = \frac{z^*}{r^*}\,, \quad \text{or} \quad z^* = r^* s^* \tag{3.57}$$

With this new dependent variable, eq 3.54 becomes:

$$\frac{d^2z^*}{dr^{*2}} = 9\,\phi_{A1}^2 \frac{z^*}{1 + s_{Rm}\dfrac{z^*}{r^*}} \tag{3.58}$$

With initial and boundary conditions:

$$(z^*)_{r^*=0} = 0\,, \quad (z^*)_{r^*=1} = 1\,, \quad \text{and} \quad \left(\frac{dz^*}{dr^*}\right)_{r^*=0} = s_0^* \tag{3.59}$$

where s_0^* denotes the dimensionless substrate concentration at the bead center ($0 < s_0^* < 1$). Equation 3.58 may then be broken down into:

$$\frac{dz^*}{dr^*} = w^* \tag{3.60}$$

$$\frac{dw^*}{dr^*} = 9\,\phi_{A1}^2 \frac{z^*}{1 + s_{Rm}\dfrac{z^*}{r^*}} \tag{3.61}$$

This set of ordinary differential equations may be integrated by any appropriate numerical ODE-solving algorithm. Applying the shooting method means that integration has to start at $(w^*)_{r^*=0} = s_0^*$, an unknown value. Therefore, a reasonable value for s_0^* has to be chosen, and after integration from $r^* = 0$ to $r^* = 1$, which should lead to $(z^*)_{r^*=1} = 1$, a correction on s_0^* must be applied until this condition is fulfilled. A combination with a Newton iteration for solving the shooting procedure is quite convenient.

If we want to calculate the effectiveness factor, we may rewrite eq 3.32:

$$\eta = \frac{\dfrac{K_m + S_R}{\dfrac{V_p}{A_p} \dfrac{k_2 E_I}{\mathcal{D}_e}} \left(\dfrac{dS}{dr}\right)_R}{S_R} \tag{3.62}$$

After we transform variables to dimensionless form, the same expression becomes:

$$\eta = \frac{1 + s_{Rm}}{3\,\phi_{A1}^2} \left(\frac{ds^*}{dr^*}\right)_{r^*=1} \tag{3.63}$$

or, after we take the new variables into account:

$$\frac{ds^*}{dr^*} = \frac{1}{r^*} \frac{dz^*}{dr^*} - \frac{1}{r^{*2}} z^* \tag{3.64}$$

it is:

$$\eta = \frac{1 + s_{Rm}}{3\,\phi_{A1}^2} \left[\left(\frac{dz^*}{dr^*}\right)_{r^*=1} - 1 \right] \tag{3.65}$$

where the gradient term can be identified with $(w^*)_{r^*=1}$ of eq 3.60.

Other numerical methods may be employed to solve the pore-diffusion problem, for example, that devised by Na and Na [34], which consists in the transformation of the boundary problem to an initial-value problem, and that given by Villadsen and Stewart [35], based on an orthogonal collocation technique, as well as a technique called invariant imbedding [36, 37].

Analytical solutions for the diffusion problem with a reaction of the Michaelis-Menten type are available only for extreme situations. Some of these will be discussed.

If the reaction is of first order ($S_R << K_m$), an analytical solution can be given for the effectiveness factor [31]:

$$\eta = \frac{1}{\phi_{A1}} \left(\frac{1}{\tanh\,(3\,\phi_{A1})} - \frac{1}{3\,\phi_{A1}} \right) \tag{3.66}$$

with ϕ_{A1} defined by eq 3.46.

The development of this formula can be found in any major textbook on the subject. The other extreme for which the reaction may be approximated by zero-order kinetics ($S_R >> K_m$) poses a more difficult problem. The reaction/diffusion equation for a single pore has been solved by Wheeler [38]. He showed that one may obtain the solution by changing the boundary conditions with respect to the concentration gradient (see eq 3.55). The effectiveness for zero-order reaction inside a particle will always be unity as long as there is substrate present at any location. For reaction orders lower than one, substrate depletion can become complete, depending on the product of catalyst activity and time. Therefore, an effectiveness factor lower than unity is obtained for zero-order reaction kinetics only when substrate depletion is complete at the end of a pore or at the center of a catalyst bead. The effectiveness factor in spherical geometry is simply the volume still occupied by substrate divided by the volume of the whole bead:

$$\eta = 1 - \left(\frac{r_c}{R}\right)^3 \tag{3.67}$$

where r_c is the radius only beyond which substrate is still present. The new boundary condition for spherical geometry is therefore:

$$\left(\frac{dS}{dr}\right)_{r=r_c} = 0 \tag{3.68}$$

Unfortunately, this leads to an unpleasant cubic equation for the radius r_c.

The problem of zero-order reaction with pore diffusion in spherical geometry was treated first by Krasuk and Smith [39] and by Weekman and Gorring [40], as well as by Onda *et al.* [41]. The only closed analytical solutions have been presented by Onda *et al.* [41], and Yamané *et al.* [42]. Translated into our nomenclature, it is:

$$\eta = 1 - \left[\frac{1}{2} + \cos\left\{\frac{2}{3}\left[\arccos\left(\sqrt{\frac{2}{3}}\,\frac{1}{\phi_{A0}}\right) + 2\pi\right]\right\}\right]^3 \tag{3.69}$$

with:

$$\phi_{A0} \geq \sqrt{\frac{2}{3}}, \quad \text{else} \quad \eta = 1 \tag{3.70}$$

This solution describes well the expected asymptotic behavior for high moduli, as will be given in eq 3.83. Other solutions presented so far do not describe this behavior due to the assumption of an incorrect boundary condition [41, 42].

In the following discussion we shall focus on some limiting cases with respect to the mass-transfer modulus. A rule of thumb says that for a first-order Aris modulus of less than 0.3, mass-transfer phenomena may be neglected. In this case the diffusional flux is much more important than the substrate consumption rate, and the effectiveness factor does not deviate much from unity. On the other hand, if the first-order Aris modulus takes values beyond 3, severe mass-transfer limitation is encountered. This is the case for highly active catalysts for which the diffusional flux limits the overall rate of reaction.

For severe mass-transfer limitation it may be assumed that the reaction takes place only in the outer layer of the particles. The effective reaction zone may be so small that the curvature of the spherical catalyst particle may be neglected. This reduces the model to:

$$\frac{d^2 S}{dr^2} = \frac{r_v^r}{\mathfrak{D}_e} \tag{3.71}$$

If we use the relation:

$$\frac{d^2 S}{dr^2} = \frac{d}{dr}\left(\frac{dS}{dr}\right) = \frac{1}{2}\frac{d}{dS}\left(\frac{dS}{dr}\right)^2 \tag{3.72}$$

we obtain:

$$\frac{d}{dS}\left(\frac{dS}{dr}\right)^2 = \frac{2}{\mathfrak{D}_e} r_v^r \tag{3.73}$$

If we further assume that the substrate concentration may drop to essentially zero beyond the effective reaction zone, we may integrate eq 3.73 simply as:

$$\left(\frac{dS}{dr}\right)_R = \sqrt{\frac{2}{\mathfrak{D}_e}\int_0^{S_R} r_v^r\, dS} \tag{3.74}$$

For Michaelis-Menten kinetics, this becomes:

$$\left(\frac{dS}{dr}\right)_R = \sqrt{\frac{2}{\mathfrak{D}_e} k_2 E_I \int_0^{S_R} \frac{S}{K_m + S}\, dS} \tag{3.75}$$

or, explicitly:

$$\left(\frac{dS}{dr}\right)_R = \sqrt{2}\sqrt{\frac{k_2 E_I}{K_m \mathfrak{D}_e}}\, K_m \sqrt{s_{Rm} - \ln(1 + s_{Rm})} \tag{3.76}$$

When the observed rate according to eq 3.30 is combined with the maximum rate according to eq 3.31, the effectiveness factor is obtained as:

$$\eta_\infty = \frac{r_v^{obs}}{r_v^R} = \frac{\sqrt{2}}{\frac{R}{3}\sqrt{\frac{k_2 E_I}{K_m \mathfrak{D}_e}}} \frac{1 + s_{Rm}}{s_{Rm}} \sqrt{s_{Rm} - \ln(1 + s_{Rm})} \tag{3.77}$$

or:

$$\eta_\infty = \frac{\sqrt{2}}{\Phi_{A1}} \frac{1 + s_{Rm}}{s_{Rm}} \sqrt{s_{Rm} - \ln(1 + s_{Rm})} \tag{3.78}$$

If we compare the solution with the definition of the Bischoff modulus of eq 3.50, we see that the effectiveness factor for high moduli in the case of Michaelis-Menten kinetics may be expressed as:

$$\eta_\infty = \frac{1}{\Phi_{BA}} \qquad \left(= \frac{3}{\Phi_{BT}} \right) \tag{3.79}$$

This shows that the Bischoff modulus leads to a single straight-line solution for high moduli, regardless of the order of magnitude of the saturation parameter (s_{Rm}).

Of further interest is how the effective observable rate would change in such a mass-transfer-limited situation. This rate is obtained again by eq 3.30, and after a small rearrangement it can be written as:

$$r_v^{obs} = \frac{3}{R} \sqrt{2 k_2 E_I \mathfrak{D}_e K_m} \sqrt{s_{Rm} - \ln(1 + s_{Rm})} \tag{3.80}$$

In the case where a zero-order approximation holds ($S_R >> K_m$), the expression is simplified because of:

$$\lim_{s_{Rm} \to \infty} [s_{Rm} - \ln(1 + s_{Rm})] = s_{Rm} \tag{3.81}$$

to give:

$$r_v^{obs} = \frac{3}{R} \sqrt{2 k_2 E_I \mathfrak{D}_e S_R} \tag{3.82}$$

For a zero-order approximation, the observed rate should change with the square root of the bulk substrate concentration. The effectiveness factor is obtained for this zero-order-approximation limiting case as:

$$\eta_\infty = \frac{\sqrt{2}}{\Phi_{A0}} \qquad \left(= \frac{3\sqrt{2}}{\Phi_{T0}} \right) \tag{3.83}$$

In the case for which the kinetics can be described by a first-order approximation ($S_R << K_m$), the term in square brackets of eq 3.81 tends to:

$$\lim_{s_{Rm} \to 0} [s_{Rm} - \ln(1 + s_{Rm})] = \frac{s_{Rm}^2}{2} \tag{3.84}$$

and the observable rate is obtained as:

$$r_v^{obs} = \frac{3}{R} \sqrt{\frac{k_2}{K_m} E_I \mathfrak{D}_e} \; S_R \tag{3.85}$$

Thus, the linear dependence of rate and substrate concentration is preserved. In consequence, the effectiveness factor depends linearly on the inverse of the Aris modulus of first order:

$$\eta_{\infty} = \frac{1}{\phi_{A1}} \left(= \frac{3}{\phi_{T1}} \right) \tag{3.86}$$

Beyond the formalism discussed, it would be of interest to arrive at a relation which contains only directly measurable quantities. This would be particularly useful since one knows the rate constants for the reaction only after the data are taken, but these data might contain diffusional effects. Therefore, it is important for us to be able to evaluate the severity of diffusion, based on rate measurements alone and not knowing the actual intrinsic kinetics. This knowledge, obviously, may be available only in rare cases.

When the gradient of the substrate concentration at the catalyst surface is expressed in terms of the observable reaction rate according to eqs 3.30 and 3.62:

$$\left(\frac{ds^*}{dr^*} \right)_{r^*=1} = \frac{R^2}{3} \frac{r_v^{obs}}{\mathcal{D}_e S_R} \tag{3.87}$$

and if this equation is introduced into eq 3.63, it yields:

$$\eta = \frac{\dfrac{R^2}{3} \dfrac{r_v^{obs}}{\mathcal{D}_e S_R}}{3\phi_{A1}^2} (1 + s_{Rm}) \tag{3.88}$$

A new modulus, originally proposed by Wagner, and Weisz and Prater, and called the *Weisz modulus* [45], may be defined (here in the form of an Aris-type modulus of first order):

$$\Phi_W = \eta \phi_{A1}^2 = \frac{R^2}{9} \frac{r_v^{obs}}{\mathcal{D}_e S_R} \tag{3.89}$$

It contains only the particle size or the specific surface area, the diffusion coefficient, and the bulk substrate concentration, in addition to the observable rate of reaction. We can calculate the Weisz modulus and estimate the effectiveness of a given catalyst by referring to graphical correlations of η and Φ_W, if we know the effective diffusion coefficient.

With multipoint measurements of the observable reaction rate, as a function of both bulk substrate concentration and particle size, even the intrinsic kinetic parameters as well as the effective diffusion coefficient may be estimated. We can also make a rough estimation of the effectiveness factor if we apply eq 3.89 together with the corresponding definition for zero-order kinetics, ϕ_{A0}. Both functions are found in a narrow range [45] (on log-log plots, maximum range is a factor of 2) so one can roughly estimate the effectiveness even without knowing the saturation parameter. For more precise evaluations, eq 3.89 has to be corrected for the saturation parameter.

In practice, the main aim will be to avoid internal mass-transfer limitations. This can be done if immobilized enzyme preparations of small particles are used. Obviously, there is a certain limit with respect to particle size reduction, due to increasing difficulties of catalyst recovery. Another method consists of designing carriers composed of a solid core and a thin external porous layer [46], in order to limit the diffusional penetration depth. We can achieve the same end by controlling the immobilization conditions in such a way that we obtain an asymmetric enzyme distribution, with maximal concentration at the surface of the particles. If a support with very small pores is chosen, enzyme may even be excluded from the inner catalyst volume.

Unfortunately, not only enzymes may be excluded from penetrating the inner pores of the support, but also macromolecular substrates. In consequence, immobilized enzymes normally exhibit very low activity with such substrates. Even macroporous catalysts are often not economical due to low diffusivities of macromolecular substrates and steric hindrance with respect to complex formation with the rigidly immobilized enzyme.

A simple test for internal mass-transfer limitation consists of performing enzymatic transformations in a discontinuously operated stirred-tank reactor with the original catalyst, as well as with a comminuted preparation. If the reduction of particle size reveals a significant effect, internal mass-transfer resistance may have to be taken into account.

14.3.3 Combined Influence of Internal and External Mass Transfer

If internal mass transfer has to be taken into account for a given immobilized enzyme preparation, it is likely that external mass transfer effects are also important—depending on the hydrodynamic conditions in the reactor. We may assess numerically the relative magnitude of influence of both external and internal mass transfer by using the *Biot number for mass transfer*, which may be defined as:

$$\mathbf{Bi} = \frac{V_p}{A_p} \frac{k_L}{\mathcal{D}_e} = \frac{\Phi_{An}^2}{\mathbf{Da}_{IIn}} \left(\frac{1}{S_{Rb}} \right)^{n-1} \tag{3.90}$$

This is not the standard definition, but one in which the characteristic length (usually R) is replaced by its shape-generalized form ($L_c = V_p/A_p$). The use of the actual nonstandard definitions for both the Damköhler number and the Biot number is obvious because of the more consistent number system obtained. At large Biot numbers, external mass transfer is fast, hence it may be neglected. At low Biot numbers, external mass transfer may be limiting, and hence internal mass transfer may be neglected.

In a model of these phenomena under steady-state conditions, the molar flux of substrate through the boundary layer has to balance that into the catalyst particle. According to eq 3.30, this can be expressed by means of the concentration gradient at the surface of the particles:

$$\frac{A_p}{V_p} D_e \left(\frac{dS}{dr}\right)_R = k_L \frac{A_p}{V_p} (S_b - S_R) \tag{3.91}$$

Or, written in dimensionless form for catalysts of spherical geometry ($V_p/A_p = R/3$):

$$\left(\frac{ds^*}{dr^*}\right)_{r^*=1} = 3\,\mathbf{Bi}\,\frac{1 - s_{Rb}}{s_{Rb}} \tag{3.92}$$

This equation must replace the boundary condition given in eq 3.55 in order to obtain the appropriate solution when eq 3.54 is solved.

General solutions of the problems are available for at least first- and zero-order reactions. The solution for the first-order approximation of Michaelis-Menten kinetics may be written in the form:

$$\frac{1}{\eta} = \frac{\phi_{A1}}{\dfrac{1}{\tanh(3\,\phi_{A1})} - \dfrac{1}{3\,\phi_{A1}}} + \frac{\phi_{A1}^2}{\mathbf{Bi}} \tag{3.93}$$

The first expression on the right-hand side may be identified with the inverse of the effectiveness factor without external mass transfer as given by eq 3.66. In consequence, eq 3.93 may be written as a sum of resistances due to both the internal and the external mass transfer:

$$\frac{1}{\eta} = \frac{1}{\eta_{int}} + \frac{1}{\eta_{ext}} = \frac{1}{\eta_{int}} + \mathbf{Da}_{II1} \tag{3.94}$$

with the resistance term due to external mass transfer identified as the Damköhler number in accordance with eq 3.11.

The solution for the zero-order approximation is given by Yamané *et al.* [42] and may be written in our nomenclature as:

$$1 - \mathbf{Da}_{II0}\,\eta = \frac{3}{2}\,\phi_{A0}^2\,\{1 - (1 - \eta)^{1/3}\}^2\,[2\,(1 - \eta)^{1/3} + 1] \tag{3.95}$$

with:

$$\phi_{A0} \geq \sqrt{\frac{(2/3)\,\mathbf{Bi}}{(2/3) + \mathbf{Bi}}}\,, \quad \text{else} \quad \eta = 1 \tag{3.96}$$

This equation has to be solved iteratively in order to obtain the effectiveness factor. A solution similar to that given in eq 3.69 may also be obtained for Bi > 2/9 [42]. The only simple solution is the particular one:

$$\eta = 1 - \left(1 - \frac{2}{9\,\phi_{A0}^2}\right)^{3/2} \tag{3.97}$$

for:

$$\mathbf{Bi} = 1/3 \quad \text{and} \quad \text{eq 3.96.} \tag{3.98}$$

In the case of a finite saturation parameter, the problem has to be solved numerically.

14.3.4 Approximate Solutions and Particular Systems

It should be added that many attempts have been made to obtain approximate or model-based semiempirical expressions for the effectiveness factor in the cases of Michaelis-Menten kinetics with and without external mass transfer. The articles of Yamané [47] and Hong [48] may serve as sources for related literature.

We have discussed only the simplest cases of reaction and mass transfer in the last sections. Inhibition by one of the reactants may considerably increase the difficulties of solving the appropriate model equations. If it may be assumed that both substrate and inhibiting reactants exhibit the same diffusivity, solutions may still be readily available by numerical methods [49]. A number of examples are known for which partitioning of the substrate between the aqueous bulk phase and the support of the immobilized enzyme has been found. If this has to be considered, the boundary condition for the surface concentration should be modified. For the solution of this problem, some data, and literature, the article of Yamané *et al.* [42] may be consulted.

Furthermore, there are some cases for which effectiveness factors larger than unity may be found in the presence of external or internal mass-transfer limitation or both. The classical example is the case of substrate inhibition [49]. If the bulk substrate concentration is chosen so that it is larger than that for which the maximum rate would have been obtained, effectiveness factors larger than unity may be found. The same may be observed in the case of enzymatic transformations accompanied by a pH change [50, 51], when the bulk pH is chosen to be suboptimal and the pH change proceeds in the direction of the pH optimum of the enzyme.

14.4 BASIC REACTOR DESIGN PRINCIPLES

Basic reactor modeling can be interpreted as a combination of mass balance and reaction kinetics. In general, open and closed systems must be distinguished. *Closed reactors* are *discontinuously operating (batch) reactors* which are charged once with reactants, and after the reaction has taken place for a given time, the reactor is discharged. *Open reactors* are *continuously operated* devices, normally charged by a constant feed flow, and generating a constant product stream. A classical combination of both types is the *fed-batch reactor* which is open for the feed but closed for the product.

The general mass balance for a reactor in which one substrate, S, is transformed may be written as:

$$\dot{V}_{in} S_{in} = \dot{V}_{out} S_{out} + V_R r_v + V_R \frac{dS}{dt} \tag{4.1}$$

The term on the left-hand side is the mass of substrate entering the reactor per unit time. The first term on the right-hand side is the mass of substrate leaving the reactor per unit time; the second term describes the rate of disappearance of substrate due to the reaction taking place; and the last term stands for substrate accumulation in the reactor. If the reactor is fed with an incompressible fluid like an aqueous phase, and volume changes due to the reaction may be neglected, the mass balance may be written as:

$$\frac{S_{in} - S_{out}}{\tau} = r_v + \frac{dS}{dt} \tag{4.2}$$

with the *space time* given as:

$$\tau = \frac{V_R}{\dot{V}} \tag{4.3}$$

The reciprocal of the space time is called the *space velocity* or, in biotechnology, the *dilution rate*.

14.4.1 Ideal Reactors

Since reactants have to be brought together in the reactor in order to be able to react, mixing obviously plays a key role in reactor design. Mixing has an ambivalent meaning when used with respect to reactor design. Mixing *per se* means that gradients, in particular concentration gradients, are destroyed. It is also connected with the term *backmixing*, which describes the fact that the space time of a volume element passing through a reactor may not be a unique value, but that it has a certain distribution called the *residence time distribution*. Backmixing, in terms of a distribution in space time or time of reaction, describes the statistics of histories the ensemble of molecules may undergo during reaction. It may be caused by the reactor or by the inherent behavior of the system with respect to flow profiles, axial and radial mixing, and diffusional phenomena. On the other hand, backmixing may be deliberately induced, by, for example, mixing or recycling.

Reactors are said to behave *ideally* if backmixing can be neglected, or if it is complete. Ideal reactors of the closed type are assumed to be ideally well-mixed, and their volume elements and molecular entities are assumed to experience the same time of reaction. Open systems can be either ideally well-mixed or not mixed at all. In an ideal *continuous stirred-tank reactor* (*CSTR*), the probability for a volume element to leave the reactor is independent of the time it has been therein. An ideal continuous reactor without backmixing is visualized usually by a tubular reactor experiencing infinite radial mixing but no axial mixing. It is therefore called a *plug-flow reactor* (*PFR*). All volume elements entering this reactor would leave it with an age precisely equivalent to the space time.

14.4.1.1 Well-Mixed Closed Reactors

Since there is no flow into or out of the reactor, the basic mass balance for a discontinuously operated stirred-tank reactor reduces to:

$$-\frac{dS}{dt} = -\frac{d}{dt}S_0(1 - X) = S_0\frac{dX}{dt} = r_v \tag{4.4}$$

according to eq 4.2, and it contains only the terms for the accumulation and the disappearance of substrate. The actual substrate concentration or substrate conversion may then be obtained by integration over the time of reaction:

$$t = S_0 \int_{X_0}^{X_f} \frac{dX}{r_v} \tag{4.5}$$

with the rate of reaction, r_v, given by the kinetic model. Expressed in terms of the catalyst-mass-referred time of reaction:

$$Et = S_0 \int_{X_0}^{X_f} \frac{dX}{r_m} \tag{4.6}$$

this is equivalent to the nomenclature we used for the integral kinetic models as given, for example, in eqs 1.21, 1.38, 1.46, etc. These equations will be reformulated in the next section.

The performance of a reactor may be characterized by its productivity with respect to its volume, which is in this case:

$$L_v = \frac{S_0}{t}(X_f - X_0) = \frac{X_f - X_0}{\int_{X_0}^{X_f} \frac{dX}{r_v}} \tag{4.7}$$

or with respect to the amount of enzyme used:

$$L_m = \frac{S_0}{Et}(X_f - X_0) = \frac{X_f - X_0}{\int\limits_{X_0}^{X_f} \frac{dX}{r_m}} \tag{4.8}$$

Thus, for a reaction which can be described by the extended integral form of the Michaelis-Menten kinetics (primes are omitted for convenience), the enzyme-referred productivity, according to eq 1.38, becomes:

$$L_m = \frac{k_2}{1 + \frac{K_m}{S_0}\frac{1}{X_f - X_0}\ln\left(\frac{1 - X_0}{1 - X_f}\right)} \tag{4.9}$$

It is obvious that the productivity in this case depends on the saturation parameter, S_0/K_m, and a maximum productivity, $L_m = k_2$, is obtained in the limit of a zero-order approximation, $S_0 >> K_m$.

14.4.1.2 Plug-Flow Reactors

Let us consider a plug-flow tubular reactor of cross section A_R, fed with a solution containing a substrate concentration of S_0. The mass balance over an incremental axial volume element spanning the cross section becomes:

$$-\dot{V}dS = \dot{V}S_0 dX = r_v A_R dz \tag{4.10}$$

The loss of substrate must be balanced by the reaction taking place in the volume $A_R dz$. If we assume that flow rate is constant, we find that a certain volume passes the reactor in a certain time, as given by:

$$A_R dz = \dot{V}dt \tag{4.11}$$

When we combine the two equations we obtain:

$$S_0 dX = r_v dt \tag{4.12}$$

With respect to the general mass balance given in eq 4.2, both eqs 4.11 and 4.12 express the local product accumulation as being equal to the disappearance of substrate. This equivalence leads to:

$$\tau = S_0 \int\limits_{X_0}^{X_f} \frac{dX}{r_v} \tag{4.13}$$

We obtain the same model as for the ideal closed reactor, and thus model the plug-flow reactor like the discontinuously operated reactor by simply exchanging the time of reaction for the space time. The solution of eq 4.13 may be written in dimensionless form either as:

$$\mathbf{Da}_{\mathrm{I0}} = \frac{k_2 E \tau}{S_0} = X_f - X_0 + \frac{1}{s_{0m}} \ln\left(\frac{1 - X_0}{1 - X_f}\right), \quad \text{or} \tag{4.14}$$

$$\mathbf{Da}_{\mathrm{I1}} = \frac{k_2 E \tau}{K_m} = s_{0m}(X_f - X_0) + \ln\left(\frac{1 - X_0}{1 - X_f}\right), \quad \text{or} \tag{4.15}$$

$$\mathbf{Da}_{\mathrm{I}} = \frac{k_2 E \tau}{K_m + S_0} = \frac{s_{0m}}{1 + s_{0m}}(X_f - X_0) + \frac{1}{1 + s_{0m}} \ln\left(\frac{1 - X_0}{1 - X_f}\right) \tag{4.16}$$

with the Damköhler numbers of the first kind for Michaelis-Menten kinetics, $\mathrm{Da_I}$, or its integer-order approximations, $\mathrm{Da_{I0}}$ or $\mathrm{Da_{I1}}$, along with the saturation parameter, $s_{0m} = S_0/K_m$. The Damköhler number of the first kind may be interpreted as the ratio of the time constant of the flow system, τ, and the time constant of the reaction, t_R, for maximum (initial) rate of reaction:

$$\mathbf{Da}_{\mathrm{I}} = \frac{\tau}{t_R} = \frac{\tau}{S_0/r_{v0}} = \frac{E\tau}{S_0/r_{m0}} \tag{4.17}$$

The different numbers defined are related by:

$$\mathbf{Da}_{\mathrm{I}} = \frac{s_{0m}}{1 + s_{0m}} \mathbf{Da}_{\mathrm{I0}} = \frac{1}{1 + s_{0m}} \mathbf{Da}_{\mathrm{I1}} \tag{4.18}$$

Equations 4.14–4.16 must be solved iteratively in order to obtain the substrate conversion as a function of the Damköhler number. However, if one of the integer-order approximations may be applied, analytical solutions result. For the zero-order approximation, $s_{0m} \to \infty$, the solution is:

$$X_f = X_0 + \mathbf{Da}_{\mathrm{I0}} \qquad (\mathbf{Da}_{\mathrm{I0}} \leq X_f - X_0) \tag{4.19}$$

and for the first-order approximation, $s_{0m} \to 0$):

$$X_f = 1 - (1 - X_0)\exp(-\mathbf{Da}_{\mathrm{I1}}) \tag{4.20}$$

At steady state there exists a concentration gradient along the axial coordinate of the plug-flow reactor. In consequence, one is less likely to observe ideal behavior in tubular reactors than in closed systems, since for closed reactors backmixing would be related to mixing on the time scale.

14.4.1.3 Well-Mixed Open Reactors

Well-mixed open reactors are usually called *steady-state mixed flow reactors* or *continuous (flow) stirred-tank reactors,* commonly abbreviated *C(F)STR*. Equation 4.2 describes the transient behavior of an ideal CSTR in terms of substrate accumulation:

$$\frac{dS}{dt} = S_0 \frac{X_f - X_0}{\tau} - r_v \tag{4.21}$$

At steady state, the term for the accumulation of reactants vanishes, and the steady-state mass balance results:

$$S_0 \frac{X_f - X_0}{\tau} = r_{vf} \tag{4.22}$$

Thus, for an ideal CSTR, the productivity is equal to the rate of reaction given by the kinetic model:

$$L_m = \frac{S_0}{E\tau}(X_f - X_0) = r_{mf} \tag{4.23}$$

It should be noted that the rate of reaction, r_{mf}, refers to the conditions inside the reactor, which are identical for a well-mixed reactor with the conditions in the exit stream, X_f. We obtain the substrate conversion for a given catalyst-mass-referred space time, $E\tau$, by solving eq 4.22, for example, for the Michaelis-Menten kinetics:

$$S_0 \frac{X_f - X_0}{\tau} = k_2 E_0 \frac{S_0(1 - X_f)}{K_m + S_0(1 - X_f)} \tag{4.24}$$

We can compare the mass balance with that of the plug-flow reactor when we write it in dimensionless form:

$$\mathbf{Da}_{\mathrm{I1}} = \frac{k_2 E\tau}{K_m} = \frac{X_f - X_0}{1 - X_f}[1 + s_{0m}(1 - X_f)] \tag{4.25}$$

It specifies the Damköhler number required in order to achieve a given substrate conversion in a CSTR. This equation may be transformed in order to obtain the substrate conversion as a function of the Damköhler number and the saturation parameter, s_{0m}:

$$X_f = \frac{1}{2}\left[X_0 + \left(1 + \frac{1}{s_{0m}}\right) + \frac{1}{s_{0m}}\mathbf{Da}_{\mathrm{I1}}\right] - \sqrt{\frac{1}{4}\left[X_0 + \left(1 + \frac{1}{s_{0m}}\right) + \frac{1}{s_{0m}}\mathbf{Da}_{\mathrm{I1}}\right]^2 - \left[\left(1 + \frac{1}{s_{0m}}\right)X_0 + \frac{1}{s_{0m}}\mathbf{Da}_{\mathrm{I1}}\right]} \tag{4.26}$$

In the limiting situation of integer-order approximations, the solution simplifies considerably. For the zero-order approximation, $s_{0m} \to \infty$, it is equal to that given by eq 4.19, and thus insensitive to backmixing; for the first-order approximation, $s_{0m} \to 0$, eq 4.26 becomes:

$$X_f = \frac{X_0 + \mathrm{Da}_{I1}}{1 + \mathrm{Da}_{I1}} \tag{4.27}$$

The mass balance for continuous stirred-tank reactors and enzyme kinetics may most conveniently be formulated as a quadratic or cubic equation in terms of the final substrate conversion:

$$a_3 X_f^3 + a_2 X_f^2 + a_1 X_f + a_0 = 0 \tag{4.28}$$

Corresponding formulae are given in Table 14.2 for some representative enzyme kinetics. These formulae apply for the case of vanishing initial substrate conversion ($X_0 = 0$). They reveal clearly the mathematical structure of the models.

TABLE 14.2 Mass Balance for Continuous Stirred-Tank Reactors with Enzyme Kinetics, Expressed As:

$$a_3 X_f^3 + a_2 X_f^2 + a_1 X_f + a_0 = 0\,, \qquad \text{for } X_0 = 0$$

Inhibition	a_3	a_2	a_1	a_0
none	0	1	$-\left(1 + \frac{1}{s_{0m}} + \mathrm{Da}_{10}\right)$	Da_{10}
product compet.	0	$1 - \frac{K_m}{K_{ic}}$	$-\left(1 + \frac{1}{s_{0m}} + \mathrm{Da}_{10}\right)$	Da_{10}
product uncompet.	$\frac{S_0}{K_{ia}}$	$1 - \frac{S_0}{K_{ia}}$	$-\left(1 + \frac{1}{s_{0m}} + \mathrm{Da}_{10}\right)$	Da_{10}
product noncompet.	$\frac{S_0}{K_{in}}$	$1 - \frac{K_m}{K_{in}} - \frac{S_0}{K_{in}}$	$-\left(1 + \frac{1}{s_{0m}} + \mathrm{Da}_{10}\right)$	Da_{10}
product mixed	$\frac{S_0}{K_{iu}}$	$1 - \frac{K_m}{K_{ic}} - \frac{S_0}{K_{iu}}$	$-\left(1 + \frac{1}{s_{0m}} + \mathrm{Da}_{10}\right)$	Da_{10}
substrate uncompet.	$-\frac{S_0}{K_{is}}$	$1 + 2\frac{S_0}{K_{is}}$	$-\left(1 + \frac{1}{s_{0m}} + \frac{S_0}{K_{is}} + \mathrm{Da}_{10}\right)$	Da_{10}

14.4.1.4 Comparison of Plug-Flow and Well-Mixed Open Reactors

The two types of ideal reactors, the plug-flow reactor (PFR) and the continuous stirred-tank reactor (CSTR) may be compared on the basis of their productivity at a given substrate conversion. According to the definition of productivity in eq 4.8, this reduces to comparing the catalyst-mass-referred space times required:

$$\frac{L_m^{\mathrm{PFR}}}{L_m^{\mathrm{CSTR}}} = \frac{E\tau\,(\mathrm{CSTR})}{E\tau\,(\mathrm{PFR})} = \frac{\mathrm{Da}_{\mathrm{II}}^{\mathrm{CSTR}}}{\mathrm{Da}_{\mathrm{II}}^{\mathrm{PFR}}} \tag{4.29}$$

The required space time in a PFR is given by eq 4.13, and that for a CSTR by eq 4.22. Thus, the performance ratio in general is given by:

$$\frac{L_m^{\mathrm{PFR}}}{L_m^{\mathrm{CSTR}}} = \frac{X_f - X_0}{r_{mf}} \, \frac{1}{\int_{X_0}^{X_f} \frac{dX}{r_m}} \tag{4.30}$$

The principal difference between the two reactors is that in the case of the PFR, the substrate concentration changes gradually from S_0 to the final value, $S_f = S_0X_f$, whereas the catalyst in a CSTR operates at the outlet substrate concentration, S_f (that is, at the lowest concentration possible). Since, for most kinetic models, the rate of reaction increases with substrate concentration, the PFR operates at higher mean rate (or productivity) than the CSTR. This can be deduced mathematically from eq 4.30, where $1/r_{mf}$ most commonly is found to be larger than the integral $\int(1/r_m)dX$, because the inverse rate as a function of substrate conversion normally is a steadily increasing function.

For Michaelis-Menten kinetics, eq 4.29 can be written explicitly:

$$\frac{L_m^{\mathrm{PFR}}}{L_m^{\mathrm{CSTR}}} = \frac{\mathrm{Da}_{\mathrm{II}}^{\mathrm{CSTR}}}{\mathrm{Da}_{\mathrm{II}}^{\mathrm{PFR}}} = \frac{X_f - X_0}{1 - X_f} \, \frac{1 + s_{0m}(1 - X_f)}{s_{0m}(X_f - X_0) + \ln\left(\dfrac{1 - X_0}{1 - X_f}\right)} \tag{4.31}$$

In the limiting case for which the zero-order approximation, $s_{0m} \to \infty$, may be applied, the performance ratio is unity, since the rate of reaction is independent of substrate concentration and hence of conversion. The same situation is found if the final conversion is only gradually different from the initial conversion. For the case of the first-order approximation, $s_{0m} \to 0$, eq 4.31 would simplify to:

$$\frac{L_m^{\mathrm{PFR}}}{L_m^{\mathrm{CSTR}}} = \frac{X_f - X_0}{1 - X_f} \, \frac{1}{\ln\left(\dfrac{1 - X_0}{1 - X_f}\right)} \tag{4.32}$$

Consequently, in almost any case the plug-flow reactor will outperform the continuous stirred-tank reactor. The inverse becomes true only for reactions exhibiting negative-order kinetics. A classical example of negative-order enzyme kinetics is the substrate inhibition model. For a pure anticompetitive substrate inhibition pattern, the model is:

$$S_0 \frac{dX}{dt} = k_2 E_0 \frac{S_0 (1 - X)}{K_m + \left(1 + \frac{S_0}{K_{is}}(1 - X)\right) S_0 (1 - X)} \tag{4.33}$$

Under certain conditions the square of the substrate concentration in the denominator leads to negative-order overall behavior. This condition is given by:

$$S_0 > \sqrt{K_m K_{is}} \tag{4.34}$$

It describes the point for which the rate of reaction as a function of substrate concentration takes a maximum value. Beyond this concentration, the reaction velocity decreases with increasing substrate concentration. Under the condition of eq 4.34, the reaction rate would first increase with substrate conversion until it reached the maximum value at:

$$X_{opt} = 1 - \frac{1}{S_0} \sqrt{K_m K_{is}} \tag{4.35}$$

and beyond this point it would decrease steadily.

A CSTR would exhibit a maximum productivity at the optimum conversion, X_{opt}, since rate of reaction and productivity are equivalent. We may derive equations 4.34 and 4.35 by setting the first derivative of the rate law with respect to substrate concentration and conversion, respectively, to zero.

With eq 4.35 the maximum productivity in a CSTR is obtained as:

$$L_m^{max}\,(\mathrm{CSTR}) = \frac{k_2}{1 + 2\sqrt{\frac{K_m}{K_{is}}}} \,, \qquad X_f = X_{opt} \tag{4.36}$$

For a PFR, unfortunately, the derivation is more difficult. The required Damköhler number for a substrate-inhibited reaction in a PFR is given by:

$$\mathrm{Da}_{\mathrm{II}}^{\mathrm{PFR}} = s_{0m}\left(1 + \frac{S_0}{K_{is}}\right)(X_f - X_0) + \ln\left(\frac{1 - X_0}{1 - X_f}\right) - \frac{1}{2}\, s_{0m} \frac{S_0}{K_{is}} (X_f^2 - X_0^2) \tag{4.37}$$

The productivity may then be calculated as:

$$L_m^{\mathrm{PFR}} = \frac{S_0 (X_f - X_0)}{\frac{K_m}{k_2} \mathrm{Da}_{\mathrm{II}}^{\mathrm{PFR}}} = k_2 s_{0m} \frac{X_f - X_0}{\mathrm{Da}_{\mathrm{II}}^{\mathrm{PFR}}} \tag{4.38}$$

The optimum productivity is obtained when the first derivative of eq 4.38 with respect to substrate conversion vanishes. This condition is written as:

$$\frac{1}{(X_{\mathrm{opt}} - X_0)(1 - X_{\mathrm{opt}})} + \frac{\ln\left(\frac{1 - X_{\mathrm{opt}}}{1 - X_0}\right)}{(X_{\mathrm{opt}} - X_0)^2} = \frac{1}{2} \frac{S_0^2}{K_m K_{is}} = \frac{1}{2} s_{0m} \frac{S_0}{K_{is}} \tag{4.39}$$

and must be solved iteratively for the optimum substrate conversion, X_{opt}. This conversion is a special point, as we demonstrate if we take the derivative of the general definition of the productivity of a PFR given in eq 4.8:

$$\frac{dL_m^{\mathrm{PFR}}}{dX_f} = \frac{d}{dX_f}\left[\frac{X_f - X_0}{\int_{X_0}^{X_f} \frac{dX}{r_m}}\right] \tag{4.40}$$

and find where it vanishes ($X_f = X_{\mathrm{opt}}$). This leads to the condition:

$$\int_{X_0}^{X_f} \frac{dX}{r_m} = \frac{X_f - X_0}{r_{mf}} , \qquad X_f = X_{\mathrm{opt}} \tag{4.41}$$

or, rearranged:

$$r_{mf} = \frac{X_f - X_0}{\int_{X_0}^{X_f} \frac{dX}{r_m}} , \qquad X_f = X_{\mathrm{opt}} \tag{4.41a}$$

This shows that at the point of maximum performance of the PFR, both types of reactors, PFR and CSTR, exhibit equal performance. Equation 4.41 may be best visualized by a graph of the inverse of the rate of reaction as a function of conversion, which gives a curve with a minimum. The space time required for a PFR is the integral under this curve, whereas that required for a CSTR is a rectangle. In consequence, the continuous stirred-tank reactor outperforms the plug-flow reactor for substrate conversions lower than that given by eqs 4.39 or 4.41. Beyond this conversion, the PFR shows better performance. In practice, however, the substrate conversion is usually limited by economic considerations. In addition, the optimum conversion is relatively high only in cases of severe substrate inhibition. Here, also, the plug-flow reactor will normally be better. The most productive configuration, however, would be a combination of a CSTR followed by a PFR. The CSTR would work at the point of optimum conversion given by eq 4.35 and the PFR would do the rest. This strategy minimizes the required overall catalyst-mass-referred space time.

Another peculiar feature is connected with the appearance of an overall negative order of reaction: if these reactions are performed in a continuous stirred-tank reactor, multiple steady states may occur. If we look at the mass balance for the substrate-inhibited reaction, as given in Table 14.2, we see that it is distinctly different from other mass balances, which can be described by a cubic equation. In fact, it may yield two or even three real solutions, depending on the parameters of the model and the reactor operating variables [52]. We can easily visualize this by plotting the mass balance according to eq 4.22 in the form:

$$\frac{S_0}{Et} X = r_m = f(X) \tag{4.42}$$

where the left-hand side, the convection term, is a straight line as a function of conversion, and intersects the rate term ($r_m = f(X)$) at at least one distinct substrate conversion. The point(s) of intersection is (are) the steady-state solution of the mass balance of the CSTR. If three real roots are obtained, the reactor may operate at either the lowest or the highest conversion obtained, with the intermediate solution being unstable. Whether the reactor will operate at the one or the other substrate conversion is fixed by the history of the reactor. Changing the space time can result in sudden steps and hysteresis effects on substrate conversion.

14.4.2 Backmixing in Nonideal Reactors

The term *backmixing* originally stems from the observation in tubular reactors that the ideal path of reactant molecules through the reactor is perturbed by diffusion, as well as by free and forced convection, in addition to the perturbation due to the establishment of velocity profiles. All these phenomena cause a flattening of the ideal axial reactant concentration profiles. In addition, they cause different reactant molecules to experience different reaction environments and different residence times. Backmixing, therefore, usually decreases the reactor performance. The only case for which backmixing has no influence on reactor performance is a reaction of zero order.

The key to understanding and measuring residence time distributions lies in tracer experiments. In almost all cases, inert tracers are employed. These are assumed not to interfere with the reaction system and the reactor construction material. Inert tracers must neither react nor adsorb on any internal structure of the reactor, and they must be homogeneously soluble in the fluid passing through the reactor. Tracer experiments and their evaluation may be performed in many different ways. Only the basic tools will be provided here. Usually, a tracer is introduced into the feed stream of the reactor and its concentration is analyzed in the exit stream. The reactor is operated at constant flow during each experiment.

Since the entrance signal must be known, the tracer is normally introduced by either of two methods: the injection of a small quantity of tracer just at the reactor entrance, or the rapid switching from a normal feed solution to one doped with a constant concentration of tracer. In the first case, the entrance signal is commonly supposed to behave like an *ideal Dirac pulse*, and like an *ideal step function* in the second case. These entrance signals are transformed in more or less characteristic ways by their passing through the reactor. The transforming function associated with the reactor, or, in other words, the flow model of the reactor, is called the *weight function* and is identical with the *E function* of the system. Since in most interesting cases in which an ideal tracer function is applied, the weight function is obtained directly from the outlet tracer signal, the individual functions will be distinguished only where necessary.

In the case of applying a tracer step function, the signal at the reactor outlet will be a function which may be zero for a lag period and then increases until it reaches the tracer concentration of the feed tracer solution, c_0. The normalized tracer response for a positive tracer step is called the *cumulative external residence time distribution*, or simply the *F function*:

$$F = \frac{c_F}{c_0}, \quad \text{with} \quad c_F = f(t) \tag{4.43}$$

According to this definition, F may take values between zero and one. The value of F for a given time t expresses the fraction of the fluid in the outflow which has a residence time in the reactor of less than t.

Associated with this probability distribution F is the probability density function E:

$$E = \frac{dF}{dt} = \frac{1}{c_0}\frac{dc_F}{dt} \tag{4.44}$$

The E function is, due to historical reasons, called the *residence time distribution* (*RTD*) or *exit age distribution*, although it represents a probability density function. According to eq 4.44, the F function may also be obtained by integration over the E function:

$$F = \int_0^t E\,dt' \tag{4.45}$$

The RTD is usually obtained from one-shot tracer experiments. Since we say the concentration of tracer at the reactor outlet is $c_E = f(t)$ after tracer injection at time zero, we can also define the E function as:

$$E = \frac{\dot{n}}{n_0} = \frac{\dot{V} c_E}{\int_0^\infty \dot{V} c_E \, dt} = \frac{c_E}{\int_0^\infty c_E \, dt} = f(t) \tag{4.46}$$

using $\dot{n}$ and n_0 as the flux of tracer at the reactor exit and the amount of tracer, respectively. The equivalence of the third and fourth expressions in eq 4.46 is correct only if tracer has been introduced proportionally to the local flow rate, in the case of velocity profiles at the injection point, and if the tracer was measured by taking the mean value over the total flow. The term *Edt* gives the fraction of the outflow having residence times in the range of t to $t+dt$. Practically, the RTD is the tracer response obtained from a one-shot, pulse, input tracer signal normalized by the amount of tracer estimated by integration over the output signal:

$$n_0 = \dot{V} \int_0^\infty c_E \, dt \tag{4.47}$$

which leads to:

$$E = \frac{c_E}{\int_0^\infty c_E \, dt} = \frac{\dot{V}}{n_0} c_E \tag{4.48}$$

Although internal age distributions may be defined [53], we shall limit our discussion to the E and F functions. Often, these functions are defined with respect to a normalized time, $\theta = t/\bar{t}$, where $\bar{t}$ is the mean residence time. The two functions are related by:

$$E_\theta = \bar{t} E \quad \text{and} \quad F_\theta = F \tag{4.49}$$

The mean residence time $\bar{t}$ is used in parallel with the space time τ because $\bar{t}$ is only an estimate, whereas τ is a value fixed by the operating variables of the reactor.

The main properties of the E function may be given as follows:

$$\int_0^\infty E \, dt = 1 \tag{4.50}$$

$$\bar{t} = \int_0^\infty t E \, dt \tag{4.51}$$

$$\sigma_t^2 = \int_0^\infty (t - \bar{t})^2 \, E \, dt \tag{4.52}$$

$$\sigma_\theta^2 = \frac{\sigma_t^2}{\bar{t}^2} \tag{4.53}$$

with the variance of the RTD, σ_t^2, and the dimensionless variance, σ_θ^2.

We may write the same relations in terms of the F function by substituting eq 4.45 into eqs 4.50–4.52, and integrating between the limits of $F = 0$ and $F = 1$. Since data are usually gathered equidistant in time and not in F, integration over time would be more convenient. This can be done by reformulation and partial integration [54]:

$$\int_0^1 dF = -\int_0^1 d(1 - F) \tag{4.54}$$

$$\begin{aligned} \bar{t} &= \int_0^1 t\,dF = -\int_0^1 t\,d(1 - F) \\ &= -[t(1 - F)]_{F=0}^{F=1} + \int_0^\infty (1 - F)\,dt = \int_0^\infty (1 - F)\,dt \end{aligned} \tag{4.55}$$

$$\begin{aligned} \sigma_t^2 &= \int_0^1 (t - \bar{t})^2 dF = \int_0^1 t^2 dF - 2\bar{t}\int_0^1 t\,dF + \bar{t}^2 \int_0^1 dF \\ &= -[t^2(1 - F)]_{F=0}^{F=1} + 2\int_0^\infty (1 - F)\,t\,dt - 2\bar{t}^2 + \bar{t}^2 \\ &= 2\int_0^\infty (1 - F)\,t\,dt - \bar{t}^2 \end{aligned} \tag{4.56}$$

or, with eq 4.55:

$$\begin{aligned} \sigma_t^2 &= 2\left[\int_0^\infty (1 - F)\,t\,dt - \frac{\bar{t}}{2}\int_0^\infty (1 - F)\,dt\right] \\ &= 2\int_0^\infty (1 - F)\left(t - \frac{\bar{t}}{2}\right)dt \end{aligned} \tag{4.57}$$

The practical procedure for data evaluation in the case of a one-shot tracer input consists in integrating the response signal over time until the tracer signal drops again to zero. This integral normalizes the response signal to obtain the experimental E function defined by eq 4.48. The E function may then be used to calculate the mean residence time and the variance of the distribution according to eqs 4.51–4.53. In the case of a positive tracer step input, we can normalize the response signal using the tracer concentration in the feed stream to obtain the F function. We obtain the mean residence time and the variances by applying eqs 4.55, 4.57, and 4.53.

What is still lacking are correlations of these mixing parameters, in particular the variance of the RTD, with reactor models for estimating the magnitude of backmixing. We shall first consider those models derived from ideally behaving reactors.

Injection of a Dirac pulse of tracer into an ideal continuous flow stirred-tank reactor at time zero would change the tracer concentration in the reactor, and thus in the exit stream, instantaneously from zero to an initial value given by the mass of tracer and the vessel volume ($c_{E0} = n_0/V_R$). The tracer response would then change according to the CSTR's mass balance:

$$-V_R \frac{dc_E}{dt} = \dot{V} c_E \tag{4.58}$$

according to eq 4.2 with a feed solution devoid of tracer. Thus, the response signal changes as:

$$c_E = c_{E0} \exp\left(-\frac{t}{\tau}\right) \tag{4.59}$$

The integral of this signal over time until infinity yields:

$$\int_0^\infty c_E \, dt = \frac{c_{E0}}{\tau} \tag{4.60}$$

The E function according to eq 4.48 is then written:

$$E = \frac{1}{\tau} \exp\left(\frac{t}{\tau}\right) \tag{4.61}$$

The F function may be obtained by integration of the E function with time:

$$F = 1 - \exp\left(\frac{t}{\tau}\right) \tag{4.62}$$

Introduction of eq 4.61 into eqs 4.51–4.53 yields the identity $\bar{t} = \tau$, and $\sigma_\theta^2 = 1$. However, the behavior of well-mixed reactors should not be investigated in this way, the reason being that the exponential decay of tracer concentration would necessitate a very long and very precise measurement. Stopping integration too early or imprecision of the baseline would lead to large errors, particularly in the estimation of the variance. The easiest and most commonly employed method for checking well-mixed reactors involves applying the mass balance given in eq 4.59 directly. A log-linear plot of tracer concentration against time would yield the space time as the slope of a straight line, and would also indicate deviations from ideality.

The RTD of tubular reactors is most commonly modeled by means of the so-called *plug-flow dispersion model*. The main assumptions of this model are that radial tracer gradients do not exist and that the axial dispersion of tracer may be described as a diffusional process. The governing partial differential equation for this situation in terms of tracer concentration is:

$$\frac{\partial c}{\partial \theta} = \frac{1}{\mathrm{Bo}} \frac{\partial^2 c}{\partial z^{*2}} - \frac{\partial c}{\partial z^*} \tag{4.63}$$

with the dimensionless coordinates $z^* = z/L_R$ and $\theta = t/\tau$. The model parameter is the *Bodenstein number* (often also called the *axial Peclet number*), defined as:

$$\mathrm{Bo} = \frac{\bar{u} L_R}{D_{ax}} = \frac{1}{\tau} \frac{L_R^2}{D_{ax}} \tag{4.64}$$

with the linear fluid velocity, $\bar{u}$, the length of the tubular reactor or the measurement section, L_R, and the axial dispersion coefficient, D_{ax}, which is a kind of effective axial diffusion coefficient. A small Bodenstein number is obtained for high degrees of backmixing, whereas a large Bodenstein number reflects the approach of plug-flow conditions. The solution of eq 4.63 depends on the boundary conditions which are applied at the point of tracer injection and measurement. Details are found in the specialized literature. We shall consider only two situations: systems which are either closed or open on both sides, the points of tracer injection and measurement.

The closed system is characterized by plug-flow conditions at the two points of reference, whereas in an open system there is no discontinuity in dispersion, including the two points of reference. This signifies that in an open system the tracer may pass the point of injection and the point of tracer detection not only by forced convection but also by diffusion. It should be noted again that the mode of tracer injection and detection, and the disturbance by probes, may influence the theoretical as well as the experimental results quite considerably, particularly in the case of open systems.

In the limit of small extents of dispersion, the RTD is a symmetrical function given by:

$$E = \sqrt{\frac{\mathrm{Bo}}{4 \pi \tau^2}} \exp\left[- \frac{\mathrm{Bo}}{4} \left(1 - \frac{t}{\tau}\right)^2 \right] \tag{4.65}$$

with mean: $\bar{t} = \tau$ and variance: $\sigma_\theta^2 = \dfrac{2}{\mathrm{Bo}}$ (4.66)

This model may be used if the Bodenstein number is larger than 100.

In the case of normal and large extents of dispersion, an analytical solution is not available for the closed system, but the mean is $\bar{t} = \tau$ and the variance:

$$\sigma_\theta^2 = \frac{2}{\mathrm{Bo}}\left\{1 - \frac{1}{\mathrm{Bo}}[1 - \exp(-\mathrm{Bo})]\right\} \tag{4.67}$$

For an open system, the normalized response signal on a Dirac tracer pulse is not identical to the RTD (weight function) of the system due to backmixing at the sites of injection and measurement. Thus, we may call the normalized observable tracer response the *C function*. It is given by [55]:

$$C = \frac{c_E}{\int_0^\infty c_E\,dt} = \frac{1}{2}\left(1 + \frac{\tau}{t}\right)\sqrt{\frac{\mathrm{Bo}}{4\pi\tau t}}\exp\left[-\frac{\mathrm{Bo}}{4}\left(1 - \frac{t}{\tau}\right)^2\frac{\tau}{t}\right] \tag{4.68}$$

with mean and variance:

$$\bar{t} = \tau\left(1 + \frac{1}{\mathrm{Bo}}\right) \quad \text{and} \quad \sigma_\theta^2 = \frac{2}{\mathrm{Bo}} + \frac{6}{\mathrm{Bo}^2} \tag{4.69}$$

The dispersion at the points of reference leads to the fact that the mean value of the RTD does not coincide with the space time, but depends on the extent of dispersion. The E function itself is written as [55, 56]:

$$E = \sqrt{\frac{\mathrm{Bo}\,\tau}{4\pi t^3}}\exp\left[-\frac{\mathrm{Bo}}{4}\left(1 - \frac{t}{\tau}\right)^2\frac{\tau}{t}\right] \tag{4.70}$$

with mean: $\bar{t} = \tau$ and variance: $\sigma_\theta^2 = 2/\mathrm{Bo}$. It may serve for parameter estimation by the two-point tracer measurement technique and the convolution method, as will be briefly discussed later.

Another very useful model of backmixing is the *tanks-in-series model*. If a cascade of equal-size, perfectly mixed vessels is assumed, the RTD of this reactor would be given by:

$$E = \frac{N}{(N-1)!}\,\frac{\left(N\frac{t}{\tau}\right)^{N-1}}{\tau}\exp\left(-N\frac{t}{\tau}\right) \tag{4.71}$$

with N the number of tanks between the point of tracer injection and the point of tracer detection, and τ the total space time between these points. The mean of this RTD is $\bar{t} = \tau$, and the variance $\sigma_\theta^2 = 1/N$. For $N = 1$, the ideal CSTR is obtained; with increasing number of tanks, the plug-flow condition is approached. In the limit of very low dispersion, we look for the relationship between the Bodenstein number and the number of tanks by using eq 4.66, and find $N = \mathrm{Bo}/2$.

Three other convenient methods for the evaluation of tracer experiments may be worth mentioning. If the normalized tracer response function from a Dirac tracer pulse is given by an analytical expression, nonlinear parameter estimation techniques may be applied in order to obtain the Bodenstein number. The mean residence time may also be estimated in order to check for model validity. This method, combined with a comparison of the experimental data and the model function by means of computer graphics, reveals many more details of the underlying phenomena than reliance on the mean value and variance of an RTD only. A second method deals with the problem that an experimental tracer pulse will never have the ideal shape of a Dirac function. We may solve this problem by introducing two tracer detection points, one for measuring the input signal and one for the output signal. A one-shot tracer injection is performed upstream of the input signal detection point. The mean and variance of both input and output signals are calculated. The characteristic variance for the section between the two points of reference is then given by:

$$\Delta \sigma_\theta^2 = \frac{\sigma^2_{t,\mathrm{out}} - \sigma_{t,\mathrm{in}}}{\left(\bar{t}_{\mathrm{out}} - \bar{t}_{\mathrm{in}}\right)^2} \tag{4.72}$$

which may be correlated with the dimensionless variance of any appropriate and known model E function.

Both techniques, two-point tracer detection and nonlinear parameter estimation, may be combined by means of the convolution integral method. In fact, one of the main assumptions in treating residence-time distributions is that the output signal is a linear transform of the input signal, due to the system weight function, which is identical with the E function. A consequence of this assumption is the additivity of variances for systems composed of series of units. The output signal may then be calculated from the input signal by means of the convolution integral:

$$\begin{aligned} C_{\mathrm{out}}(t) = C_{\mathrm{in}} * E &= \int_0^t C_{\mathrm{in}}(t - t')\, E(t')\, dt' \\ &= \int_0^t C_{\mathrm{in}}(t')\, E(t - t')\, dt' \end{aligned} \tag{4.73}$$

where the time in parentheses refers to the corresponding value of the associated functions. The convolution integral may be visualized from the last expression in eq 4.73: a slice of tracer passing the first detector will appear at the second at a time, and with a distribution, prescribed by the E function of the system. Summing these partial response signals over all slices of tracer entering the system gives the tracer response. We can break equation 4.73 into time-equidistant steps in order to compute the integral. By nonlinear regression methods, we may obtain the model parameters of a given weight function by comparing the experimental response signal with the theoretical value calculated by convolution in the real-time domain. Although this method needs a lot more computation time than the other methods discussed, the advantage of not having to rely on ideal pulse injection and the possibility of graphical representation of the result are obvious.

In most of the cases in which tracer experiments are performed, the main questions are whether the reactor deviates considerably from ideal behavior, and whether there is any malfunction of the system. In other cases, for example, fluidized beds, the tracer experiment should lead to a method which permits determination of the reactor performance. Strictly this is possible only if the rates of reaction are related linearly to the reactant concentrations. For enzymatic reactions this holds only in the limit of the first-order approximation. In this case the final conversion may be calculated by means of the experimentally obtained RTD as:

$$X_f = \int_0^1 X_t \, dF = \int_0^\infty X_t E \, dt \tag{4.74}$$

where X_t is the conversion which would result after time t in a batch reactor for the kinetics given. This conversion is weighted by the fraction of the outflow having the appropriate residence time ($dF = E\,dt$). Summation over all time slices yields the substrate conversion at the reactor exit. We can solve the partial differential equation resulting for a first-order reaction in a tubular reactor, the flow pattern of which may be described by the dispersion model. The solution may be given as:

$$X_f = 1 - \frac{4a \exp\left(\frac{\mathrm{Bo}}{2}\right)}{(1+a)^2 \exp\left(a\frac{\mathrm{Bo}}{2}\right) - (1-a)^2 \exp\left(-a\frac{\mathrm{Bo}}{2}\right)} \tag{4.75}$$

with the parameter:

$$a = \sqrt{1 + 4\frac{\mathrm{Da}_{I1}}{\mathrm{Bo}}} \tag{4.76}$$

containing the Damköhler number of the first-order approximation for enzyme kinetics.

For reactions with nonlinear kinetics, the residence-time distribution itself is generally insufficient for evaluating the reactor performance. The application of eq 4.74 yields a lower bound to conversion, and no exact solution. The general problem may be illustrated by the fact that a system consisting of a continuous-flow stirred-tank reactor and a tubular reactor connected in series would yield the same RTD regardless of how they are arranged. When we recall the earlier discussion of reactor performance in the presence of substrate inhibition (Section 14.4.1.4), we see that this may be quite misleading. The problem is related to the influence of early/late mixing. Therefore, a multiparameter RTD model composed of different compartments in series, in parallel, or both, may be necessary to obtain a reliable description of the actual reactor behavior.

Nevertheless, a very useful reactor model is the tanks-in-series model, because it simplifies evaluation of the reactor performance. In the case of Michaelis-Menten kinetics, the mass balance of a CSTR, as given by eq 4.26, would merely have to be solved a number of times equal to the number of equivalent stirred tanks obtained by tracer experiments.

14.5 NOMENCLATURE

14.5.1 Roman-Letter Symbols

Symbol	Unit	Meaning
a		general constant or model parameter
$\mathbf{a}$		vector of model parameters
a_p	m^{-1}	volume-specific surface area of catalyst particle (A_p/V_p)
A_p	m^2	total external surface area of catalyst particles
A_R	m^3	cross-sectional area of tubular reactors
b		general constant
Bi	-	Biot number for mass transfer $[(V_p/A_p)(k_L/\mathcal{D}_e)]$
Bo	-	Bodenstein number $[\overline{u}L_R/D_{ax}]$
c	mol m^{-3}	concentration
c_0	mol m^{-3}	tracer concentration in feed
c_E	mol m^{-3}	tracer concentration response signal from a Dirac tracer pulse
c_F	mol m^{-3}	tracer concentration response signal from a tracer step function
C	mol m^{-3}	coenzyme concentration
C	s^{-1}	normalized tracer response after Dirac tracer pulse
C_{ij}	-	covariance coefficients
d_R	m	diameter of tubular reactors
D_{ax}	$m^2\ s^{-1}$	axial dispersion coefficient
$\mathcal{D}$	$m^2\ s^{-1}$	diffusion coefficient in solution
$\mathcal{D}_e$	$m^2\ s^{-1}$	effective diffusion coefficient (in catalyst particles)
Da_I	-	Damköhler number of the first kind $[\tau/t_R = E\tau/(S_0/r_{m0})]$
Da_{In}	-	same, for n^{th}-order reaction
Da_{IIn}	-	Damköhler number of the second kind for n^{th}-order reaction $[(V_p/A_p)(kS_b^n)/(k_L S_b)]$

$\mathbf{e}$	kg m^{-3}	vector of concentration of enzyme species
E	s^{-1}	exit age distribution function, E function, $f(t)$
E_θ	-	same, $f(\theta)$
E	kg m^{-3}	enzyme concentration
E_0	kg m^{-3}	total concentration of enzyme species
E_I	kg m^{-3}	enzyme concentration inside catalyst particle $[m_E/V_p]$
ES	kg m^{-3}	concentration of enzyme-substrate complex (same for inhibitors EI, etc.)
Et	kg s m^{-3}	catalyst-referred time of reaction $[E_0t]$
$E\tau$	kg s m^{-3}	catalyst-referred space time $[E_0\tau = m_E/\dot{V}]$
$\mathbf{f}$		vector of residuals
$\mathbf{f}^T\mathbf{f}$		residual sum of squares
F	-	cumulative external residence time distribution function, F function, $f(t)$
F_θ	-	same, $f(\theta)$
$F(a)$		general objective function
$\mathbf{g}$		gradient vector
$\mathbf{G}$		Hessian matrix, second partial derivatives
$\mathbf{H}$		inverse of Hessian matrix
ΔH_R	kJ mol^{-1}	reaction enthalpy
$\mathbf{I}$		identity matrix
I	mol m^{-3}	inhibitor concentration
$\mathbf{j}$		vector of partial derivatives
$\mathbf{J}$		Jacobian matrix, first partial derivatives
k		general rate constant
k_2	mol kg^{-1} s^{-1}	rate constant for rate-limiting step of product formation by decomposition of the enzyme-substrate complex
k_i^+, k_i^-		rate constant for forward or backward reaction step i
k_L	m s^{-1}	mass-transfer coefficient (liquid/solid)
K'	m^3 mol^{-1}	lumped parameter
K_{eq}		equilibrium constant of reaction
K_{ic}	mol m^{-3}	inhibitor constant for a competitive inhibitor
K_{ij}	mol m^{-3}	same, for reaction step j
K_{in}	mol m^{-3}	same, noncompetitive inhibition
K_{is}	mol m^{-3}	same, substrate inhibition
K_{iu}	mol m^{-3}	same, uncompetitive inhibition

K_m	mol m^{-3}	Michaelis-Menten (Briggs-Haldane) constant
K_s	mol m^{-3}	equilibrium constant for the dissociation of the enzyme-substrate complex (true Michaelis-Menten constant)
L_c	m	characteristic length, in general $[V/A]$
L_m	mol kg^{-1} s^{-1}	catalyst-mass-referred productivity
L_v	mol m^{-3} s^{-1}	reactor-volume-referred productivity (space time yield)
L_R	m	length of tubular reactor
m	kg	mass
m_E	kg	mass of active enzyme
$\mathbf{M}$		coefficient matrix
n	-	order of reaction
n_0	mol	injected amount of tracer
$\dot{n}$	mol s^{-1}	molar flow (of tracer)
N	-	number of equivalent stirred vessels in the tanks-in-series model
P	mol m^{-3}	product concentration
q	-	lumped parameter
r	m	radial position (in catalyst particle)
r_c	m	limiting radius for zero-order reaction-diffusion problem
r_m	mol kg^{-1} s^{-1}	enzyme-mass-referred rate of reaction
r_v	mol m^{-3} s^{-1}	reactor-volume-referred rate of reaction
r^*	-	dimensionless radial position $[r/R]$
R	kJ mol^{-1} K^{-1}	general gas constant $(8.3144 \cdot 10^{-3})$
R	m	radius (of catalyst particle)
s_{0m}	-	normalized substrate concentration $[S_0/K_m]$
s_{bm}	-	same, $[S_b/K_m]$
s_{Rb}	-	same, $[S_R/S_b]$
s_{Rm}	-	same, $[S_R/K_m]$
s^*	-	dimensionless substrate concentration $[S_r/S_R]$
s_0^*	-	same, at bead center
S	mol m^{-3}	substrate concentration
$\mathbf{S}$		direction of search (vector)
t	s	time
$\bar{t}$	s	mean residence time

t_R	s	time constant of reaction $[S_0/r_{v0}]$
t_S	-	percentile point of Student's t distribution
T	K	absolute temperature
$\overline{u}$	m s^{-1}	superficial fluid velocity
V	m^3	volume
V_p	m^3	effective volume of catalyst particles
V_R	m^3	reactor volume
$\dot{V}$	m^3 s^{-1}	volumetric flow rate
w^*	-	dimensionless gradient (dz^*/dr^*)
$\mathbf{w}$		gradient (vector)
X	-	substrate conversion
X_t	-	same, in a discontinuously operating reactor at operating time t
y_c ; $\mathbf{y}_c$		calculated dependent variable; vector of
y_i ; $\mathbf{y}$		observed dependent variable; vector of
Y	-	yield
z	m	general length coordinate
z^*	-	dimensionless variable $[r^* s^*]$

14.5.2 Greek-Letter Symbols

α		step width, iteration parameter
δ	m	concentration boundary layer thickness
ε_p	-	porosity of catalyst particle
η	-	(catalyst) effectiveness factor
η_∞	-	same, for high Thiele moduli
θ	-	dimensionless operating time $[t/\tau]$
σ_{ij}^2		elements of variance-covariance matrix
σ_{obs}^2		variance of observations
σ_t^2	s^2	variance of E function over time

σ_M^2		variance-covariance matrix
σ_θ^2	-	dimensionless variance of E function
τ	s	space time $[V_R/\dot{V}]$
τ_p	-	tortuosity factor
ϕ_{An}	-	shape-generalized Thiele modulus, "Aris modulus"
$\phi_{BT,A}$	-	Bischoff modulus defined according to Thiele/Aris
$\phi_{OT,A}$	-	order-generalized Thiele/Aris modulus
ϕ_{Tn}	-	Thiele modulus for reaction order n
φ	-	enantiomeric excess
Φ	-	ratio of specificity constants
Φ_W	-	Weisz modulus

14.5.3 Additional Subscripts

b	bulk
C	coenzyme
eq	equilibrium
ext	external
f	final
i	counter of species or iteration step counter
in	feed; incoming
int	internal
j	counter
0	initial; feed
opt	optimal
out	outflowing
PL, PD	referring to product of form L or D
r	at radial position r
R	at the particle surface
S	substrate
SL, SD	referring to substrate of form L or D

14.5.4 Additional Superscripts

b	bulk
C	coenzyme
CSTR	referring to an ideal continuous-flow stirred-tank reactor
D	chiral substance of form D
f	referring to final conversion X_f

$_L$	chiral substance of form L
$_{max}$	maximal
$_{obs}$	observed
$_P$	product
$_{PFR}$	referring to a plug-flow reactor
$_r$	at radial position r
$_R$	at surface of particle
$_S$	substrate
^{T}	transpose (vector, matrix)
$'$	lumped (parameter)

14.6 REFERENCES

1. King, E. L., Altman, C., *J. Phys. Chem. 60*, 1375-1378 (1956).
2. Wong, J. T.-F., Hanes, C. S., *Can. J. Biochem. Physiol. 40*, 763-804 (1962).
3. Seshagiri, N., *J. Theor. Biol. 34*, 469-486 (1972).
4. Kuo-Chen, C., Carter, R., Forsén, S., *Chemica Scripta 18*, 82-86 (1981).
5. Hurst, R. O., *Can. J. Biochem. 47*, 941-944 (1969).
6. Fromm, H. J., *Methods Enzymol. 63*, 84-103 (1979).
7. Kullmann, W., *Biochem. J. 220*, 405-416 (1984).
8. Chen, C.-S., Fujimoto, Y., Girdaukas, G., Sih, C. J., *J. Am. Chem. Soc. 104*, 7294-7299 (1982).
9. Sih, C. J., Chen, C.-S., Girdaukas, G., Zhou, B.-N. In *Basic Biology of New Developments in Biotechnology*; Hollaender, A., Laskin, A. I., Rogers, P., Eds.; Plenum Press, New York, 1983; pp 215-230.
10. Rosenbrock, H. H., *Computer J. 3*, 175-184 (1960).
11. Nelder, J. A., Mead, R., *Computer J. 7*, 308-313 (1965).
12. Levenberg, K., *Q. Appl. Math. 2*, 164-168 (1944).
13. Marquardt, D. W., *Soc. Ind. Appl. Math. J. 11*, 431-441 (1963).
14. Powell, J. D., *Computer J. 7*, 303-307 (1965).
15. Fletcher, R., Reeves, C. M., *Computer J. 7*, 149-154 (1964).
16. Huang, H. Y., *J. Optim. Theory Appl. 5*, 405-423 (1970).
17. Dennis, J. E., Moré, J., *SIAM Rev. 19*, 46-89 (1977).
18. Forsman, E.-S., Heikonen, M., Kiviniemi, L., Kreula, M., Linko, P., *Milchwissenschaft 34*, 618-621 (1979).
19. Lineweaver, H., Burk, D., *J. Am. Chem. Soc. 56*, 658-666 (1934).
20. Eadie, G. S., *J. Biol. Chem. 146*, 85-93 (1942).
21. Hofstee, B. H. J., *Nature 184*, 1296-1298 (1959).
22. Hanes, C. S., *Biochem. J. 26*, 1406-1421 (1932).
23. Eisenthal, R., Cornish-Bowden, A., *Biochem. J. 139*, 715-720 (1974).
24. Wichmann, R., Wandrey, C., *Enzyme Engineering 5*, 259-261 (1980).

25. Walker, A. C., Schmidt, C. L. A., *Arch. Biochem. 5*, 445-467 (1944).
26. Foster, R. J., Niemann, C., *Proc. Natl. Acad. Sci. 39*, 999-1003 (1953).
27. Jennings, R. R., Niemann, C., *J. Am. Chem. Soc. 77*, 5432-5433 (1955).
28. Halwachs, W., *Biotechnol. Bioeng. 20*, 281-285 (1978).
29. Schwerdt, G. W., *J. Biol. Chem. 244*, 1278-1284 & 1285-1290 (1969).
30. Thiele, E. W., *Ind. Eng. Chem. 31*, 916-920 (1939).
31. Aris, R., *Chem. Eng. Sci. 6*, 262-268 (1957).
32. Bischoff, K. B., *AIChE J. 11*, 351-355 (1965).
33. Lapidus, L. *Digital Computation for Chemical Engineers*; McGraw-Hill, New York, 1962; p 106.
34. Na, H. S., Na, T. Y., *Math. Biosci. 6*, 25-35 (1970).
35. Villadsen, J. V., Stewart, W. E., *Chem. Eng. Sci. 22*, 1483-1501 (1967).
36. Lee, E. S., *Ind. Eng. Chem. 60*, 55-64 (1968).
37. Allen, D. W., Gerhard, E. R., Likins, M. R., *AIChE J. 22*, 593-594 (1976).
38. Wheeler, A., *Adv. Catalysis 3*, 249-327 (1951).
39. Krasuk, J. H., Smith, J. M., *Ind. Eng. Chem. Fundam. 4*, 102-106 (1965).
40. Weekman, V. W., Gorring, R. L., *J. Catalysis 4*, 260-270 (1965).
41. Onda, K., Murakami, Y., Kobayashi, T., Hattori, T., *Kagaku Kogaku 32*, 801-807 (1968) (cited in [49]).
42. Yamané, T., Araki, S., Sada, J. E., *Ferment. Technol. 59*, 367-374 (1981).
43. Kargi, F., Moo-Young, M. In *Comprehensive Biotechnology*; Moo-Young, M., Cooney, C., Humphrey, A. E., Eds.; Pergamon, New York, 1985; Vol. 2, p 44.
44. Bailey, J. E., Ollis, D. F. *Biochemical Engineering Fundamentals*, 2nd ed.; McGraw-Hill, New York, 1986; p 215.
45. Weisz, P. B., *Science 179*, 433-440 (1973).
46. Horwath, C., Engasser, J.-M., *Ind. Eng. Chem. Fundam. 12*, 229-235 (1973).
47. Yamané, T., *J. Ferment. Technol. 59*, 375-381 (1981).
48. Hong, J., *AIChE J. 30*, 526 (1984).
49. Moo-Young, M., Kobayashi, T., *Can. J. Chem. Eng. 50*, 162-167 (1972).
50. Bailey, J. E., Chow, M. T. C., *Biotechnol. Bioeng. 16*, 1345-1357 (1974).
51. Halwachs, W., Wandrey, C., Schügerl, K., *Biotechnol. Bioeng. 20*, 541-554 (1978).
52. O'Neill, S. P., Lilly, M. D., Rowe, P. N., *Chem. Eng. Sci. 26*, 173-175 (1971).
53. Robinson, B. A., Tester, J. W., *Chem. Eng. Sci. 41*, 469-483 (1986).
54. Klinkenberg, I. A., *Trans. Inst. Chem. Eng. 43* , T141-T156 (1965).
55. Pallaske, U., *Chem.-Ing.-Tech. 56*, 46-47 (1984).
56. See [B2], p 107.

14.7 LITERATURE FOR FURTHER READING

The following lists of references summarize the literature of more general interest for specific topics than those given above. They contain textbooks and papers used for writing the chapter as well as literature for those readers interested in more details.

14.7.1 Further Reading in Enzyme Kinetics

K1. Dixon, M., Webb, E. C. *Enzymes*, 3rd ed.; Longman Group Ltd., London, 1979.

K2. Laidler, K. J., Bunting, P. S. *The Chemical Kinetics of Enzyme Action*, 2nd ed.; Clarendon Press, Oxford, 1973.

K3. Wong, J. T.-F. *Kinetics of Enzyme Mechanisms*; Academic Press, London, 1975.

K4. *Enzyme Kinetics and Mechanism*; Purich, D. L., Ed.; *Methods Enzymol. 63/64* (1979/80).

K5. Fromm, H. J. *Initial Rate Enzyme Kinetics*; *(Molec. Biol. Biochem. Biophys. 22)*, Springer-Verlag, Berlin, 1975.

K6. Bisswanger, H. *Theorie und Methoden der Enzymkinetik*; Verlag Chemie, Weinheim, 1979.

K7. Hayashi, K., Sakamoto, N. *Dynamic Analysis of Enzyme Systems*; Japan Scientific Society Press and Springer-Verlag, Tokyo/Berlin, 1986.

K8. Segel, I. H. *Enzyme Kinetics*; John Wiley & Sons, New York, 1975.

14.7.2 Further Reading in Parameter Estimation Methods

P1. *Numerical Methods*; Churchhouse R. F., Ed.; Vol. 3 In *Handbook of Applicable Mathematics*; Ledermann, W., Ed.; John Wiley & Sons, Chichester, 1981.

P2. Wolfe, M. A. *Numerical Methods for Unconstrained Optimization*; Van Nostrand Reinhold, London, 1978.

P3. Bard, Y. *Nonlinear Parameter Estimation*; Academic Press, New York, 1974.

P4. Hoffmann, U., Hofmann, H. *Einführung in die Optimierung*; Verlag Chemie, Weinheim, 1971.

P5. Himmelblau, D. M. *Process Analysis by Statistical Methods*; John Wiley & Sons, New York, 1970.

P6. Rice, J. R. *Numerical Methods, Software, and Analysis*, 3rd ed.; McGraw-Hill, Singapore, 1987.

14.7.3 Further Reading in Mass-Transfer Phenomena

T1. Satterfield, C. N. *Mass Transfer in Heterogeneous Catalysis*; M.I.T. Press, Cambridge, MA, 1970.

T2. Aris, R. *The Mathematical Theory of Diffusion and Reaction in Permeable Catalysts*; Vol. 1, *The Theory of the Steady State*; Vol. 2, *Questions of Uniqueness, Stability and Transient Behaviour*; Clarendon Press, Oxford, 1975.

T3. Doraiswamy, L. K., Sharma, M. M. *Heterogeneous Reactions*; John Wiley & Sons, New York, 1984; Vol. 1, Chapters 3-7 & 18.

T4. Emig, G., Hofmann, H., *Chem.-Ing.-Tech. 47*, 889-892 (1975).

T5. Roberts, G. W., Satterfield, C. N., *Ind. Eng. Chem. Fundam. 4*, 288-293 (1965).

14.7.4 Further Reading in Reactor Design

14.7.4.1 General Reaction Engineering

R1. Levenspiel, O. *Chemical Reaction Engineering*, 2nd ed.; John Wiley & Sons, New York, 1972.

R2. Froment, G. F., Bischoff, K. B. *Chemical Reactor Analysis and Design*; John Wiley & Sons, New York, 1979.

R3. Baerns, M., Hofmann, H., Renken, A. *Chemische Reaktionstechnik*; Georg Thieme Verlag, Stuttgart, 1987.

R4. Trambouze, P., van Landeghem, H., Wauquier, J.-P. *Les Réacteurs Chimiques*; Édition Technip, Paris, 1984.

14.7.4.2 Backmixing

B1. See [R1], Chapters 9 & 10.

B2. Nauman, E. B., Buffham, B. A. *Mixing in Continuous Flow Systems*; John Wiley & Sons, New York, 1983.

B3. Wen, C. Y., Fan, L. T. *Models for Flow Systems and Chemical Reactors*; Marcel Dekker, New York, 1975.

INDEX